Design of
CONCRETE
MIXES

FIFTH EDITION

Design of
CONCRETE
MIXES

FIFTH EDITION

N Krishna Raju
BE, MSc (Engg), PhD MI Struct. E, C. Eng, MIE

Emeritus Professor of Civil Engineering
MS Ramaiah Institute of Technology
Bengaluru

CBS Publishers & Distributors Pvt Ltd

New Delhi • Bengaluru • Chennai • Kochi • Kolkata • Lucknow • Mumbai
Hyderabad • Jharkhand • Nagpur • Patna • Pune • Uttarakhand

Design of
Concrete Mixes
FIFTH EDITION

ISBN: 978-81-239-2467-0

Copyright © Author and Publisher

Fifth Edition: 2014
  Reprint: 2015, 2016, 2018, 2023
Fourth Edition: 2002
Third Edition: 1988
Second Edition: 1983
First Edition: 1974

Published by Satish Kumar Jain and produced by Varun Jain for

CBS Publishers & Distributors Pvt Ltd

4819/XI Prahlad Street, 24 Ansari Road, Daryaganj, New Delhi 110 002, India
Ph: 011-23289259, 23266861   Website: www.cbspd.com
                                    e-mail: delhi@cbspd.com
Corporate Office: 204 FIE, Industrial Area, Patparganj, Delhi 110 092
Ph: 011-4934 4934   Fax: 011-4934 4935   e-mail: publishing@cbspd.com; publicity@cbspd.com

Branches

- **Bengaluru:** Seema House 2975, 17th Cross, K.R. Road, Banasankari 2nd Stage, Bengaluru 560 070, Karnataka, India
  Ph: +91-80-26771678/79   Fax: +91-80-26771680   e-mail: bangalore@cbspd.com
- **Chennai:** 7, Subbaraya Street, Shenoy Nagar, Chennai 600 030, Tamil Nadu, India
  Ph: +91-44-26680620, 26681266   Fax: +91-44-42032115   e-mail: chennai@cbspd.com
- **Kochi:** 42/1325, 1326, Power House Road, Opp KSEB, Power House, Ernakulam 682 018, Kerala, India
  Ph: +91-484-4059061-65   Fax: +91-484-4059065   e-mail: kochi@cbspd.com
- **Kolkata:** 147, Hind Ceramics Compound, 1st Floor, Nilgunj Road, Belghoria, Kolkata-700056, West Bengal, India
  Ph: 033-25633055, 033-25633056   e-mail: kolkata@cbspd.com
- **Lucknow:** Basement, Khushnuma Complex, 7-Meerabai Marg (Behind Jawahar Bhawan) Lucknow 226001, India
  Ph: 0522-4000032   e-mail: tiwari.lucknow@cbspd.com
- **Mumbai:** PWD Shed. Gala no. 25/26, Ramchandra Bhatt Marg, Next to JJ Hospital Gate no. 2, Opp. Union Bank of India Noorbaug Mumbai-400009, Maharashtra, India
  Ph: 022-66661880/89   e-mail: mumbai@cbspd.com

Representatives

- **Hyderabad** 0-9885175004   • **Jharkhand** 0-9811541605   • **Nagpur** 0-9421945513
- **Patna** 0-9334159340   • **Pune** 0-9923910676   • **Uttarakhand** 0-9716462459

Printed at Chaman Interprises, Daryaganj, Delhi, India

Preface to the Fifth Edition

Revolutionary developments have taken place in the technology of cements, admixtures and concretes during the last two decades in keeping with the advances in the construction industry.

The fourth edition of this book has seen several reprints to meet the ever increasing demands of civil engineering students, teachers, practising civil, structural and highway design and construction engineers.

During the last decade, concretes of different types have made inroads into the construction industry. Phenomenal developments in the field of nanotechnology has resulted in the production of nanoconcrete with superior properties. The normal concrete of the twentieth century is being gradually replaced by high-performance concrete. New innovative concretes have been developed using the waste products of coal-based thermal stations.

Utilization of waste materials, like pond ash from thermal stations has paved the way for the production of pond ash concrete. Chapter 23 covers the design aspects of pond ash concrete and its properties. Chapter 24 deals with a systematic study of the production, properties and various applications of nanoconcrete. Chapter 25 covers the production, properties and uses of high-performance concrete which is replacing the conventional concrete for major infrastructure projects.

The author is grateful to his ex-colleagues Dr Manamohan R Kalgal and Dr RN Pranesh for providing valuable experimental source material on pond ash concrete compiled in Chapter 23. The author is also grateful to many of his colleagues in his profession and practising engineers for providing useful data on the latest developments in the field of nano and high-performance concretes. Sincere thanks are also due to CBS Publishers & Distributors Pvt Ltd for their continued encouragement and effective cooperation in bringing out the fifth edition in a record time.

N Krishna Raju

Preface to the First Edition

The primary object of studying the various properties of concrete and its ingredients is to design concrete mixes most suited for given job requirements. As a result of extensive research in the field of concrete over the past fifty years, it is possible in the present state of the art to produce concretes of low, medium and high strength and varying densities.

The object of this book is to provide a broad outline of the various empirical methods generally used for the design of concrete mixes of different types. Very little of the book is new but the material has been consolidated and streamlined from a wide variety of sources in order to make it accessible to the student as well as to the practising engineer.

The salient properties of concrete and its constituents, which form the background material for proportioning the concrete mixes, are covered in the first five chapters. The statistical concepts essential for producing concrete of uniform quality in the field are presented in Chapter 4. The well established British and American methods of designing the normal density concrete, form the subject matter of Chapters 6 and 7.

The data presented in the subsequent chapters are useful in the design of special types of concrete such as high alumina cement concrete, lightweight aggregate concrete, mass concrete, fly-ash cement concrete and high density concrete used for radiation shielding of nuclear reactors. The methods of producing ultra-high-strength concrete are covered in Chapter 17 and a brief introduction to the use of computers in mix design is presented in the chapter.

The design of concrete mixes of different types is illustrated with a number of examples at the end of each chapter. Metric and international system of units are used throughout the book in presenting design graphs and tables which are based on published information. The material requirements presented in the book generally conform to the Indian, British and ASTM specifications.

The book is based on courses of lectures and laboratory work given by the author at the Karnataka Regional Engineering College, University of Mysore, for undergraduate and postgraduate courses in the civil engineering stream.

N Krishna Raju

Acknowledgements

The author gratefully acknowledges the following institutions, journals and societies in respect of the reproduction of certain tables and figures mentioned throughout the text.

American Concrete Institute; American Society for Testing Materials; British Standards Institution; Cement and Concrete (India); Cement and Concrete Association, London; Civil Engineering and Concrete (UK); Institution of Civil Engineers, London; Indian Standards Institution; International Association for Bridge and Structural Engineering; Rilem, Paris; Magazine of Concrete Research, London; Portland Cement Association; Road Research Laboratory, DSIR; Structural Engineer.

The complete details of the sources are presented in the form of references at the end of the book.

The author is deeply indebted to Dr EW Bennett for introducing him to his fascinating and useful subject and for his constant encouragement. In the preparation of this work, the author has the advantage of the generous help of many, but in particular wishes to acknowledge with thanks the encouragement provided by Prof BS Basavarajaiah and colleagues of the civil engineering faculty of the Karnataka Regional Engineering College, Srinivasnagar.

The author is deeply indebted to Shri MS Ramaiah, Founder Chairman, Shri MR Jayaram, Chairman, MS Ramaiah Institute of Technology, Bangalore, and many of his colleagues for their constant encouragement and timely help during the preparation of the revised edition of the book. The author is grateful to CBS Publishers and Distributors for their cooperation in the production and release of this edition.

N Krishna Raju

Contents

1

Principles of Concrete Mix Design

1.1 INTRODUCTION

Concrete is a widely used structural material consisting essentially of a binder and mineral filler. It has the unique distinction of being the only construction material actually manufactured on the site, whereas other materials are merely shaped to use at the worksite. Good or bad concrete is made from the same discrete materials like, grains of sand, gravel or pieces of crushed rock and the innmerable fine particles of cement powder mixed with water. Ever since the time of Romans, there has been a continuous effort by research workers in the field of cement and concrete technology to produce better cements resulting in concretes of overall improved quality. The introduction of reinforced concrete as an alternative to steel construction, in the beginning of 20th century, necessitated the development and use of low and medium strength concretes.

In 1930, when prestressing was introduced by Freyssinet, the importance of using high-strength concrete to withstand the high compressive stresses developed due to prestressing was recognised. This naturally led to the development of concrete with compressive strength ranging from 30 to 40 N/mm², by proper selection of the ingredients and improved methods of manufacture. Recent developments in the field of concrete mix design have indicated that it is possible in the present state of the art to produce ultra-high-strength concrete, which has any desired 28-day cube compressive strength ranging from 70 to 100 N/mm², without recourse to unusual material or processing and without incurring any significant technical difficulties. The upper limit for the range of strength is considered to be about 110 N/mm², since this appears to be the ceiling above which, it becomes necessary to resort to special materials and sophisticated manufacturing techniques, like polymerization. In keeping with the demands of nuclear age, high-density concrete has been successfully used for radiation shielding of nuclear reactors. Considerable progress has been achieved in the design and use of structural lightweight concretes, which have the dual advantage of reduced density coupled with increased thermal insulation. With the present state of knowledge in the field of concrete mix design, it is possible to select and design concrete capable of resisting heat, sea water, frost and chemical attack arising out of industrial effluents.

1.2 EARLY MIX DESIGN METHODS

When concrete was first adopted as a structural material during the nineteenth century, compressive strength was perhaps the only criterion in the proportioning of a concrete mix. The concepts of workability, durability and other factors influencing the mix proportions, as they are understood now are of comparatively recent origin. The strength of concrete was supposed to increase with the increase in quantity of cement and with better compaction. The role of mixing water was not clearly understood except in so far as it helped concrete to become plastic for easy compaction. It was also realised that use of aggregates having less voids resulted in stronger concrete. Some of the earlier mix design methods are based on the principles of minium voids and maximum density.

   (a) *Minimum voids method:* The principle of this method is to proportion the ingredients so that the resulting concrete is dense with the minium percentage of voids. The void contents of the fine and coarse aggregates are predetermined. The quantity of the aggregate in the mix should be sufficient enough to fill up the voids in the coarse aggregate and similarly the cement content of the mix is governed by the voids in the fine aggregate. The quantity of water should be sufficient enough to render the mix workable.

   During mixing, the presence of water will apparently increase the void content in the aggregates and the method does not yield dense and strong concrete since graded aggregates are not used and there is no control over the water content in the mix in relation to the quantity of cement used.

(b) *Fuller's maximum density method:* Fuller advocated the maximum density theory which assumes that, the greater the amount of solid particles that can be packed in a given volume of concrete, the higher its strength. The method implies that the aggregates should be graded so that the mixture has the maximum density. Fuller and Thompson suggested ideal grading curves based on the equation given by

$$p = 100 \, (d/D)^{1/2}$$

where,

p = percentage of material smaller than size 'd'

D = maximum particle size

If it is required to grade a mix with 20 mm maximum size coarse aggregate and 4.75 mm fine aggregate, the value of the percentage of material (p) finer than 4.75 mm is given by

$$p = 100 \, (4.75/20)^{1/2} = 50 \text{ per cent}$$

Hence the fine aggregates including cement and coarse aggregates are combined in equal proportions and in addition, the quantity of particles of various intermediate sizes should correspond with Fuller's ideal curve. A major drawback of the method is that the ideal curve is based on the assumption that the aggregate is carefully packed to achieve maximum density and the effect of particle interference is ignored. In practice, it was found that the aggregate graded to give maximum density results in a harsh and unworkable mix since workability is improved only when there is excess of paste above that required to fill the voids in the sand and also an excess of mortar above that required to fill the voids in the coarse aggregate.

The concept of an ideal grading curve is now discredited and concrete can be made successfully from aggregates having grading much different from the ideal curve.

(c) *Talbot-Richart method:* The method suggested by Talbot and Richart[1] is based on the experimental investigations in which the compressive strength of concrete was found to depend upon cement space ratio which is given by $\dfrac{(c)}{(c+v)}$,

where,

c = solid volume of cement

v = volume of water plus voids in a unit volume of freshly made concrete

The compressive strength of concrete was also found to depend on the 'Basic Water Content', which represents the volume of water corresponding to the minimum volume of mortar. A relative water content in the range of 1.2 to 1.4 which implies 20 to 40 per cent more water than the Basic Water Content is recommended for the mixes. The method requires the computations of cement space ratios for several combinations of cement and sand at the same relative water content. If the resultant mixes are not satisfactory, new values of relative water content are chosen and the calculations are to be repeated.

The method is not popular since it requires more laboratory work than that can be justified for moderate sized jobs. In addition, the method does not take into account the water cement ratio law as the basis of strength and involves tedious calculations and trial mixes to achieve workable mixes.

(d) *Fineness modulus method:* Significant advances were made in concrete mix design when Duff Abrams[2] formulated the water/cement ratio law in 1918. The law states that when concrete is fully compacted, its compressive strength is inversely proportional to the water/cement ratio. For the purpose of applying this law, concrete can be considered as fully compacted, if it contains less than 2 per cent of air voids. Feret's law[3], established in 1892, appears to have been similar except that he used the ratio of cement to water plus air voids. According to Abrams, the 28 days compressive strength

of concrete is given by $S = \dfrac{K_1}{K_2^x}$

where, K_1 and K_2 are empirical constants equal to 700 and 3.5 respectively, x is the water/cement ratio by volume and S is the compressive strength expressed in kg/cm^2.

A new method of mix proportioning was suggested by Abrams, in which the workability was assessed in relation to a new property of the aggregate, designated as 'Fineness Modulus' which is a measure of the mean size of particles in the entire body of aggregates.

The fineness modulus is obtained as the sum of the cumulative percentage retained on a set of standard sieves. The proportion of fine to coarse aggregate is dependent on the richness of cement content in the mix and is given by the formula:

$$P = 100\frac{(A - B)}{(A - C)}$$

where,

P = the percentage of fine aggregate by weight of the total aggregate

A = the fineness modulus of the coarse aggregate

C = the fineness modulus of the fine aggregate

B = the maximum fineness modulus permissible depending upon the quantity of cement used and the maximum size of aggregate, as specified in Table 1.1.

Table 1.1: Maximum permissible value of fineness modulus

Amount of total aggregate per 50 kg bag of cement (m^3)	Fineness modulus for maximum aggregate size of	
	20 mm	40 mm
0.1132	5.1	5.8
0.1415	4.9	5.6
0.1698	4.8	5.5
0.1981	4.7	5.4
0.2264	4.6	5.3
0.2547	4.5	5.2

If the density of coarse aggregate is assumed as 1600 kg/m³, the aggregate/cement ratio by weight for the recommended range varies from 3.6 to 8.1. The water/cement ratio for the required strength of concrete is evaluated from the water/cement ratio law. The major drawback of the method is that it is possible for aggregates to have substantially different gradings and yet have the same fineness modulus, requiring number of trial mixes to achieve the desired workability.

The early mix design procedures discussed above may be considered to have paved the way for better insight into the field of concrete mix design and they have also indicated the importance of the various parameters influencing the concrete mix. The subsequent investigators realised the importance of the practical approach in formulating the mix design procedures since the parameters involved are such that they are not easily amenable for theoretical postulations. Most of the mix design methods which are popular at the present day are invariably based on extensive experimental investigations.

1.3 FACTORS INFLUENCING THE CHOICE OF MIX PROPORTIONS

The fundamental requirement of a concrete mix is that it should be satisfactory both in the fresh as well as in the hardened state, possessing certain minimum desirable properties, like workability, strength and durability. Besides these requirements, it is essential that the concrete mix is prepared as economically as possible by using the least possible amount of cement content per unit volume of concrete, with due regard to the strength and durability requirements. Since concrete is produced by mixing several discrete materials, the number of variables governing the choice of mix proportions are necessarily large. However, continuous research work in this field by various investigators has helped us to identify the significant parameters controlling the proportions of ingredients in the mix.

1.3.1 Compressive Strength

The usual primary requirement of good concrete is a satisfactory compressive strength in its hardened state. Many of the desirable properties, like durability, impermeability, abrasion resistance, are highly influenced by the strength of concrete. For purposes of mix design, the strength of concrete can be considered to be solely dependent on the water/cement ratio for low and medium strength concrete mixes. In the case of high-strength concrete mixes, the aggregate/cement ratio, workability of the mix and the type and maximum size of aggregate influence the selection of water/cement ratio for a desired strength of concrete.

The difference between the design strength and the minium site strength depends upon the degree of quality control to be exercised. The strength of concrete also depends upon the type of cement used and, the method of curing employed, since the rate of hardening of cements of different types varies considerably. However, the strength of concrete made with different cements is approximately the same after one year according to the investigations of Gonnerman and Lerch[4]. In most of the mix design methods, the water/cement ratio required to produce the design compressive strength is determined by curves or tables which are based on the water/cement ratio law developed by Abrams.

1.3.2 Workability

The workability of a concrete mix is mainly determined to suit the type of construction, placing conditions and the means of compaction available at site. The properties of fresh concrete, amount and condition of reinforcement and the shape and size of mould are important factors which control workability. For heavily reinforced sections, more workable concrete should be used than in massive construction. The main factor affecting workability is the water content in the mix. Other parameters influencing workability are the maximum size of aggregate, its grading, texture and shape and the mix proportions.

The proportion of the finer fractions of the aggregate controls cohesiveness of the mix. It is essential that the mix is cohesive so that compacting will result in a uniform and void free mass without segregation. For purposes of mix design, the degree of workability may be specified as slump, compacting factor or Vebe time of the fresh concrete mix. In most of the mix design procedures, one of these factors is invariably used as a measure of the workability of the mix. Air entrainment significantly improves the workability of the mix and this aspect has been recognised and incorporated in the method of design recommended by the American Concrete Institute[5]. The degree of workability and the corresponding magnitude of slump, compacting factor and Vebe time to be used for different types of construction and placing conditions are discussed in Chapter 3.

1.3.3 Type, Size and Grading of Aggregate

Good concrete can be made by using different types of aggregates, like rounded and irregular gravel and crushed rock which is mostly angular in shape. The maximum nominal size of the aggregate to be selected for a particular job depends upon the width of section and the spacing of reinforcement. According to the Indian Standard Code of Practice IS: 456[6], the maximum size of the aggregate is restricted to 5 mm less than the minimum clear distance between the main bars for heavily reinforced concrete members, such as ribs of main beams. It is generally advantageous to use as large a maximum size of aggregate as possible and experimental investigations by Bloem[7] have indicated that the improvement in the properties of concrete with an increase in the size of aggregate does not extend beyond about 40 mm.

The grading of aggregate is a major factor, influencing the workability of a concrete mix. The grading of the aggregates should be such as to ensure that the voids between the larger aggregates are filled with smaller fractions and mortar so as to achieve maximum density and strength. The coarser and finer fractions of aggregates available at site can be suitably combined to obtain the desired standard grading. It is important to note that good concrete can be made by using gap graded or continuously graded aggregate within a permissible limiting range.

1.3.4 Aggregate/Cement Ratio

The various factors involved in selecting the aggregate/cement ratio of a mix are, the desired workability, size, shape, texture and overall grading of the aggregates. The choice of the aggregate/cement ratio is generally made from charts or tables prepared from comprehensive laboratory investigations. It is also possible to reduce the combined grading curve to a numerical value in terms of the specific surface or fineness modulus, and then the workability can be linked mathematically by an equation or graphically by a series of curves to the aggregate/cement ratio, the water/cement ratio and the angularity.

The aggregate/cement ratio affects the strength of concretes in the high strength range to a significant degree and this is one of the reasons for considering the design of high strength concretes separately. It is important to note that mixes with very low water/cement and aggregate/cement ratios, having an extremely high cement content of the order of 450 to 550 kg/m^3, exhibit retrogression of strength, especially when large size aggregates are used. The reduction of strength is attributed to the loss of aggregate-cement bond due to stresses induced by shrinkage[8]. For a constant water/cement ratio, a leaner mix leads to a higher strength according to the investigations of Singh[9]. This pattern of behaviour is due to the absorption of water by aggregates leading to a reduction in the effective water/cement ratio of the mix.

1.3.5 Durability

Generally, concrete made from suitable ingredients, with proper compaction is durable under ordinary conditions of exposure. In such cases, the mix is designed by selecting the water/cement ratio on the basis of strength and workability rather than durability criterion. If the conditions of exposure are such that high durability is essential, the mix has to be designed by limiting the values of the water/cement ratio depending upon the type of exposure. In case of severe exposure to cycles of freezing and thawing, concrete mixes have to be designed with suitable air entrainment. It is also essential that frost-resistant concrete should have a low water content, low absorption and low permeability.

The choice of the type of cement is important, when concrete is exposed to chemical attack by sea water or industrial effluents. For concrete to be used in spillways and airport runways, resistance to erosion and abrasion is important, and this can be controlled by using hard coarse aggregates, like flint gravel or granite, and compacting the mix to achieve maximum density. The resistance of concrete to fire can be improved by using natural aggregates, like dolerites, basalts, limestones, and the manufactured lightweight aggregates, like foamed slag and sintered clay. In cases where heat-resistant concrete[10] is required, use of high alumina and cement with crushed rock or fire brick aggregates is recommended.

1.4 GENERALISED FORMAT OF CONCRETE MIX DESIGN

The various methods of mix proportioning, generally used for the design of ordinary concrete, are all based on the relation between strength and water/cement ratio as well as on the relation between workability, water/cement and aggregate/cement ratios. The basic factors influencing the sequential decision to be made in the course of mix design are compiled in the flow chart shown in Fig. 1.1[11]. In view of the large number of variables involved in the design of a concrete mix, trial mixes are invariably made using the designed proportions of ingredients. The desired properties of the trial mixes are checked and suitable adjustments are made and the process repeated until a satisfactory mix suitable for a particular job is obtained. Mix design methods being mostly empirical, minor variations exist in the process of selecting the mix proportions by using different methods. The widely used American Concrete

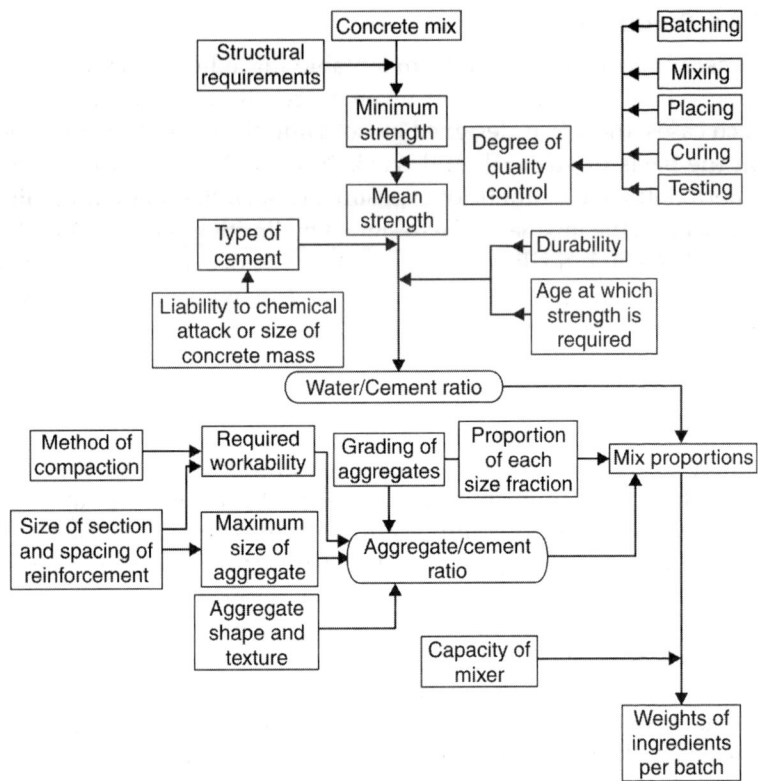

Fig. 1.1: Flowchart for concrete mix design

Institute[5] methods and the procedure recommended in Road Note No. 4[12] differ especially in the estimation of water content in the mix. In the design of mixes based on surface and angularity factors[13], the water/cement ratio is first fixed on the basis of desired strength while the aggregate/cement ratio is estimated using the aggregate properties, workability and the water content in the mix. In the case of lightweight concrete mixes, the net water/cement ratio would depend upon the amount of water absorbed by the lightweight aggregates. Consequently, the water/cement ratio law cannot be directly used in the mix design computations and generally proportioning on the basis of cement content related to the strength is preferred. Trial mixes of required workability and a given cement content are usually made before finalising the mix proportions suitable for the job in hand.

2

Material Requirements

2.1 JOB SUITABILITY AND STANDARD REQUIREMENTS OF CEMENTS

The different types of cements commonly used for making concrete have been developed to suit the strength, durability and other special requirements needed for a variety of jobs. The most widely used types of Portland cements in India, United Kingdom and United State of America are compiled in Table 2.1.

Table 2.1: Types of Portland cements

India	UK	USA
Ordinary Portland	Ordinary Portland	Type I
–	–	Type II
Rapid hardening Portland	Rapid hardening Portland	Type III
Low heat Portland	Low heat Portland	Type IV
–	Sulphate resisting Portland	Type V
Portland blast furnace	Portland blast furnace	Type IS
Portland Pozzolana	Portland Pozzolana	Type IP

There are other types, like high-alumina cement, super sulphate cement, expanding cements and special cements, mainly used for specific situations.

The physical and chemical properties of Portland cements covered by the Indian[14] and British Standard[15] specifications include the lime saturation factor, magnesia and sulphur content, fineness, soundness and setting time of cements and the tensile and compressive strength of standard mortar cubes. The maximum and minimum values specified for the different physical and chemical requirements are compiled in Tables 2.2 and 2.3. The values in brackets refer to the British Standards. The corresponding requirements of the ASTM specification[16] are given in Tables 2.4 and 2.5.

Table 2.2: Physical requirements of Indian and British Standard specifications for cements

Test	Types of cement			
	Ordinary Portland	Rapid hardening Portland	Low heat Portland	Portland blast furnace
1	2	3	4	5
Fineness of grinding				
Specific surface (cm^2/gm) minimum	2250 (2250)	3250 (3250)	3200 (3200)	2250 (2250)
Setting time				
Initial set not less than (mins.)	30 (45)	30 (45)	60 (60)	30 (45)
Final set not more than (mins.)	10 (10)	10 (10)	10 (10)	10 (10)
Soundness				
Le Chatelier test expansion after one hour's boiling (mm)	10 (10)	10 (10)	10 (10)	10 (10)
Compression strength (kg/cm^2) 1 : 3 sand mortar cubes (minimum values)				
1 day	–	–	160	–	–
3 days	–	160 (155)	275 (210)	100 (77)	160 (112)
7 days	–	220 (238)	– (280)	160 (140)	220 (210)
28 days	–	–	–	350	–
	–	–	(280)	–	
Heat of hydration (cal/gm.)				
at 7 days not more than	–	–	65	–
	–	–	(60)	–
at 28 days not more than	–	–	(75)	–

Table 2.3: Chemical requirements of Indian and British Standard specifications for cements

	Types of cement		
Requirement	*Ordinary and rapid hardening portland*	*Low heat Portland*	*Portland blast furnace slag*
Maximum permissible values (percentage)			
(a) Insoluble residue	1.5 (1.5)	1.5	1.5
(b) Magnesia (MgO)	6.0 (4.0)	6.0	8.0
(c) Sulphur as SO_3	2.75 (2.50)	2.75	3.0
(d) Loss on ignition	4.0 (4.0)	4.0	4.0
(e) Lime saturation factor	0.66–1.02 (0.66–1.02)	1.0	1.0

Table 2.4: Physical requirement of ASTM specifications for cements

	Types of cement		
Test	*Type-I*	*Type-III*	*Type-IV*
Fineness of grinding			
Specific surface (cm^2/gm) minimum			
(a) Turbidometer test –	1600	–	1600
(b) Air permeability test –	2600	–	2600
Time of setting (alternative methods)			
Gillmore test			
Initial set not less than (minutes)	60	60	60
Final set not more than (hours)	10	10	10
Vicat test			
Set, minutes not less than	45	45	45
Soundness			
Autoclave expansion—maximum (per cent)	0.50	0.50	0.50
Compressive strength (kg/cm^2)			
1 : 2.75 sand mortar cubes (minimum values)			
1 day –	–	120	–
3 days –	84	210	–
7 days –	147	–	56
28 days –	245	–	140

Table 2.5: Chemical requirement of ASTM specifications for cements

Requirement	Types of cement		
	Type-I	Type-III	Type-IV
(a) Insoluble residue			
Maximum per cent	0.75	0.75	0.75
(b) Loss of ignition			
Maximum per cent	3.0	3.0	2.30
(c) Sulphur trioxide (SO_3)			
Maximum per cent	2.5	3.0	2.30
(d) Magnesium oxide (MgO)			
Maximum per cent	5.0	5.0	5.0
(e) Ferric oxide (Fe_2O_3)			
Maximum per cent	–	–	6.5
(f) Tricalcium silicate			
($3CaO\ SiO_2$)			
Maximum per cent	–	–	35
(g) Dicalcium silicate			
($2CaO\ SiO_2$)			
Maximum per cent	–	–	40
(h) Tricalcium aluminate			
($3CaO\ Al_2O_3$)			
Maximum per cent	–	15	7

For determination of the fineness of cement, the sieve test is not entirely satisfactory because, cements giving the same result of residue by this test can have very different distribution of finer particles. The air permeability methods[17] are now preferred.

The Wagner turbidometer method, used in America, is based on the opacity of suspension of cement in kerosene. The air permeability method gives results which are 1.6 to 1.8 times those obtained with the Wagner turbidometer. The salient properties which are useful in selecting a particular type of cement for a specified job are the rate of strength development and heat evolution, drying shrinkage, resistance to cracking and inherent resistance to chemical deterioration. A comparative analysis of these properties for different cements is shown in Table 2.6 which is due to Lea[18].

Ordinary Portland cement is by far the most common type which is admirably suited for use in all general concrete construction, when there is no exposure to sulphates in the soil or in ground water. Low heat Portland cements are preferred in the construction of massive structures, like dams, retaining walls and bridge abutments mainly to control the amount of heat generated at tolerable levels. Rapid hardening Portland cements are of great advantage in the production of precast structural concrete elements, while the sulphate-resisting and high-alumina cements are ideally suited to resist sulphate attack in marine constructions.

Table 2.6: Salient properties of different types of cements

Types of cement	Rate of strength development	Rate of heat evolution	Drying shrinkage	Resistance to cracking	Resistance to chemical deterioration
Ordinary Portland	Medium	Medium	Medium	Medium	Low
Rapid hardening Portland	High	High	Medium	Low	Low
Low heat Portland	Low	Low	Somewhat higher	High	Medium
Portland blast furnace	Medium	Medium	Medium	Medium	Medium
Sulphate resisting Portland	Low to medium	Low to medium	Medium	Medium	High
High alumina	Very high	Very high	Medium	Low	Very high
Super sulphate	Medium	Very low	Medium	–	High
Pozzolanic cements	Low	Low to medium	Somewhat higher	High	High

2.2 GRADING REQUIREMENTS FOR AGGREGATES

The grading or particle size distribution of aggregates is a major factor determining the workability, segregation, bleeding, handling, placing and finishing characteristics of the concrete. The grading of fine aggregate has been found to influence the properties of fresh concrete more than that of the coarse aggregate. Although good workable concrete can be made with various gradings of aggregate, there are identifiable limits within which a grading must lie to produce a satisfactory concrete, but these depend upon the shape, surface texture and type of aggregate and the amount of flaky or elongated material. According to Neville[19], grading is of vital importance in the proportioning of concrete mixes, but its exact role in mathematical terms is not fully known, and the behaviour of this type of semi-liquid mixture of granular materials is still imperfectly understood. Grading requirements recommended by the Indian[20] British[21] and American[22] Standards for fine and coarse aggregates separately and for all in aggregates are presented in Tables 2.7 to 2.9, respectively.

Table 2.7: IS, BS, and ASTM grading requirements for fine aggregate

IS sieve designation	*Percentage passing, (by weight)*				
	Grading zone I	*Grading zone II*	*Grading zone III*	*Grading zone IV*	*ASTM C.33–57*
10 mm	100	100	100	100	100
4.75 mm	90–100	90–100	95–100	95–100	95–100
2.36 mm	60–95	75–100	85–100	95–100	80–100
1.18 mm	30–70	55–90	75–100	90–100	50–85
600 micron	15–34	35–59	60–79	80–100	25–60
300 micron	5–20	8–30	12–40	15–50	10–30
150 micron	0–10	0–10	0–10	0–15	2–10

The combined grading of coarse and fine aggregate mixture should be such that a reasonable workability with minimum segregation is obtained in the concrete mix. The standard continuous grading curves, which form the basis of the design method recommended by road note number 4[12] are shown in Figs 2.1 and 2.2 for aggregates of maximum size 40 and 20 mm, respectively. Similar curves for aggregates of 10 mm maximum size, suggested by McIntosh and Erntroy[23], are reproduced in Fig. 2.3. The outer curves nos. 1 and 4 represent the limits for normal continuous gradings. The coarsest grading curve no. 1 is suitable for dry mixes with a low water/cement ratio, while the finest grading curve no. 4 is suited for wet mixes, where high workability is required.

In practice, fine and coarse aggregates of several sizes are stock piled separately and they can be combined to conform to a desired grading either by analytical or graphical methods detailed in Chapter 6. Due to the presence of over and undersize aggregates and because of variations with any fraction size, the combined practical gradings are likely to lie in a zone rather than coincide exactly with the type grading curves.

Table 2.8: IS and BS grading requirements for coarse aggregate

IS sieve designation	Percentage passing for single sized aggregate of nominal size						Percentage passing for graded aggregate of nominal size			
	63 mm	40 mm	20 mm	16 mm	12.5 mm	10 mm	40 mm	20 mm	16 mm	12.5 mm
80 mm	100	–	–	–	–	–	–	–	–	–
63 mm	85–100	100	–	–	–	–	–	–	–	–
40 mm	0–30	85–100	100	–	–	–	95–100	100	–	–
20 mm	0–5	0–20	85–100	100	–	–	30–70	95–100	100	100
16 mm	–	–	–	85–100	100	–	–	–	90–100	–
12.5 mm	–	–	–	–	85–100	100	–	–	–	90–100
10 mm	–	0.5	0–20	0–30	0–45	85–100	10–35	25–55	30–70	40–85
4.75 mm	–	–	0–5	0–5	0–10	0–20	0–5	0–10	0–10	0–10
2.36 mm	–	–	–	–	–	0–5	–	–	–	–

Table 2.9: IS and BS grading requirements for all-in aggregates

IS sieve designation	Percentage passing for all-in aggregate of	
	40 mm nominal size	20 mm nominal size
80 mm	100	–
40 mm	95–100	95–100
20 mm	45–75	95–100
4.75 mm	25–75	30–50
600 micron	8–30	10–35
150 micron	0–6	0–6

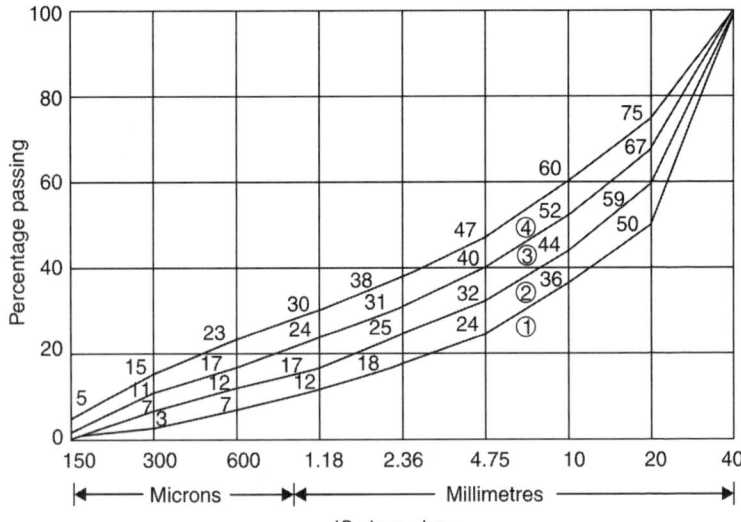

Fig. 2.1: Curves of four gradings of 40 mm aggregates

2.3 SPECIFIC GRAVITY, BULK DENSITY AND VOID CONTENT OF AGGREGATES

In the process of mix design, a knowledge of the specific gravity and bulk density of aggregates used is essential to set out the proportions of material by weight and volume and also to compute the quantity of aggregate required in a unit volume of concrete. Standard tests[24] are available to ascertain these properties for a given representative sample of aggregate to be used in the mix. Table 2.10 gives typical

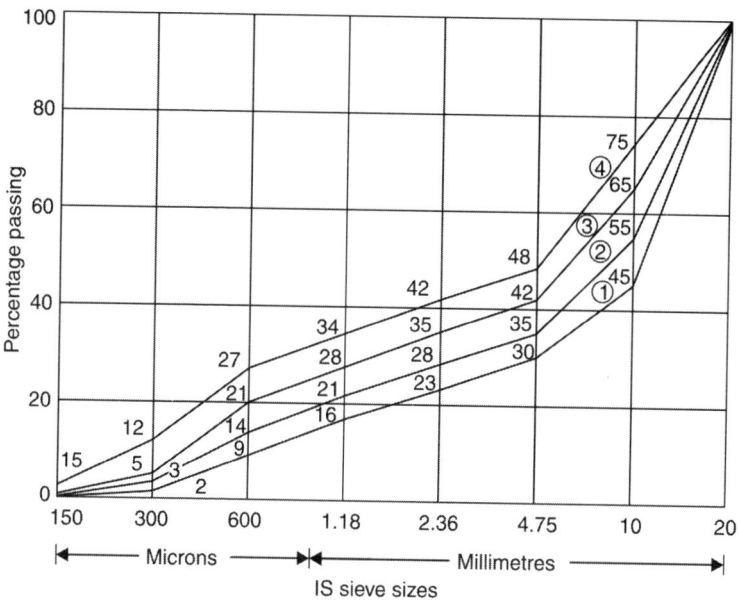

Fig. 2.2: Curves of four gradings of 20 mm aggregate

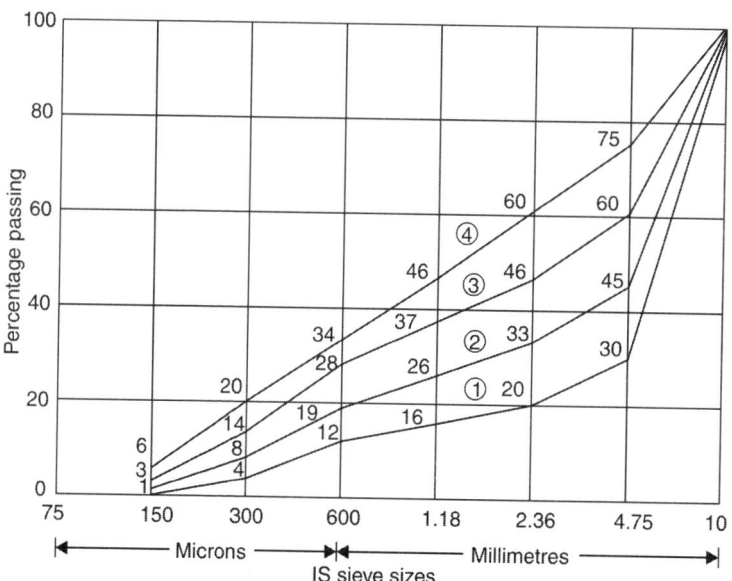

Fig. 2.3: McIntosh and Erntroy's type grading curves for 10 mm aggregate

values of these properties for the commonly used aggregates as reported by the Concrete Association of India[25].

Table 2.10: Specific gravity, bulk density and voids of different types of aggregates

Specific gravity		
Trap	–	2.90
Granite	–	2.80
Gravel	–	2.66
Sand	–	2.65
Bulk density (kg/l)		
River sand:		
Fine	–	1.44
Medium	–	1.52
Coarse	–	1.60
Beach or river shingle	–	1.60
Broken stone	–	1.60
Stone screenings	–	1.44
Broken granite	–	1.68
Voids (per cent)		
River sand:		
Fine	–	43
Coarse	–	35
Mixed and moist	–	38
Mixed and dry	–	30
Broken stone, graded:		
25 mm maximum size	–	46
50 mm maximum size	–	45
63 mm maximum size	–	41
Stone screenings	–	48

The extent to which an aggregate can be compacted to produce a minimum void content is dependent on the size distribution and shape of the aggregate particles. Rounded aggregates when compacted contain less voids than irregular and flaky aggregates of the same nominal size. A reduction in the void content of the coarse aggregate will in turn reduce the quantity of mortar required for the mix, resulting in marginal savings of cement and sand.

2.4 BULKING OF FINE AGGREGATE

The presence of moisture in fine aggregates to the extent of 3 to 10 per cent increases the volume of the aggregate by as much 20 to 30 per cent. The magnitude of bulking depends to a large extent on the fineness of the grading. For a given moisture content, bulking of fine sand may be nearly 2 to 3 times that of coarse sand. Table 2.11[25] gives the details of percentage bulking of different types of sands at various moisture contents.

Table 2.11: Bulking of sand for various moisture contents

Moisture per cent	Percentage bulking in		
	Fine sand	Medium sand	Coarse sand
1	16	8	6
2	26	16	12
3	32	22	15
4	36	27	17
5	38	29	18
6	37	28	18
8	35	26	16
10	32	22	12
12	28	19	8
15	22	12	2
17	18	7	0
20	9	0	0
27	0	0	0

The mix proportions are usually designed in terms of the weight or volume of dry aggregates and during mixing operations, adjustments have to be made for the amount of water contained in the aggregates. If weigh-batching is adopted, the weight of aggregates have to be increased to allow for the weight of water contained in them and the amount of water added to the mix, should be correspondingly reduced.

In the case of volume batching, bulking of fine aggregate should also be considered together with the adjustments for the actual quantity of water contained in the aggregate. The methods of adjustments of mix proportion for moisture content in aggregates is illustrated in the mix design example of Chapter 6.

3

Workability of Fresh Concrete

3.1 SIGNIFICANCE OF WORKABILITY

The earlier investigators, Glanville, Collins and Mathews[26], defined workability of concrete as the amount of useful internal work necessary to produce full compaction. However, recent developments have indicated the workability can best be described as a collective term to include all the essential properties of the concrete in the plastic condition. According to Newman[27], these properties comprise the following:

(a) *Compactability:* The property representing the ease with which the concrete is compacted by expelling the air voids. This is similar to the definition of Glanville, Collins and Mathews of the Road Research Laboratory.

(b) *Stability:* The property of the concrete which resists segregation of its ingredients in transit or during compaction.

(c) *Mobility:* The property which determines the ease with which concrete can flow round the reinforcement and fill out the angles etc., peculiar to the particular sections being cast.

(d) *Finishability:* The property which helps to provide a smooth surface finish by trowelling or other means.

A knowledge of the workability is most essential for the production of a well-designed concrete mix, which can be easily placed and compacted with the minimum of effort. The factors controlling the workability of a concrete mix are too many, and the prominent among them are, the overall grading, maximum size, shape and surface texture of the aggregates; the ratio of coarse to fine aggregates and their bulk densities; the quantity of water in the mix; the absorption capacity of the aggregate and the richness of the mix.

3.2 MEASUREMENT OF WORKABILITY

The various tests developed over the years correlate workability with some easily determinable physical quantity. The slump test which is used extensively in site work, does not measure workability as such but is very useful in detecting variations in the uniformity of a mix of given nominal proportions. The compacting factor test developed by Glanville et al[26] was intended to measure the compactability of concrete. However, Cusens[28] has shown that the test is very misleading for dry mixes which tend to stick in the hoppers. It is found that the implied assumption that all mixes with the same compacting factor require the same amount of useful work is not always justified. The compacting factor test is mainly applicable to mixes, where hand compaction may possibly be used, and for control purposes.

The Kelly ball test[29] developed in USA is an alternative to the slump test. This test has the unique advantage that it can be applied to concrete in a wheel barrow or actually in the form and the test is simpler and quicker to perform than the slump test. The flow test covered by the ASTM Standard[30] does not measure workability, as concretes having the same flow may differ considerably in their workability. However, the test gives a good assessment of consistence of concrete and its proneness to segregation.

The Vebe Consistometer test, developed in Sweden by V. Bahrner[31], is basically a remoulding test of the type developed earlier by Powers[32]. In the case of controlled concrete mixes, where vibration is invariably used, the Vebe test represents a good reproduction of practical conditions and indicates a combination of both compactability and mobility of vibrated concrete. Although there is no ideal test for workability, the various test results which indicate the different aspects of workability should be judiciously used in designing a concrete mix suitable for a particular job.

3.3 DEGREE OF WORKABILITY

In the design of concrete mixes, the desired degree of workability is fixed based on the conditions of placing and the compacting methods available at the worksite. A guide to the degree of workability recommended in Road Note no. 4 is reproduced in Table 3.1 in which the values of compacting factor and slump suitable for different conditions of placing are specified. These values are applicable to concrete made with 20 or 40 mm maximum size of aggregate. The recommendations of Taylor[33] shown in Table 3.2 include three different sizes of aggregates of 10, 20 and 40 mm.

Table 3.1: Uses of concrete of different degrees of workability according to Road Note no. 4

Degree of workability	Slump[†] (mm)	Compacting factor Small apparatus	Large apparatus	Use for which concrete is suitable
Very low	0 to 25	0.78	0.80	Roads vibrated by power-operated machines. At the more workable end of this group, concrete may be compacted in certain cases with hand-operated machines.
Low	25 to 50	0.85	0.87	Roads vibrated by hand-operated machines. At the more workable end of this group, concrete may be manually compacted in roads using aggregate of rounded or irregular shape. Mass concrete foundations without vibration or lightly reinforced sections with vibration.
Medium	50 to 100	0.92	0.935	At the less workable end of this group, manually compacted flat slabs using crushed aggregates. Normal reinforced concrete manually compacted and heavily reinforced sections with vibration.
High	100 to 180	0.95	0.96	For sections with congested reinforcement. Not normally suitable for vibration.

† The slump is not definitely related to the workability or the compacting factor. The figures given must, therefore, be regarded as providing a rough indication of the order of the slump and nothing more.

The compacting factor figures have been obtained by means of the compacting factor test for workability described in Road Research Technical Paper no. 5[26].

Vibration is invariably used as a means of compaction, in the case of high quality concrete mixes. In such cases, a suitable Vebe time ensures in general satisfactory compactability, mobility and finishability. Table 3.3 shows the suggested mean values of compactability

Table 3.2: Selected values of compacting factors and slump for different conditions of placing

Placing condition	Degree of workability	Nominal maximum size of aggregate (mm)	Compacting factor	Slump (mm)
Section subject to extremely intensive or prolonged vibration or to vibration accompanied by pressure	Extremely low	10 20	0.65 0.68	0 0
Small section subject to intensive vibration, large sections with normal vibration	Very low	10 20 40	0.75 0.78 0.78	0 0–12 0–25
Simply reinforced sections with vibration, large sections without vibration	Low	10 20 40	0.83 0.85 0.85	0–6 12–25 25–50
Simply reinforced sections without vibration, heavily reinforced sections with vibration	Medium	10 20 40	0.90 0.92 0.92	6–25 25–50 50–100
Heavily reinforced sections without vibration. Sections not normally suitable for vibration	High	10 20 40	0.95 0.95 0.95	25–100 50–125 100–175

for different applications and equipment according to the Hughes[34]. At the lower end of the table where the workability range is extremely low and where table vibrators are essential, Vebe time is the only means of ascertaining the degree of workability as the compacting factor and slump of very dry mixes cannot be measured.

3.4 WORKABILITY AIDS

Finely ground mineral powders, like lime, bentonite, kaolin, silica flour, chalk and diatomaceous earth, are used as workability aids. These materials, when ground as fine as cement, increase the

Table 3.3: Suggested mean values of compactability for different applications and equipment

Application	Vebe (secs)	Compacting factor	Slump (mm)
Heavily reinforced sections with light immersion vibrators. Simply reinforced sections without vibration	3	0.93	40–100
Heavily reinforced sections with immersion vibrators. Concrete suitable for pumping	4	0.91	25–75
Reinforced sections with immersion vibrators	6	0.87	13–50
Reinforced sections with heavy immersion vibrators. Road slabs with hand-operated machines	8	0.83	0–25
Mass concrete and other large sections with heavy immersion vibrators. Road slabs with power-operated machines	12	0.78	–
Table vibration of heavy precast units, etc.	16	–	–
Table vibration of light precast units, etc.	24	–	–
Heavily vibrated or mechanically compacted concrete in precast unit, etc.	32	–	–
Extremely intensive vibration, pressure may also be necessary	Above 40	–	–

amount of mortar in the mix and prevent segregation of harsh mixes, without any extra water being added to the mix. If they are used in large quantities, the amount of water has to be increased and there will be consequent loss of strength.

Air entrainment besides increasing the workability of harsh mixes improves the resistance of concrete to weathering and in particular to the action of frost. However, the main disadvantage of entrained air is the reduction in the strength of concrete, which can be compensated by reducing the water/cement ratio. The effect of the percentage of air entrained on the workability of concrete, measured as compacting factor, has been reported by Wright[35]. If the air content is increased by five per cent, the corresponding increase in the compaction factor was observed to be as much as 0.07, which corresponds to an increase in slump of nearly 40 mm. A greater increase in the workability is possible with wet rather than dry mixes and also with lean mixes than with rich mixes.

4

Statistical Quality
Control of Concrete

4.1 VARIABILITY OF CONCRETE STRENGTH

Concrete produced at site is likely to have variability of strength from batch to batch and also within the batch. The magnitude of this variation depends on several factors, such as the quality of materials, method of batching, proportioning, mixing, placing and the overall workmanship and supervision at site.

The purpose of controlling the quality of concrete using the statistical means is to produce concrete of uniform quality, which can be achieved by good workmanship and plant maintained at peak efficiency. Quality control of concrete will be of immense value of considerable size or importance where the specifications insist on certain minimum requirements and the effort which goes to the testing and analysis could more than offset by the resulting savings in the overall concreting operations. The compressive strength test results of cubes from random sampling of a mix exhibit variations, which are inherent in the various operations involved in making and testing of concrete. If a large number of cube strength test results are plotted on a histogram, the results are found to follow a bell-shaped curve: termed as "Normal Distribution Curve". The results are said to follow a normal distribution, if they are equally spaced about the mean value and if the largest number of cubes have a strength close to the mean value, the number falling off as the results are much greater or lesser than the value. However, some divergence from the smooth curve is only to be expected, particularly, if the number of results available is relatively small. The arithmetic mean or the average value of a number of test results gives no indication of the extent of variation of strength. However, this can be ascertained by relating the individual strengths to the mean strength and determining the variation from the mean with the help of the properties of the normal distribution curve.

Standard Deviation

The root mean square deviation of the whole consignment is termed as the "Standard Deviation" and is defined numerically as:

$$\sigma = \sqrt{\frac{\Sigma(x - \bar{x})}{n-1}}$$

where,

σ = the standard deviation of the population
x = any value in the set of numbers
$\bar{x}$ = the arithmetic means of the set
n = the number values in the set

The standard deviation increases with increasing variability and is expressed in the same units as the numbers. A unique property of the standard deviation is that the proportion of all the results falling within or outside certain limits can be related to it when the concerned results occur at random. In the case of concrete work, this assumption can generally be made without serious loss of accuracy as long as the techniques of random sampling[36] are followed.

The characteristics of the normal distribution curve are fixed by the average value and the standard deviation. The spread of the curve along the horizontal scale is governed by the standard deviation, while the position of the curve along the vertical scale is fixed by the average value. It is possible to fix the limits below or above which the proportion of results can be expected to fall: the limits are set out as $(\bar{x} \pm k\sigma)$, where k is the probability factor having the value shown in Fig. 4.1. For different values of k, the percentage of results falling above and below a particular value is illustrated in Fig. 4.2, in relation to the area bounded by the normal probability curve.

The values of standard deviation suggested by Himsworth[37] for different control conditions are compiled in Table 4.1.

Coefficient of Variation

An alternative method of expressing the variation of results about the mean is by coefficient of variation, which is a non-dimensional measure of variation obtained by dividing the standard deviation by the average and is expressed as:

$$V = \frac{\sigma}{\bar{x}} \times 100$$

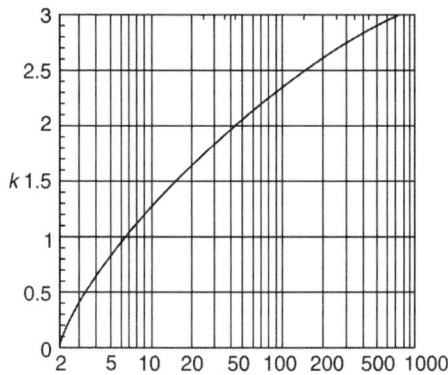

Fig. 4.1: Relation between the factor k and the proportion of results expected to be below the minimum strength

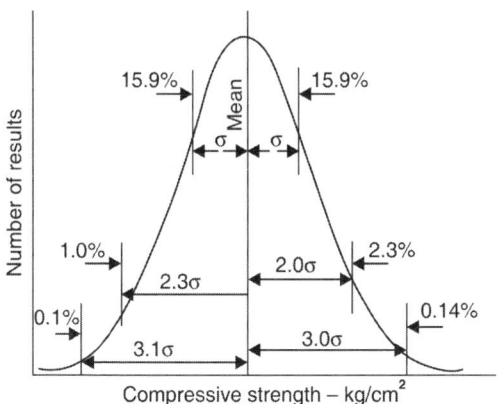

Fig. 4.2: Normal probability curve

With a constant coefficient of variation, the standard deviation increases with strength and is larger for high strength concrete. The values of coefficient of variation suggested by Stanton Walker[38] for different degrees of control are shown in Table 4.2. The standard

Table 4.1: Values of standard deviations for different types of control (according to Himsworth)

Type of control	Standard deviation	
	kg/cm²	N/mm²
Excellent	28	2.8
Very good	35	3.5
Good	42	4.2
Fair	56	5.6
Poor	70	7.0
Uncontrolled	84	8.4

Table 4.2: Values of coefficient of variation for different degrees of control (according to Stanton Walker)

Type of control	Coefficient of variation (per cent)
Attainable only in well-controlled laboratory tests	5
Excellent, approching laboratory precision	10
Excellent	12
Good	15
Fair	18
Fair minus	20
Bad	25

deviation and the coefficient of variation are useful in the design and control of strength of concrete mixes. The method of computing the standard deviation and coefficient of variation for a given test data is illustrated by the following example:

Example of statistical analysis

Cube number	Compressive strength (kg/cm²) x	$(x - \bar{x})$	$(x - \bar{x})^2$
1	165	−35	1225
2	155	+45	2025
3	225	+25	625
4	185	−15	225
5	210	+10	100

(Contd...)

(*Contd...*)

Cube number	Compressive strength (kg/cm²) x	$(x - \bar{x})$	$(x - \bar{x})^2$
6	265	+65	4225
7	195	−5	25
8	250	+50	2500
9	180	−20	400
10	230	+30	900
11	260	+60	3600
12	195	−5	25
13	160	−40	1600
14	175	−25	625
15	150	−50	2500
$\Sigma x =$	3000	$\Sigma(x - \bar{x})^2 =$	20600

Mean strength, $\qquad \bar{x} = \dfrac{\Sigma x}{n} = \dfrac{3000}{15} = 200 \text{ kg/cm}^2$

Standard deviation, $\quad \sigma = \sqrt{\dfrac{\Sigma(x - \bar{x})^2}{n-1}} = \sqrt{\dfrac{20600}{14}} = 38.5 \text{ kg/cm}^2$

Coefficient of variation, $V = \dfrac{\sigma}{\bar{x}} \times 100 \qquad = \dfrac{38.5}{200} \times 200$

$$= 19.25\%$$

4.2 APPLICATION OF STATISTICS TO MIX DESIGN

The advantages of applying the principles of statistics to the design of concrete mixes and to control the quality of concrete at site was first realised during the late 1940s by the early investigators[39,40]. In the first symposium on Mix Design and Quality Control of Concrete[41] held in 1954, many research workers have emphasized the importance of statistical techniques with relevance to mix design.

Relation between Mean and Minimum Strength

In the design of concrete mixes, the average strength to be aimed should be appreciably higher than the minimum, if the quality of the concrete is to comply with the requirements of the specifications. If from previous experience, the expected variation in the compressive strength is represented by a certain standard deviation or coefficient of variation, it is possible to compute the average design strength of the mix, which would carry with it a predetermined chance or results falling below a specified minimum strength using the following relations:

$$\sigma_{mean} = \sigma_{min} + k\,\sigma$$
$$= \sigma_{min}/(1 - k.V)$$

The standard deviation or the coefficient of variation method can be used to estimate the mean design strength of a mix. However, for a given degree of control, the standard deviation method yields higher mean strength than the coefficient of variation method for low and medium strength concretes as shown by a comparative analysis made in Table 4.3. For high strength concretes, the coefficient of variation method yields higher values of the mean strength. The values shown in the table are computed for a constant probability (k) of 2.33, corresponding to 1 per cent of results falling below the minimum.

Table 4.3: Mean strength from standard deviation and coefficient of variation method

Type of control	Minimum strength range (kg/cm²) Method used	150	300	400
		Computed mean strength (kg/cm²)		
Excellent	Standard deviation: 28 kg/cm²	215	365	515
	Coefficient of variation: 5 per cent	167	335	500
Good	Standard deviation: 42 kg/cm²	248	398	548
	Coefficient of variation: 15 per cent	202	405	607
Fair	Standard deviation: 56 kg/cm²	280	430	580
	Coefficient of variation: 18 per cent	213	425	638

The cost of production being dependent on the mean strength of concrete, the method of evaluation should be consistent with the observed trend of results for the different ranges of strengths.

Murdock[42] and Erntroy[43] have shown that the coefficient of variation more nearly represents a particular standard of control at relatively low strengths, while the standard deviation more nearly represents the standard at high strengths. The variability in the compressive strength of concrete made on site is related more or less directly to the variability of the water/cement ratio and it has been reported by Erntroy[44] that a particular standard of quality control can be associated with a Control Ratio, which is defined as the ratio of the water/cement value required to produce the mean strength to that of minimum strength. The recommended values of the control ratio for different types of supervision and standards of quality control are compiled in Table 4.4. The standard of control which is economically justified on a particular contract depends upon both the quantity and quality of the concrete involved. The economical control ratios, shown in Table 4.5, are based on the work of Erntroy[45] and the Joint Committee of the Institution of Civil and Structural Engineers[46].

Table 4.4: Control ratios

Proportion of test result expected below minimum	Type of supervison	Control ratio Standards of control			
		A	B	C	D
	Poor	0.78	0.75	0.72	0.70
1 in 25	Normal	0.82	0.79	0.77	0.75
	Good	0.86	0.83	0.82	0.80
	Poor	0.76	0.71	0.69	0.66
1 in 40	Normal	0.80	0.76	0.74	0.72
	Good	0.84	0.81	0.79	0.78
	Poor	0.71	0.67	0.63	0.60
1 in 100	Normal	0.76	0.72	0.69	0.67
	Good	0.81	0.77	0.75	0.74
Control A	Batching of cement and aggregates by weight with servo operation				
Control B	Batching of cement and aggregates by weight with manual operation				
Control C	Batching cement by weight and aggregates by volume				
Control D	Batching cement and aggregate by volume				

4.3 USE OF CONTROL CHARTS

Quality control charts[47,48] can be prepared by plotting any variable against a measure of progress over a period of time. The charts are helpful to detect the presence of extraneous variations in a series of experimental observations and thus to determine at an early stage any deviations of the characteristics of the distribution of the results from the requirements of the specifications.

Table 4.5: Economical control ratios

Minimum specified 28 days cube strength kg/cm² (N/mm²)	Size of contact (m³)			
	Very large (over 40,000)	Large (over 8,000)	Medium (over 800)	Small (under 800)
280 (28.0)	0.82+	0.82	0.82	0.82
175–280 (17.5–28.0)	0.82	0.82	0.82–0.79	0.79
140–175 (14.0–17.5)	0.82	0.79	0.79	0.77
70–140 (7.0–14.0)	0.82	0.79	0.77	0.77
Unspecified	0.70	0.70	0.70	0.70

A typical example[49] of a quality control chart is shown in Fig. 4.3, in which the compressive strength of cubes are plotted against the progress of work. Where it is desired to verify the assumptions made in the mix design, the design mean strength and the control levels, corresponding to any desired proportion of test results falling below and above the specified minimum are drawn in the control chart and the results are plotted as and when they are obtained. If the quality is as expected, the proportion of the plotted results falling below any selected lower control limit and above the corresponding upper control limit will be equal and appropriate to the levels considered.

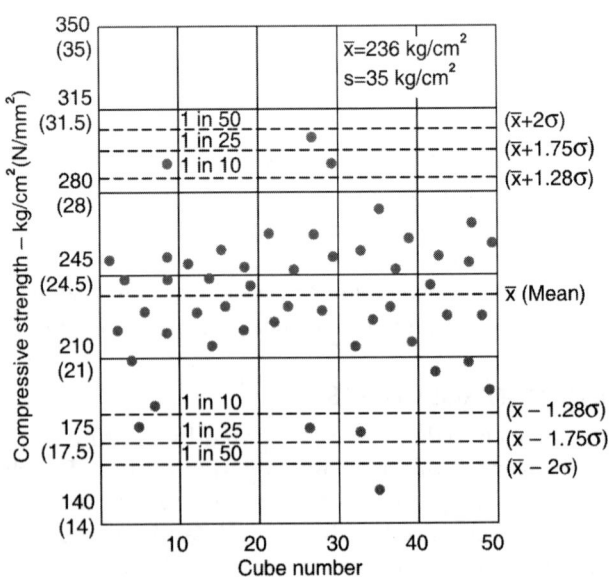

Fig. 4.3: Quality control chart

By visual examination of the charts, we can interpret whether the average strength of the concrete being produced is below the required average by counting the number of points occurring on either side of the control level for the average. If the average strength appears to be low, this should be checked by calculation and if confirmed suitable, modifications are made in the mix proportions so that a higher average strength is obtained.

4.4 RATE OF SAMPLING AND TESTING

It is essential to keep a good control on the quality of concrete produced by casting the required number of specimens from random samples and testing them at suitable intervals to obtain results as quickly as possible to enable the level of control to be established with reasonable accuracy in a short time. The cost of implementing a suitable system of sampling and testing is not negligible and hence this should be fixed, based on the volume of concrete involved and the type of work for which the concrete is used.

The recommendations of the new British Euro Code BS EN 1:2004[50], regarding the sampling of concrete is shown in Table 4.6 in which three different rates are specified. Sampling within rate 1, would be suitable for concrete used in highly stressed structural elements, while that in rate 3, involving sampling at less frequent intervals is suitable for mass concrete work not subjected to high stresses. For most of the ordinary structural concrete work, rate 2 would be sufficient. In the early stages of a job, it is preferable to resort to frequent sampling to establish the general level of control. The Indian Standard Code IS: 456–2000 prescribes sampling for every 150 m³ of concrete or part thereof, the samples being drawn

Table 4.6: Rates of sampling and testing

Rate of sampling	Rate 1	Rate 2	Rate 3
	Sample from one batch selected randomly to represent and average volume of not more than		
	10 m³ or	20 m³ or	50 m³ or
	10 batches	20 batches	50 batches
	whichever is the lesser volume		
Maximum quantity of concrete at risk under any one decision	30 m³	60 m³	150 m³

on each day for the first four days of concreting and thereafter atleast once in seven days of concreting.

The exact requirements of specifications with regard to the acceptance criteria for concrete generally vary from one code to the other while the British Code BS EN 1992–1–1:2004 stipulates that not more than 5 per cent of the test results to fall below the 28 day characteristics cube strength, the Indian Standard Code IS: 456–2000 requires that only one out of five consecutive tests may give a value less than the specified strength. The corresponding requirement according to the American Concrete Institute Standard ACI: 214–77[51], is that not more than 10 per cent of the test results are permitted to be below the specified design strength.

5

Strength and Durability of Concrete

5.1 COMPRESSIVE STRENGTH OF CONCRETE

In the design of concrete mixes, the compressive strength of concrete is generally the main target since it usually represents an overall picture of the quality of the concrete. For fully compacted concrete, water/cement ratio is the principal parameter governing the strength of concrete, according to the law established by Abrams as early as in 1919. The compressive strength is the maximum load per unit area sustained by a concrete specimen before failure under compression. Although the compression test on concrete is simple to carry out, the tests results are difficult to interpret in terms of actual strength which is influenced by many factors. Many of the important properties of concrete, like the modulus of elasticity, resistance to shrinkage, the creep and durability, improve with the increase in compressive strength. In the elastic design of structural concrete members, the permissible bending, direct and shear stresses in concrete are invariably related to the compressive strength of concrete. The introduction of limit state philosophy[52,53] in the design of structural concrete elements makes the strength characteristics of concrete more important.

A major portion of the structural concrete used is proportioned to have a strength of 20–30 N/mm^2 at 28 days. High strength concrete used in prestressed work is designed to have a 28 days compressive strength of 35–50 N/mm^2. The compressive strength of mass concrete may be as low as 75–150 kg/cm^2, while that of dry lean concrete only 3.5–7.0 N/mm^2. The strength of concrete increases with time and temperature, according to the relations developed by Plowman[54] based on experimental investigations. The relation between water/cement ratio and compressive strength of 15 cm by 30 cm cylinders is shown in Fig. 5.1 which is based on Taylor's observations.

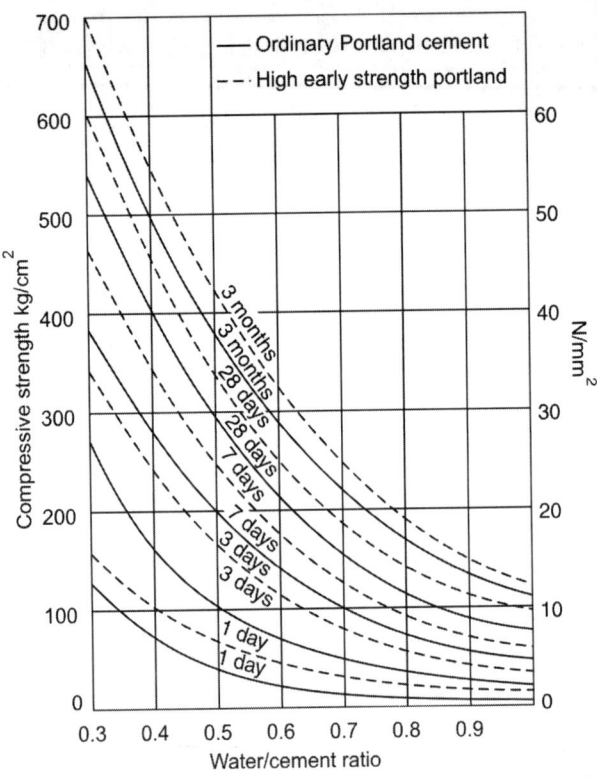

Fig. 5.1: Water/cement ratio and compressive strength relationship (15 cm dia × 30 cm height cylinder)

5.2 EFFECT OF SPECIMEN SIZE AND SHAPE ON COMPRESSIVE STRENGTH

The ultimate compressive strength of concrete and other brittle materials depends to a large extent on the size of specimen tested. For the same relative shape of specimen, the apparent strength of concrete and the variation of results decrease as the physical size of test specimen is increased[56]. This phenomenon is best explained in the light of Griffith's theory of Critical Flaws[57] and the weakest link theory[58].

At present, different types of specimens are used for determining the compressive strength of concrete in different countries. The common types of specimens used being cubes, cylinders and prisms. For crushing strength tests, cubes are used in United Kingdom, Germany, India and many other countries in Europe. Cylinders are the standard specimens in United States, Canada, Australia and New Zealand, while prismatic specimens are preferred in France. It is well established by various investigators[59–62] that for specimens with the same cross-section, the apparent strength of concrete decreases as the height of the test specimen is increased, up to a ratio of height to the least lateral dimension of about 2, where a lower limit appears to be reached. In test specimens like cubes where this ratio is unity, the apparent strength is highly influenced due to the frictional restraint between the ends of specimen and the platens of the testing machine[63].

The American cylinder which has a height/diameter ratio of 2 has been found to give a more accurate assessment of the uniaxial compression strength of concrete, than the cube, although the lateral stresses may have a small effect. The ratio of cylinder to cube strength depends primarily on the level of strength of concrete and is found to be higher for high strength concrete according to the data of Evans[64]. According to the British Standard, BS EN 12390–3:2009[65] and the Indian Standard IS: 456–2000, the cylinder strength may be taken as equal to 75 and 80 per cent of the cube strength, respectively. However, experiments have shown that there is no unique relation between the strengths of the specimens of the two shapes. Fig. 5.2 shows the relation between the cube and cylinder compressive strengths as recommended by Taylor[66]. The ratio of cylinder to cube strength approaches almost unity for very high strength concrete having a 28-day strength of 70 N/mm^2.

5.3 EFFECT OF AIR-ENTRAINMENT ON COMPRESSIVE STRENGTH OF CONCRETE

The strength of concrete is a direct function of its density ratio and the voids caused by entrained air will affect the strength in the same way as voids of any other origin. According to the data of Wright[35], when the air content is increased from zero to eight per cent the compressive strength of concrete is reduced linearly by as much as 40 per cent. However, air entrainment improves the workability of the concrete mix so that water/cement ratio may be correspondingly

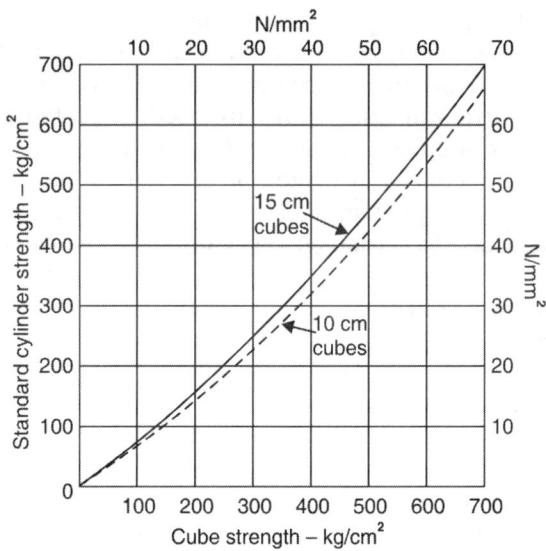

Fig. 5.2: Relation between cube and cylinder strength

reduced. For lean mixes with aggregate/cement ratio in range of 8 or more and particularly when angular aggregate is used, the improvement in workability due to air entrainment is such that the possible decrease in the water/cement ratio compensates fully for the loss of strength due to the presence of voids. The main purpose of air entrainment is to produce frost-resistant concrete and to improve its resistance to the destructive action of de-icing agents.

The exposure conditions are very important in the design of concrete mixes and in cases of severe exposure involving frequent alterations of freezing and thawing, the mixes have to be designed invariably with entrained air to improve its durability. The American Concrete Institute Standard, ACI 211.1–1991[67], recommends an average total air content varying from 1.0 to 7.5 per cent depending upon the exposure conditions and also the maximum size of aggregate used in the mix. The combined relation between compressive strength, water/cement ratio and percentage of entrained air is shown in Fig. 5.3, which is based on the work or Akroyd[68]. An air entrained mix is also less liable to segregate during mixing, transporting, placing and compacting and bleeding is considerably reduced resulting in an improved frost resistance of the top layer of a slab or lift.

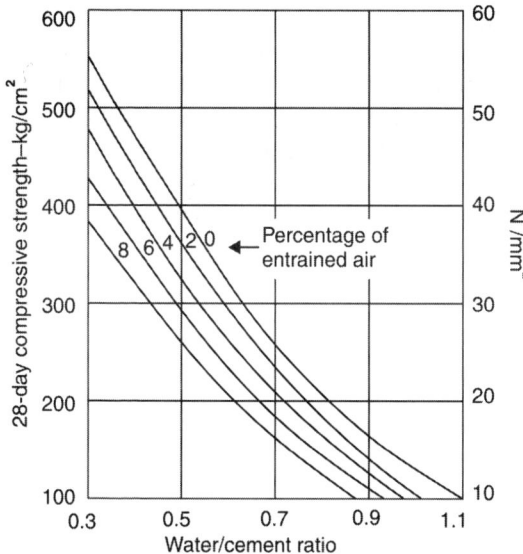

Fig. 5.3: Effect of air entrainment on compressive strength

The most dependable and accurate method of determining the air content is by the pressure method using commercial air metres. The method is ideally suited for site use and the test is covered by the ASTM Standard C: 231–10[69] and the corresponding Indian Standard IS: 1199–1959[70].

5.4 FLEXURAL STRENGTH OF CONCRETE

In the design of concrete mixes to be used in the construction of roads or airport pavements, the criterion of flexural strength is more important than the compressive strength. The flexural strength of concrete is determined by subjecting a plain concrete beam to flexural under transverse loads. The theoretical maximum-tensile stress reached in the bottom fibre of a standard test beam is often referred to as the modulus of rupture, the magnitude of which depends on the dimensions of the beam and the type of loading. The test data reported by Wright[71] indicates that two-point loading will yield a lower value of the modulus of rupture than when a single point load is applied at the centre of test beam. The modulus of rupture test is covered by the Indian Standard IS: 516–1959[72], the British Standard BS EN 12390–4:2009[65] and the ASTM Standard C: 78–10[73].

Several empirical formulas relating to the compressive and flexural strength of concrete have been suggested by various investigators. However, the number of parameters influencing the ratio of compressive to flexural strength being numerous, there is no simple relation which is generally applicable. The ratio of modulus of rupture to compressive strength has been found to decrease with increasing compressive strength of concrete according to the data reported by Price[74].

A knowledge of the flexural strength of concrete is useful in ascertaining the limit state of cracking in structural concrete elements since visible cracks are most likely to develop at stresses corresponding to that of the modulus of rupture.

5.5 DURABILITY CRITERION IN CONCRETE MIX DESIGN

In general, durability must always be defined in relation to definite exposure conditions and this is recognized in current codes which allow different depths of cover and different qualities of concrete for varying exposures. In the case of concrete which is to be exposed to severe wide range of adverse weather conditions, it is the durability requirement that will determine the water/cement ratio to be used in the mix since it is well established by various investigators[75-77], that water/cement ratio is an important factor in durability. The American Concrete Institute Standard ACI 211.1–1991[67] recommends certain maximum permissible water/cement ratios for use in different types of structures and exposure conditions (refer Table 7.3). It is widely accepted that concrete which will be exposed to a combination of moisture and freezing should be made with entrained air.

An excellent summary by Nurse[78] dealing with the assessment of concrete durability indicates the importance of a minimum cement content in the mix. The recent Indian Standard IS: 456–2000[6] Code[50] for structural concrete specifies minimum cement content required in Portland cement concretes to ensure durability under varying degrees of exposure classified as severe, moderate and mild. These recommendations are reproduced in Table 5.1 which gives the minimum cement content for plain and reinforced concrete work.

5.6 PRESCRIBED MIXES FOR ORDINARY CONCRETE

Nominal or prescribed mixes are generally preferred in the case of small jobs where the cost involved in exercising the quality control

Table 5.1: Minimum cement content, maximum W/C ratio, and minimum grade of concrete for different exposure conditions with normal weight aggregates of 20 mm nominal maximum size. (Table 5 of IS: 456–2000)

Exposure	Plain concrete		Reinforced concrete		Minimum grade of concrete	
	Minimum cement content (kg/m^3)	Maximum free W/C ratio	Minimum cement content (kg/m^3)	Maximum free W/C ratio	P.C.C	R.C.C
Mild	220	0.60	300	0.55		M-20
Moderate	240	0.60	300	0.50	M-15	M-25
Severe	250	0.50	320	0.45	M-20	M-25
Very severe	260	0.45	340	0.45	M-20	M-35
Extreme	280	0.40	360	0.40	M-25	M-40

Adjustments to minimum cement contents for aggregates other than 20 mm nominal maximum size. (Table 6 of IS: 456–2000)

Nominal maximum aggregate size (mm)	Adjustments to minimum cement contents
10	+40
20	0
40	−30

measures required for controlled concrete is not justifiable. The mix proportions are specified for a desired strength on the basis of weight. The nominal mix proportions for ordinary concrete recommended by the Indian Standard Code, IS: 456–2000[6] is shown in Table 5.2, which covers the compressive strength range of concrete from 5 to 20 N/mm² at 28 days. The nominal mixes came into use with the growth of empirical knowledge or concrete, for it was found that if concrete was mixed with sufficient water to make it plastic and easy to handle, then it could be expected to achieve the required strength.

The new British Euro Code, BS EN: 1992–1–1:2004[50] also recommends prescribed mixes for ordinary structural concrete. The code recognizes 5 grades of concrete ranging from a minimum of 70 to a maximum of 300 kg/cm². The weights of ingredients required to produce approximately one cubic metre of fully compacted concrete, for different nominal maximum sizes of aggregates are

Table 5.2: Proportions for nominal mix concrete

Grade of concrete	Total quantity of dry aggregates by mass per 50 kg of cement, to be taken as the sum of the individual masses of fine and coarse aggregates (kg) max	Proportion of fine aggregate or coarse aggregate (by mass)	Quantity of water per 50 kg of cement (max) (litres)
1	2	3	4
M-5	800	Generally 1:2 but	60
M-7.5	625	subject to an upper	45
M-10	480	limit of 1:1½ and a	34
M-15	330	lower limit of 1:2½	32
M-20	250		30

reproduced in Table 5.3. No recommendations are made regarding the quantity of water required for the mixes. However, the degree of workability and the limits to slump that may be expected are specified for each of the nominal maximum sizes of aggregates varying from 10 to 40 mm.

The ACI-Standard ACI: 211.1–1991 recommends arbitrary mix proportions for small jobs where time and personnel are not available to determine the proportions in accordance with the recommended procedures. The mixes specified in Table 5.4 are based on concrete that has just enough water in it to permit ready working into forms without any segregation. For each maximum size of aggregate ranging from 12.5 to 50 mm, three different mixes are specified which help in deciding upon a suitable mix for a small job. In contrast to the Indian and British specifications, the American Standard does not indicate the range of compressive strength of concrete that could be expected with the specified mixes.

A survey of the various nominal mixes indicates that they result in a wastage of materials, for only part of the potential strength of the concrete is utilized. The strength specified is not very high considering the richness of the mixes. In this connection, the British Code recommendations are by far the most comprehensive since they not only specify a wide range of strengths but also include the workability, aggregate/cement ratio and grading characteristics of

Table 5.3: Prescribed mixes for ordinary structural concrete

Concrete grade kg/cm² (N/mm²)		Nominal max. size of aggregate (mm)							
		Weights of ingredients to produce approximately one cube metre of fully compacted concrete (kg)							
		40		20		14		10	
		Medium	High	Medium	High	Medium	High	Medium	High
	Workability limits to slump that may be expected (mm)	50–100	100–150	25–75	75–125	10–50	50–100	10–25	25–50
70	Cement (kg)	180	200	210	230	—	—	—	—
(7)	Total aggregate (kg)	1950	1850	1900	1800	—	—	—	—
	Fine aggregate (%)	30–45	30–45	30–45	30–50	—	—	—	—
100	Cement (kg)	210	230	240	260	—	—	—	—
(10)	Total aggregate (kg)	1900	1850	1850	1800	—	—	—	—
	Fine aggregate (%)	30–45	30–45	45–50	35–50	—	—	—	—
150	Cement (kg)	250	270	280	310	—	—	—	—
(15)	Total aggregate (kg)	1850	1800	1800	1750	—	—	—	—
	Fine aggregate (%)	30–45	30–45	30–50	35–70	—	—	—	—
200	Cement (kg)	300	320	320	350	340	380	360	410
(20)	Total aggregate (kg)	1850	1750	1800	1750	1750	1700	1750	1650
	Sang: Zone 1	35	40	40	45	45	50	50	55
	(%) Zone 2	30	35	35	40	40	45	45	50
	Zone 3	30	30	30	35	35	40	40	45

(Contd...)

(Contd...)

Weights of ingredients to produce approximately one cube metre of fully compacted concrete (kg)

Concrete grade kg/cm² (N/mm²)	Nominal max. size of aggregate (mm)	40		20		14		10	
	Workability limits to slump that may be expected (mm)	Medium 50–100	High 100–150	Medium 25–75	High 75–125	Medium 10–50	High 50–100	Medium 10–25	High 25–50
250 (25)	Cement (kg)	340	360	360	390	380	420	400	450
	Total aggregate (kg)	1800	1750	1750	1700	1700	1650	1700	1600
	Sand: Zone 1	35	40	40	45	45	50	50	55
	(%) Zone 2	30	35	35	40	40	45	45	50
	Zone 3	30	30	30	35	35	40	40	45
300 (30)	Cement (kg)	370	390	400	430	430	470	460	510
	Total aggregate (kg)	1750	1700	1700	1650	1700	1600	1650	1550
	Sand: Zone 1	35	40	40	45	45	50	50	55
	(%) Zone 2	30	35	35	40	40	45	45	50
	Zone 3	30	30	30	35	35	40	40	45

Table 5.4: Concrete mixes for small jobs*

Maximum size of aggregate (mm)	Mix desig- nation	Approximate quantity of cement per m³ of concrete (kg)	Aggregate, kg per 1-bag batch (50 kg)			
			Sand†		Gravel or crushed stone	Iron blast furnace slag
			Air- entrained concrete††	Concrete without air		
12.5	A	395	125	130	91	77
	B	390	120	125	101	88
	C	385	120	125	109	96
19	A	373	120	125	120	104
	B	361	120	125	130	114
	C	345	109	114	154	136
25	A	361	120	125	130	112
	B	350	114	120	146	128
	C	345	109	114	154	136
38	A	340	120	125	154	130
	B	327	114	120	170	146
	C	322	109	114	183	160
50	A	322	120	125	176	144
	B	316	114	120	192	160
	C	305	109	114	202	170

Note: Air-entrained concrete should be used in all structures which will be exposed to alternate cycles of freezing and thawing

* May be used without adjustment

† Weights are for dry sand. If damp sand is used, increase weight of sand by 5.31 kg for 1-bag batch sand, if very wet and is used, add 10.692 kg for 1 bag batch

†† Air-entrained concrete can be obtained by the use of an air-entraining cement or by adding an air-entraining agent. If an agent is used, the amount recommended by the manufacturer will, in most cases, produce the desired air content

Procedure: Select the proper maximum size of aggregate and then using mix B, add just enough water to produce a sufficiently workable consistency. If the concrete appears to be undersanded, use mix A, and if it appears to be oversanded, use mix C

the aggregates. Nominal mixes are still considered suitable for proportioning concrete for small jobs as long as the concrete is mixed with only sufficient water to produce a plastic workable mix. However, for important jobs requiring large quantities of concrete of high quality, only controlled concrete should be used and the design of such concretes will be discussed in the subsequent chapters.

Design of Low and Medium Strength Concrete Mixes According to Road Note No. 4

6.1 BASIS OF THE METHODS

The method of designing concrete mix proportions according to the Road Note No. 4 is mainly based on the experimental investigations by Glanville et al[26], dealing with the effect of aggregate grading on the strength and workability of concrete. It is important to note that designed concrete must be satisfactory both in the plastic, as well as in the hardened state and necessarily the choice of mix proportions is governed by both these conditions. The required water/cement ratio to produce the design characteristic strength depends upon the characteristics of the cement. By using the same water/cement ratio, different strengths are produced by ordinary Portland, rapid hardening Portland and high alumina cements. Since the strength is, more or less, proportional to water/cement ratio for a given type of cement, the Road Note No. 4 presents a set of curves relative water/cement ratio and compressive strength at different ages and for different types of cements.

The workability of concrete is categorised as very low, low, medium and high, and this can be selected to suit the placing conditions, type of construction and method of compaction. A rough indication of the degree of workability is obtained by the corresponding values of the slump and compacting factor of the concrete measured at the worksite. Extensive laboratory investigations were conducted at the Road Research Laboratory to study the relation among the various parameters, like aggregate/cement ratio, water/cement ratio, degree of workability, grading, and shape and size of aggregates. Based on the result, design tables are presented in Road

Note No. 4 to select suitable aggregate/cement ratios for a concrete mix, after fixing up the other influencing parameters. If the locally available aggregate does not conform to the standard grading, the finer and coarser fractions of the aggregate can suitably be combined to obtain the desired standard grading.

6.2 MIX DESIGN PROCEDURE

The sequences of operations to be followed in designing a concrete mix of low and medium strengths according to Road Note No. 4 are detailed below:

(a) The average compressive strength of the concrete mix to be designed is obtained by applying the control factors to the minimum compressive strength. The relation between the minimum and average compressive strengths depends upon the degree of quality control and is listed in Table 6.1.

Table 6.1: Estimated relation between the minimum and mean compressive strengths of site cubes (Road Note No. 4)

Conditions	Minimum strength as a percentage of average strength
Very good control with weigh batching. Use of graded aggregates, moisture determination of aggregate, etc. Constant supervision	75
Fair control with weigh batching. Use of two sizes of aggregate only. Water content left to mixer driver's judgement. Occasional supervison	60
Poor control, inaccurate volume batching of all-in aggregates. No supervison	40

(b) The water/cement ratio needed to give the necessary average compressive strength, at the desired age and for the type of cement used, can be read off directly from Fig. 6.1.

(c) The proportion of combined aggregate to cement is determined from Tables 6.2 and 6.3 for aggregates of maximum size 40 and 20 mm, respectively. The tables provide for four different degrees of workability detailed in Table 3.1 and these correspond to the four standard gradings discussed in section 2.2. Aggregate/cement ratios for aggregates of maximum size 10 mm, compiled in Table 6.4, are based on the work of McIntosh and Erntory[23]. The data provided is applicable to gravel and crushed rock aggregates.

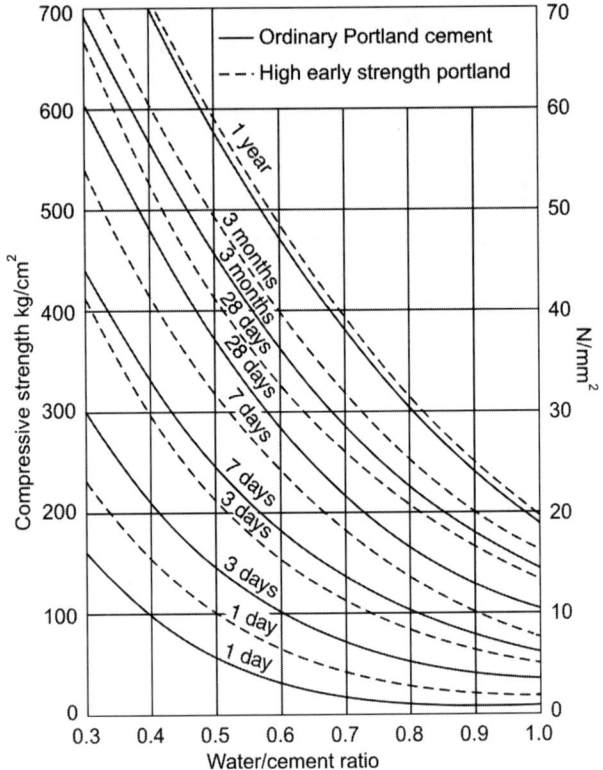

Fig. 6.1: Relation between compressive strength and water/cement ratio for 10 cm cubes of fully compacted concrete

(d) If the aggregate available at the works site differs from the standard gradings, the coarse and fine aggregates must be combined so as to produce one of the standard gradings.

(e) The proportion of cement, water, fine and coarse aggregates is determined, having a knowledge of the water and aggregate/cement ratios of the mix.

(f) The quantities of ingredients required to produce one cubic metre of concrete are calculated by the absolute volume method, using the specific of cement and aggregates.

6.3 METHOD OF COMBINING AGGREGATES

Generally, the aggregates quarried and stockpiled at works site do not conform to the standard gradings. In such cases, it is desirable to

Table 6.2: Aggregate/cement ratio required to give four degrees of workability with different water/cement ratios and gradings

(a) 40 mm Rounded gravel aggregate

Degree of workability (Table 3.1)	Very low				Low				Medium				High			
Grading curve number (Fig. 2.1)	1	2	3	4	1	2	3	4	1	2	3	4	1	2	3	4
0.35	5.0	4.5	3.9	3.4	4.3	3.9	3.5	3.1	3.4	3.1	2.9	2.7	—	—	—	—
0.40	7.6	6.5	5.7	4.9	5.9	5.6	5.0	4.4	4.7	4.6	4.3	3.8	4.1	4.0	3.9	3.5
0.45	8.9	8.6	7.7	6.5	7.6	7.4	6.7	5.8	6.0	6.1	5.7	5.0	5.2	5.3	5.0	4.6
0.50				8.0			8.2	7.2	7.5	7.6	7.1	6.3	6.3	6.5	6.2	6.7
0.55								8.4		8.9	8.1	7.3	S	7.7	7.4	6.7
6.0																7.6

Total water/cement ratio by weight

S Indicates that the mix would segregate

Note: These values have been obtained by extrapolation of other data and are not based directly on the results of trial mixes

(Contd...)

(Contd...)

(b) 40 mm Irregular gravel aggregate

Total water/cement ratio by weight	Degree of workability (Table 3.1)															
	Very low				Low				Medium				High			
Grading curve number (Fig. 2.1)	1	2	3	4	1	2	3	4	1	2	3	4	1	2	3	4
0.35	4.0	3.9	3.6	3.2	3.4	3.3	3.2	2.9	3.8	3.8	3.7	3.4	3.4	3.5	3.3	3.1
0.40	5.3	5.3	4.8	4.3	4.5	4.5	4.2	3.8	4.6	4.7	4.6	4.3	4.1	4.4	4.3	4.0
0.45	6.5	6.5	6.0	5.3	5.6	5.6	5.3	4.8	5.5	5.7	5.5	5.1	4.8	5.2	5.1	4.8
0.50	7.7	7.7	7.1	6.3	6.6	6.6	6.3	6.7	6.2	6.5	6.3	5.9	S	5.9	5.9	5.5
0.55			8.1	7.3	7.6	7.6	7.2	6.6	7.0	7.3	7.1	6.6	S	S	6.7	6.3
0.60								7.4	7.8	8.1	7.8	7.3			7.3	6.9
0.65								8.1				7.9				7.4
0.70																8.0
0.75																

S indicates that the mix would segregate

(Contd....)

(Contd...)

(c) 40 mm Crushed rock aggregate

Degree of workability (Table 3.1)	Very low				Low				Medium				High			
Grading curve number (Fig. 2.1)	1	2	3	4	1	2	3	4	1	2	3	4	1	2	3	4
0.35	3.4	3.4	3.2	2.9												
Total 0.40	4.9	4.6	4.2	3.8	4.0	3.8	3.6	3.3	3.3	3.3	3.2	3.0	3.1	3.1	2.9	2.7
water/ 0.45	6.0	5.7	5.2	4.7	4.9	4.7	4.4	4.2	4.1	4.1	3.9	3.8	3.7	3.8	3.7	3.4
cement ratio 0.50	7.2	6.8	6.2	5.6	5.8	5.6	5.3	5.0	4.8	4.8	4.7	4.6	4.4	4.5	4.5	4.2
by weight 0.55	8.1	7.7	7.1	6.4	6.6	6.4	6.1	5.8	5.5	5.5	5.4	5.3	S	5.2	5.2	4.8
0.60		8.6	8.0	7.2	7.4	7.2	6.9	6.6	6.1	6.2	6.1	6.0		S	5.9	5.6
0.65			8.8	7.9	8.1	7.9	7.6	7.3	S	6.9	6.8	6.6			6.5	6.2
0.70				8.6		8.5	8.3	7.9		7.5	7.5	7.3			7.1	6.8
0.75				8.5			8.1	7.8								7.4

S indicates that the mix would segregate

Note: These values have obtained by extrapolation of other data and are not based directly on the results of trial mixes

Table 6.3: Aggregate/cement ratio required to give four degrees of workability with different water/cement ratios and gradings

(a) 20 mm Rounded aggregate

Degree of workability (Table 3.1)	Very low				Low				Medium				High			
Grading curve number (Fig. 2.2)	1	2	3	4	1	2	3	4	1	2	3	4	1	2	3	4
Water/cement ratio by weight 0.35	4.5	4.5	3.5	3.2	3.8	3.6	3.2	3.1	3.1	3.0	2.8	2.7	2.8	2.8	2.6	2.5
0.40	6.6	6.3	5.3	4.5	5.3	5.1	4.5	4.1	4.2	4.2	3.9	3.7	3.6	3.7	3.5	3.3
0.45	8.0	7.7	6.7	5.8	6.9	6.6	5.9	5.1	5.3	5.3	5.0	4.5	4.6	4.8	4.5	4.1
0.50	–	–	8.0	7.0	8.2	8.0	7.0	6.0	6.3	6.3	5.9	5.4	5.5	5.7	5.3	4.8
0.55	–	–	–	8.1	–	–	8.2	6.9	7.3	7.3	7.4	6.4	6.3	6.5	6.1	5.5
0.60	–	–	–	–	–	–	–	7.7	–	–	8.0	7.2	x	7.2	6.8	6.1
0.65	–				–	–	–	8.3	–	–	–	7.8	x	7.7	7.4	6.6
0.70					–	–	–	–	–	–	–	–	x	–	7.9	7.2
0.75					–				–	–	–	–	x	–	–	7.6
0.80													x	–	–	–
0.85													x	–	–	–
0.90																

– Indicates that the mix was outside the range tested

x Indicates that the mix would segregate

Note: These proportions are based on specific gravities of approximately 2.5 for the coarse aggregate and 2.6 for the fine aggregate

(Contd...)

(Contd...)

(a) 20 mm Irregular aggregate

Degree of workability (Table 3.1)	Very low				Low				Medium				High			
Grading curve number (Fig. 2.2)	1	2	3	4	1	2	3	4	1	2	3	4	1	2	3	4
Water/cement ratio by weight																
0.35	3.7	3.7	3.5	3.0	3.0	3.0	3.0	2.7	2.6	2.6	2.7	2.4	2.4	2.5	2.5	2.2
0.40	4.8	4.7	4.7	4.0	3.9	3.9	3.8	3.5	3.3	3.4	3.5	3.2	3.1	3.2	3.2	2.9
0.45	6.0	5.8	5.7	5.0	4.8	4.8	4.6	4.3	4.0	4.1	4.2	3.9	x	3.9	3.9	3.5
0.50	7.2	6.8	6.5	5.9	5.5	5.5	5.4	5.0	4.6	4.8	4.8	4.5	x	4.4	4.4	4.1
0.55	8.3	7.8	7.3	6.7	6.2	6.2	6.0	5.7	x	5.4	5.4	5.1	x	4.8	4.9	4.7
0.60	9.4	8.6	8.0	7.4	6.8	6.9	6.7	6.2	x	6.0	6.0	5.6	x	x	5.4	5.2
0.65	–	–	–	8.0	7.4	7.5	7.3	6.8	x	x	6.4	6.1	x	x	5.8	5.6
0.70	–	–	–	–	8.0	0.8	7.7	7.4	x	x	6.8	6.6	x	x	6.2	6.1
0.75					–	–	–	7.9	x	x	7.2	7.0	x	x	6.6	6.5
0.80					–	–	–	–	x	x	7.5	7.4	x	x	x	7.0
0.85									x	x	7.8	7.8	x	x	x	7.4
0.90									x	x	x	8.1	x	x	x	7.7
0.95									x	x	x	–	x	x	x	8.0
1.00									x	x			x	x	x	x

(Contd...)

(Contd...)

(a) 20 mm Angular aggregate

Water/cement ratio by weight (Grading curve number (Fig. 2.2))	Very low				Low				Medium				High			
Degree of workability (Table 3.1)	1	2	3	4	1	2	3	4	1	2	3	4	1	2	3	4
0.35	3.2	3.0	2.9	2.7	2.7	2.7	2.5	2.5	2.4	2.4	2.3	2.2	2.2	2.3	2.1	2.1
0.40	4.5	4.2	3.7	3.5	3.5	3.5	3.2	3.0	3.1	3.1	2.9	2.7	2.9	2.9	2.8	2.6
0.45	5.5	5.0	4.6	4.3	4.3	4.2	3.9	3.7	3.7	3.7	3.4	3.3	3.5	3.5	3.2	3.1
0.50	6.5	5.8	5.4	5.0	5.0	4.9	4.5	4.3	4.2	4.2	3.9	3.8	x	3.9	3.8	3.5
0.55	7.2	6.6	6.0	5.6	5.7	5.4	5.0	4.8	4.7	4.7	4.5	4.3	x	x	4.3	4.0
0.60	7.8	7.2	6.6	6.3	6.3	6.0	5.6	5.3	x	5.2	4.9	4.8	x	x	4.7	4.4
0.65	8.3	7.8	7.2	6.9	6.9	6.5	6.1	5.8	x	5.7	5.4	5.2	x	x	5.1	4.9
0.70	8.7	8.3	7.7	7.5	7.4	7.0	6.5	6.3	x	6.2	5.8	5.7	x	x	5.5	5.3
0.75	—	—	8.2	8.0	7.9	7.5	7.0	6.8	x	x	6.2	6.1	x	x	5.8	5.7
0.80			—	—	—	—	7.4	7.2	x	x	6.6	6.5	x	x	6.1	6.0
0.85					—	—	7.8	7.6	x	x	7.1	6.9	x	x	6.4	6.3
0.90					—	—	—	—	x	x	7.5	7.3	x	x	x	6.7
0.95									x	x	8.0	7.6	x	x	x	7.0
1.00									x	x	—	—	x	x	x	7.3

These proportions are based on specific gravities of approximately 2.7 for both coarse and fine aggregates

Table 6.4: Aggregate/cement ratio required to give four degrees of workability with different water/cement ratios and gradings

(a) 10 mm Rounded aggregate

Degree of workability (Table 3.1)	Very low				Low				Medium				High			
Grading curve number (Fig. 2.3)	1	2	3	4	1	2	3	4	1	2	3	4	1	2	3	4
0.40	5.6	5.0	4.2	3.2	4.5	3.9	3.3	2.6	3.9	3.5	3.0	2.4	3.5	3.2	2.8	2.0
0.45	7.2	6.4	5.3	4.1	5.5	4.9	4.1	3.2	4.7	4.3	3.7	3.0	4.2	3.9	3.4	2.9
0.50		7.8	6.4	4.9	6.5	5.8	4.9	3.8	5.4	5.0	4.3	3.5	4.8	4.5	4.0	3.4
0.55			7.5	5.7	7.4	6.7	5.7	4.4	6.1	5.7	4.9	4.0	5.3	5.1	4.5	3.9
0.60				6.5		7.5	6.4	5.0	6.7	6.3	5.5	4.5	5.8	5.6	5.0	4.3
0.65				7.2			7.1	5.6	7.3	6.9	6.1	5.0	S	6.1	5.5	4.7
0.70							7.7	6.2	7.9	7.5	6.7	5.5		6.6	6.0	5.1
0.75								6.7			7.2	5.9		7.1	6.5	5.5
0.80								7.2			7.7	6.3		7.6	6.9	5.9
0.85												6.8			7.3	6.3
0.90												7.2			7.7	6.7
0.95																7.0
1.00																7.3

S indicates that the mix would segregate

(Contd...)

(Contd...)

(b) 10 mm Irregular gravel aggregate

Degree of workability (Table 3.1)	Very low				Low				Medium				High			
Grading curve number (Fig. 2.3)	1	2	3	4	1	2	3	4	1	2	3	4	1	2	3	4
0.40	4.1	3.8	3.3	2.8	3.3	3.1	2.8	2.3	3.5	3.4	3.2	2.8	3.2	3.1	3.0	2.7
0.45	5.1	4.8	4.3	3.6	4.1	3.9	3.5	3.0	4.2	4.1	3.8	3.4	S	3.8	3.6	3.2
0.50	6.1	5.8	5.2	4.4	4.8	4.6	4.2	3.7	S	4.7	4.4	4.0		4.4	4.2	3.7
0.55	7.0	6.7	6.1	5.2	5.5	5.3	4.9	4.3		5.3	5.0	4.5		4.9	4.7	4.2
0.60	7.9	7.6	7.0	6.0	S	6.0	5.6	4.9		5.9	5.6	5.0		5.4	5.2	4.6
0.65			7.8	6.8		6.6	6.2	5.5		6.4	6.1	5.5		5.9	5.7	5.0
0.70						7.2	6.8	6.1		6.9	6.6	6.0		6.4	6.1	5.4
0.75						7.8	7.4	6.7		7.4	7.1	6.4		6.8	6.5	5.8
0.80							8.0	7.3		7.9	7.5	6.8		7.2	6.9	6.2
0.85											8.0	7.2		7.6	7.3	6.6
0.90														S	7.7	6.9
0.95															8.0	7.2
1.00																

Water/cement ratio by weight

S indicates that the mix would segregate

(Contd...)

(Contd...)

(c) 10 mm Crushed rock aggregate

Degree of workability (Table 3.1)	Very low				Low				Medium				High			
Grading curve number (Fig. 2.3) → / Water/cement ratio by weight ↓	1	2	3	4	1	2	3	4	1	2	3	4	1	2	3	4
0.40	3.7	3.3	2.8	2.0	3.8	3.6	3.0	2.2	3.3	3.1	2.7	2.1	S	3.2	2.9	2.4
0.45	4.5	4.1	3.5	2.6	4.4	4.2	3.6	2.7	3.8	3.7	3.2	2.6		3.7	3.4	2.8
0.50	5.2	4.9	4.2	3.2	4.9	4.8	4.2	3.2	S	4.2	3.7	3.0		4.2	3.8	3.2
0.55	5.9	5.6	4.9	3.8	S	5.3	4.7	3.7		4.7	4.2	3.4		4.6	4.2	3.6
0.60	6.6	6.3	5.5	4.3		5.8	5.2	4.2		5.1	4.6	3.8		5.0	4.6	4.0
0.65	7.3	7.0	6.1	4.8		6.3	5.7	4.6		5.6	5.1	4.2		5.4	5.0	4.4
0.70	7.9	7.6	6.7	5.3		6.8	6.2	5.0		6.0	5.5	4.6		5.8	5.4	4.7
0.75			7.3	5.8		7.2	6.6	5.5		6.4	5.9	5.0		6.1	5.8	5.1
0.80			7.8	6.3		7.6	7.1	6.0		6.7	6.3	5.4		6.4	6.1	5.4
0.85				6.8			7.5	6.4		7.1	6.7	5.8		6.7	6.5	5.7
0.90				7.3			7.9	6.8		7.5	7.1	6.1		7.0	6.7	6.1
0.95								7.2		7.8	7.5	6.5		7.3	7.0	6.4
1.00											7.8	6.9		7.6	7.3	6.7
1.05												7.2		S	7.6	7.0
1.10															7.9	7.3
1.15																
1.20																

S indicates that the mix would segregate

proportion the available materials in a such a way that the grading of the combined aggregate corresponds to one of the four standard grading curves. This can be done by analytical calculations or graphically using the method suggested in Road Note No. 4.

6.3.1 Analytical Method

This method is best illustrated by a numerical example. The gradings of fine and coarse aggregates available at work sites are detailed in Table 6.5. The fine and coarse aggregates of 20 mm maximum size have to be combined so as to approximate to the coarsest grading of the standard grading curve no. 1 (Fig. 2.2), the most economical mix having the highest permissible aggregate/cement ratio. On this curve, 30 per cent of the total aggregates passes the 4.75 mm, IS sieve.

If $x : y$ are the proportions of fine and coarse aggregates in the combined state, then to satisfy the condition that 30 per cent of the combined aggregate passes the 4.75 mm IS sieve, we have

$$1.0\, x + 0.07\, y \ = \ 0.3\,(x + y)$$

or $\qquad x : y \ = \ 1 : 3$

Hence the fine and coarse aggregates have to be combined in the proportions of 1:3. The grading of the resulting combined aggregates is determined by multiplying columns (a) and (b) of Table 6.5 by 1.0 and 3.0 respectively and dividing the sum of these products by 4.0. The resulting combined grading is shown in column (f), the values being rounded off to the nearest per cent. In comparison with the standard grading curve shown in column (g), the combined grading is found to be in good agreement, except for minor deviations.

6.3.2 Graphical Method

The graphical method of combining the coarse and fine aggregates to conform to a standard grading is illustrated in Fig. 6.2 for the same aggregates used in the previous section. In this method, the gradings of the fine and coarse aggregates are marked off along the vertical axis by marking points and numbering them with the sieve size or number such that the ordinate of each point represents the percentage of material passing the corresponding sieve. Points representing the sieve size or number of the left hand and right hand axes corresponding to the fine and coarse aggregates respectively are joined by straight lines. A vertical line is drawn through the point where the sloping line representing 4.75 mm IS sieve intersects the horizontal line representing the percentage of material passing the

Table 6.5: Analytical method of combining aggregates to obtain a type grading

IS sieve size	Percentage passing Fine aggregate	Coarse aggregate	Col. (a) X 1.00	Col. (b) X 3.00	Col. (c) + Col. (d)	Grading of combined aggregate Col. (e) ÷ 4	Standard grading curve no (Fig. 2.2)
	(a)	(b)	(c)	(d)	(e)	(f)	(g)
20 mm	100	100	100	300	400	100	100
10 mm	100	31	100	93	193	48	45
4.75 mm	100	7	100	21	121	30	30
2.36 mm	92	0	92	0	92	23	23
1.18 mm	76	0	76	0	76	19	16
600 micron	48	0	48	0	48	12	9
300 micron	20	0	20	0	20	5	2
250 micron	3	0	3	0	3	1	0

4.75 mm IS sieve required in the combined grading. In the present problems, this value is 30 per cent for the coarsest standard grading curve no. 1 (Fig. 2.2). The ordinates of the intersections of the combined aggregates line with the sloping lines represent the grading of the combined aggregate as shown by the dotted line in Fig. 6.2. The values are found to be more or less the same as the obtained by the analytical method and it should be observed that both the methods are approximations based on quantities passing two specific sieve sizes.

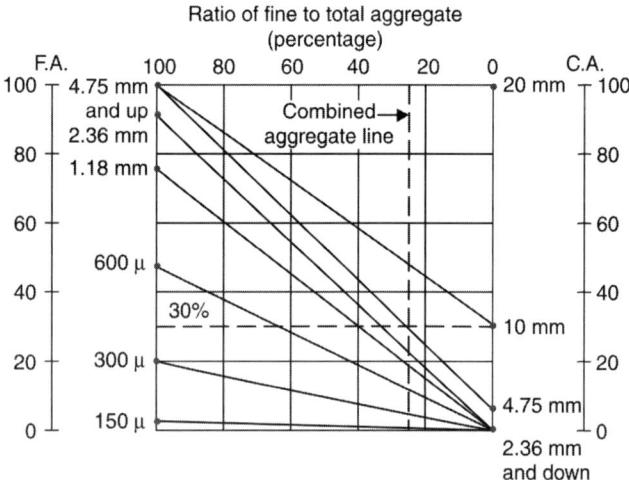

Fig. 6.2: Graphical method of combining coarse and fine aggregates

6.4 DETERMINATION OF THE WEIGHTS OF INGREDIENTS BY ABSOLUTE VOLUME METHOD

When the mix proportions have been determined based on the knowledge of water/cement, aggregate/cement ratios and the relative proportions of the aggregates of various sizes, the weight of cement required to produce one cubic metre of concrete can be calculated using the absolute volume method. The principle underlying the method being that the volume of compacted concrete is equal to the sum of the absolute volumes of all ingredients.

If C = weight of cement required per cubic metre of concrete

ρ_c = specific gravity of cement

ρ_{fa} = specific gravity of fine aggregate

ρ_{ca} = specific gravity of coarse aggregate

v = percentage of entrained air in concrete

ρ_w = density of water

and the proportions by weight of materials;

Cement : Fine aggregate : Coarse aggregate : Water is as

1 : N_f : N_c : w

then the weight of cement 'C' is evaluated by the relation:

$$\frac{C}{\rho_c \, \rho_w} + \frac{N_f.C}{\rho_{fa} \, \rho_w} + \frac{N_c.C}{\rho_{ca} \, \rho_w} + \frac{W.C.}{\rho_w} + 0.01 \, v = 1 \, m^3$$

The weights of water, and fine and coarse aggregates required for one cubic metre of concrete are obtained as the product of cement C and the corresponding proportion by weight of the individual materials. If any admixtures are used as additional ingredients in the concrete mix, similar terms may be added to the equation.

6.5 LIMITATIONS OF ROAD NOTE NO. 4 METHOD

The design tables of aggregate/cement ratios recommended in Road Note No. 4 cover only three shapes of aggregates and four types of grading, but in practice many aggregates having different shapes, sizes, and properties have to be used for the mixes. At best, the method should be considered as a guide to select the mix proportions since it is strictly applicable only to the actual aggregates used in their derivation. In addition, the data of Tables 6.2 and 6.3 are based on water in excess of that absorbed by the aggregate, while that in Table 6.4, refers to the total water added to dry air aggregate. In view of this, due care must be exercised in using the tabulated data and suitable adjustments are made for water contained in the aggregates. The design tables are only applicable for aggregates of given specific gravity ρ indicated in the tables. If aggregates having a specific gravity ρ_1 are used, the aggregate/cement ratios should be multiplied by the ratio ρ_1/ρ, since the relations developed are based on the gross apparent volume of the solid particles.

It is important to note that in particular jobs, the water/cement ratio selected for the mix should be suitable from the stand-point of durability rather than strength. Due care must be exercised in such cases since no recommendations are made in this regard in Road Note No. 4. The design tables refer to mixes in which the coarse and fine aggregates are of the same shape. In cases where rounded river sand is used with irregular gravel or crushed rock coarse aggregates, interpolation between the values of the relevant tables will be necessary.

The Road Note No. 4 data cannot be used directly for the design of air-entrained concrete. However, the nominal mix designed by using the data can be suitably adjusted for the desired air content in the mix.

6.6 MIX DESIGN EXAMPLES

6.6.1 Design a Concrete Mix to Suit the Following Data Using the Method of Road Note No. 4

Specified works cube strength: 200 kg/cm^2 at 28 days.

Degree of control: Very good with weigh batching and constant supervision.

Degree of workability: High, since concrete is required for casting the junctions of columns and beams with congested reinforcement.

Type of cement: Ordinary Portland

Type of fine aggregate: Natural sand

Type of coarse aggregate: Crushed granite

(Angular aggregate) of 20 mm maximum size. The aggregates available at works site have the following grading:

I.S. sieve size	Percentage passing	
	Coarse aggregate	Fine aggregate
20 mm	100	–
10 mm	31	–
4.75 mm	7	–
2.36 mm	–	100
1.18 mm	–	92
600 micron	–	76
300 micron	–	48
150 micron	–	3

The specific gravity and bulk-density of the various ingredients of the mix are as follows:

Material	Specific gravity	Dry bulk-density (kg/m^3)
Cement	3.15	1472
Coarse aggregate	2.60	1520
Fine aggregate	2.60	1680

Design the concrete mix and set out field mix proportions for weigh batching and volume batching. Also calculate the quantities of materials required 1 cubic metre of concrete.

    (a) By weight

    (b) By volume

Design of concrete mix

For the degree of control used (Table 6.1), minimum strength is 75 per cent of the average strength.

$$\text{Mean design strength} = \frac{200}{0.75} = 270 \text{ kg/cm}^2$$

Water/cement ratio (from Fig. 6.1) = 0.62

For high workability, the required

aggregate/cement ratio (from Table 6.3) = 4.8

The given aggregates do not conform to any particular standard grading curve. They should be combined in suitable proportions to correspond to the nearest practical grading, which in this case is grading 3. The fine and coarse aggregates are combined graphically as shown in Fig. 6.3.

The proportion of fine to total aggregate = 40 per cent

Hence the proportions by weight of ingredients are given by

Cement	:	Fine aggregate	:	Coarse aggregate
1	:	$\dfrac{4.8 \times 60}{100}$	:	$\dfrac{4.8 \times 60}{100}$
2	:	1.92	:	2.88

Proportion by volume

1	:	$\dfrac{1.92 \times 1472}{1680}$	:	$\dfrac{2.88 \times 1472}{1520}$
or 1	:	1.68	:	2.78

Quantity of materials required for 1 m³ of concrete

If C = weight of cement required per m³ of concrete

Then by absolute volume method, we have

$$\frac{C}{3.15 \times 10^3} + \frac{1.92\,C}{2.6 \times 10^3} + \frac{2.88\,C}{2.6 \times 10^3} + \frac{0.62\,C}{10^3} = 1$$

∴ C   = 355 kg

Cement = 355 kg

Water   = 0.62 × 355   = 680 kg

FA $\quad = \quad 1.92 \times 355 \quad = \quad 1020$ kg

Density of fresh concrete $\quad = \quad 2275$ kg/m^3

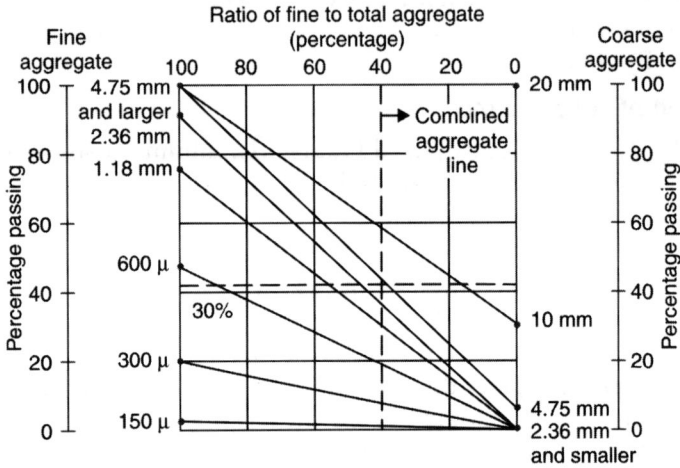

Fig. 6.3: Graphical method of combining aggregates

Material Quantities by Volume

				m^8
Cement	=	355/1472	=	0.231
Water	=	220/1000	=	0.220
FA	=	680/1680	=	0.405
CA	=	1020/1520	=	0.674

6.6.2 Adjustment of Mix Proportions for Water Contained in Aggregates

In the previous example, the quantities of ingredients required for the concrete mix were computed based on dry aggregates. Generally, in the field, the aggregates contain some moisture depending upon the storing conditions, which can be ascertained by tests. Having a knowledge of the moisture content in the aggregates, it is possible to adjust the mix proportions to yield one cubic metre of concrete using wet aggregates. If the coarse and fine aggregates of the previous example contain 3 and 6 per cent of moisture by weight, respectively and the bulking of fine aggregate is 25 per cent, the adjustments necessary for mix proportions based on dry aggregates are detailed in Tables 6.6 and 6.7 for weigh batching and volume batching, respectively.

Table 6.6: Adjustments for weigh batching

1	2	3	4	5	6	7	8
Fine aggregate			Coarse aggregate			Water	
Percent of water by dry weight	Total amount of water contained in aggregate kg	Weight of wet aggregate required kg	Percent of water by dry weight	Total amount of water contained in aggregate kg	Weight of wet aggregate required kg	Total weight of water in combined aggregate kg	Weight of water require to be added kg
6	$\dfrac{680\times6}{100}$ $=41$	$(680+41)$ $=721$	3	$\dfrac{1020\times3}{100}$ $=31$	$(1020+31)$ $=1051$	$(41+31)$ $=72$	$(220-72)$ $=148$

Quantities of materials required for 1 m³ of concrete

Dry materials (kg)	Wet materials (kg)
Cement	355	355
Water	220	148
FA	680	721
CA	1020	1051

Table 6.7: Adjustments for volume batching

1	2	3	4	5	6	7	8	9
Fine aggregate				*Coarse aggregate*			*Water*	
Percent of water by weight	Amount of water contained in aggregate kg	Dry density of aggregate kg/m³	Volume of wet aggregate required m³	Percent of water by weight	Amount of water contained in aggregate kg	Total weight of water in combined aggregate kg	Weight of water required to be added kg	Volume of water required to be added m³
6	$\dfrac{6 \times 680}{100}$ $= 41$	$\dfrac{(1680 \times 100)}{125}$ $= 1340$	$\dfrac{680}{1340}$ $= 0.51$	3	$\dfrac{3 \times 1020}{100}$ $= 31$	$(41 + 31)$ $= 72$	$(220-72)$ $= 148$	$\dfrac{148}{1000}$ $= 0.148$

Quantities of materials required for 1 m³ of concrete

	Dry materials (m³)	Wet materials (m³)
Cement	0.231	0.231
Water	0.220	0.148
FA	0.405	0.510
CA	0.674	0.674

In the case of weigh batching, it is merely a case of increasing the weight of aggregates to allow for the weight of water contained in them and also reducing the amount of water added to the mix. In the case of volume batching, allowance has to be made for the bulking of fine aggregate caused by the water as well as the adjustments for the actual quantity of water contained in the aggregate.

6.6.3 A Concrete Mix is Required having a Minimum Compressive Strength of 300 kg/cm², at 28 days for Use in a Road Slab

Degree of control: Very good with weigh batching and constant supervision

Degree of cement: Very low since compacting will be effected by vibration

Type of cement: Ordinary Portland (sp. gravity = 3.15)

Type of fine aggregate: Natural sand (sp. gravity = 2.60)

Type of coarse aggregates are: Irregular aggregate (sp. gravity = 2.50)

The aggregates are stockpiled at works site in three heaps and have the grading shown below:

I.S. sieve	Cumulative percentage passing		
size	Fine aggregate	Coarse aggregate	
		4.75 to 20 mm	20 to 40 mm
40 mm	–	100	100
20 mm	–	99	13
10 mm	100	33	8
4.75 mm	99	5	2
2.36 mm	76	0	0
1.18 mm	58	–	–
600 micron	40	–	–
300 micron	12	–	–
150 micron	2	–	–

Design the concrete mix and calculate the quantities of ingredients required for a cubic metre of concrete. What is the density of fresh concrete?

Mix Design

Mean design strength $= \dfrac{300}{0.75} = 400 \text{ kg/cm}^2$ (Table 6.1)

Water/cement ratio (Fig. 6.1) = 0.48

For 'Very Low' workability and for water/cement ratio of 0.48, the aggregate/cement ratio for the coarsest grading, which is the most economical, is found to be 7.2 (Table 6.2).

Since the gradings of the aggregates do not conform to the standard grading curves, they should be combined to achieve the desired grading.

The two coarse aggregates are combined first, using the percentage passing the 20 mm sieve as a criterion as shown in Fig. 6.4.

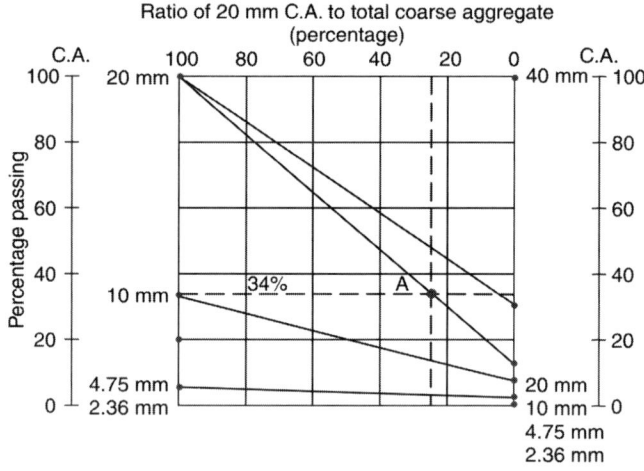

Fig. 6.4: Graphical method of combining coarse aggregates of two different sizes

A vertical line is drawn through the point where the line joining the 20 mm values intersects the horizontal line representing the correct percentage of aggregate smaller than 20 mm. For the standard grading curve no. 1, for the present problem, (50–24) = 26 parts of aggregate coarser than 4.75 mm are to pass the 20 mm sieve, while 50 parts are to be retained. The ratio amounts to 26 : (60+26) or 34 per cent of all coarse aggregate. A horizontal line is drawn through the 34 per cent point to intersect the 20 mm line at A. A vertical line drawn through A gives the quantity of material between the sieves 4.75 mm and 20 mm, as a percentage of the total coarse aggregate. Fig. 6.4 indicates this value to be 24 per cent. The vertical line represents the grading of the combined coarse aggregate and this is combined with the fine aggregate in Fig. 6.4, yielding 22 parts of the fine aggregate to be combined with 78 parts of the aggregate coarser than 4.75 mm sieve. Hence the aggregates are to the combined in the ratio of

$$(22): \frac{24}{100} \times 78 : \frac{76}{100} \times 78 \text{ or } 1:0.85:2.69$$

The vertical line through B in Fig. 6.5 represents the combined grading.

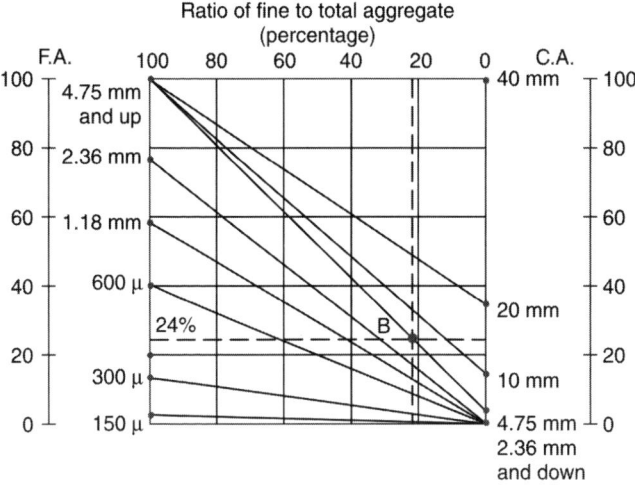

Fig. 6.5: Graphical method of combining fine and coarse aggregates

Since the aggregate/cement ratio is 7.2, the proportions by weight of ingredients are given by:

Cement	:	Fine aggregate	:	Coarse aggregate (4.75–20 mm)	:	Coarse aggregate (20–40 mm)
1	:	$\dfrac{1 \times 7.2}{4.54}$	:	$\dfrac{0.85 \times 7.2}{4.54}$	:	$\dfrac{2.69 \times 7.2}{4.54}$
1	:	1.59	:	1.35	:	4.25

If C is the weight of cement required per cubic metre of concrete, then by absolute volume method, we have the relation

$$\frac{C}{3.15 \times 10^3} + \frac{1.59\,C}{2.60 \times 10^3} + \frac{1.35\,C}{2.50 \times 10^3} + \frac{4.25\,C}{2.50 \times 10^3} + \frac{0.48\,C}{10^3} = 1$$

$\therefore$ C $\quad = \quad 275$ kg

Water $= \; 4.8 \times 275 \qquad = \; 132$ kg

FA $\quad = \; 1.59 \times 275 \qquad = \; 437$ kg

CA $\quad = \; 1.35 \times 275 \qquad = \; 372$ kg

(4.75–20 mm)

CA $\quad = \; 4.25 \times 275 \qquad = \; 1170$ kg

(20–40 mm)

Density of fresh concrete $= \; 2386$ kg/m^3

7

Design of Concrete Mixes According to American Concrete Institute Standards

7.1 GENERAL FEATURES OF THE ACI METHOD

The American Concrete Institute recommends a method of mix design[5], considering the most economical use of available materials to produce concrete of desirable workability, durability and strength. The design tables incorporating the basic relationships between the parameters are useful in selecting the optimum combinations of the ingredients of concrete mixes. The ACI mix proportioning method is suitable for normal and heavy weight concretes having a maximum 28 days cylinder compressive strength in the range of 15 to 45 N/mm^2 with workability (slump) in the range of 25 to 100 mm.

The ACI method presumes that the workability of a mix with given maximum size of well-graded aggregates is dependent upon the water content, the amount of entrained air and certain chemical admixtures, but is largely independent of mix proportions particularly the amount of cementing material. The method also assumes that the optimum ratio of bulk volume of coarse aggregate to the total volume of concrete depends only on the maximum size of coarse aggregate and on the grading of the fine aggregate expressed as fineness modulus. The water/cement ratio is selected based on the strength and durability requirements. Knowing the volumes of water, coarse aggregate and cement content, the quantity of fine aggregate required is determined by the absolute volume method, allowing for the quantity of air entrained in the mix. However, the final mix proportions should be established by trial and necessary adjustments required for the field mixes.

73

7.2 DESIGN OF CONCRETE MIXES ACCORDING TO ACI 211.1-1991

The procedure to be followed in designing a concrete mix is detailed in the following steps:

(a) *Consistency of the mix expressed as slump*: Depending upon the type of construction and compacting equipment available at site, the consistency expressed as slump is selected based upon the placing conditions. The generally used slump recommended for specific applications are compiled in Table 7.1.

Table 7.1: Recommended slumps for various applications

	Slump (mm)	
Type of construction	*Maximum*	*Minimum*
Reinforced foundation walls and footings, plain footings, caissons and substructure walls	75	25
Beams and reinforced walls, building columns	100	25
Pavements and slabs	75	25
Mass concrete	50	25

(b) *Maximum aggregate size*: The maximum size of the coarse aggregate is selected based on considerations of the thickness of the member. The maximum size is restricted to one-third of the minimum thickness of the member and three-fourths of the minimum clear space between the reinforcing bars. Larger size may result in improper compaction rendering the concrete honeycombed with large air pockets. The maximum size of the aggregate is determined by the grading analysis. The maximum sizes of aggregate generally recommended for various type of construction are compiled in Table 7.2.

(c) *Target means strength*: The mean compressive strength of concrete from trial batch tests must equal or exceed the target means strength expressed as

$$f'_{cr} = (f'_c + 1.34\,S) \text{ and } f'_{cr} = (f'_c + 2.33\,S - 3.45)$$

where, f'_c = specified compressive strength of concrete (N/mm²)
S = standard deviation (N/mm²)
In the absence of field data, the standard deviation may be selected from Table 7.3.

Table 7.2: Maximum sizes of aggregate recommended for various types of construction

Minimum dimension of section (mm)	Maximum size of aggregate (mm)			
	Reinforced walls, beams	un-reinforced walls and columns	Heavily reinforced slabs	Lightly reinforced or un-reinforced slabs
62.5–125	12.5–20	20	20–25	20–40
150–275	20–40	40	40	40–80
300–600	40–80	80	40–80	80
750 or more	40–80	150	40–80	80–150

Table 7.3: Target mean compressive strength

Specified compressive strength f'_c (N/mm²)	Target mean compressive strength f'_{cr} (N/mm²)
Less than 21	$f'_c + 7.0$
21 to 35	$f'_c + 8.5$
Over 35	$f'_c + 10.0$

(d) *Water and air content*: The desirable approximate mixing water and air contents are selected from Table 7.4 for the desired slump and maximum size of the aggregate with due regard for the type of exposure ensuring the durability.

(e) *Water/cement ratio*: The water/cement ratio is selected as the minimum from Tables 7.5 and 7.6 for the dual criterion of desired target mean strength and durability considerations based upon the type of exposure.

(f) *Cement content*: The cement content in the mix is computed from the selected mixing water content and the water/cement ratio should also conform to the maximum and minimum specified water/cement ratios based on durability considerations.

(g) *Coarse aggregate content*: The quantity of coarse aggregate in the mix is estimated from Table 7.7 for the selected maximum size of coarse aggregate and the fineness modulus of fine aggregate.

(h) The fine aggregate content is determined by subtracting the sum of the volumes of coarse aggregate, cement, water and air content from the unit volume of concrete.

    (i) If the aggregates contain excessive moisture, suitable adjustments are made in the field to account for the water contained in aggregates to determine the field mix quantities.

Table 7.4: Approximate mixing water and air contents for different slumps and maximum sizes of aggregates

Slump (mm)	Maximum quantity of water (kg/m³) for specified nominal maximum size of aggregate@							
	10	14	20	28	40	56*	80*	150*
1. Non-air-entrained concrete								
Stiff-plastic (25–50)	207	199	190	179	166	154	130	113
Plastic (75–100)	228	216	285	193	181	169	145	124
Flowing (150–175)	243	228	216	202	190	178	160	–
Approximate								
Entrapped air (%)	3.0	2.5	2.0	1.5	1.0	0.5	0.3	0.2
2. Air-entrained concrete								
Stiff-plastic (25–50)	181	175	168	160	150	142	122	107
Plastic (75–100)	202	193	184	175	165	157	133	119
Flowing (150–175)	216	205	197	184	174	166	154	–
3. Recommended average total air content (%)								
Mild exposure	4.5	4.0	3.5	3.0	2.5	2.0	1.5	1.0
Moderate exposure	6.0	5.5	5.0	4.5	4.5	4.0	3.5	3.0
Severe exposure	7.5	7.0	6.0	6.0	5.5	5.0	4.5	4.0

Note: @ Table gives the maximum water content for reasonably well-shaped aggregates.
* Slump values are based on tests made after removal of particles larger than 40 mm

7.3 MIX DESIGN PROCEDURE FOR NO-SLUMP CONCRETE ACCORDING TO ACI 211.3–75 (REVISED IN 1987 AND REAPPROVED IN 1992)

The ACI standard recommended practice[79] for selecting proportions for No-slump concrete of drier consistencies adopts the compacting factor[26] measurement of workability. It is important to note that concrete having a consistency in the range from 'extremely dry' to 'stiff' which does not have any slump should be compacted by vibration.

    The sequential operations to be followed in selecting the mix proportions for No-slump concrete are detailed below:

    (a) The maximum size of the aggregate is determined based on the sectional dimensions and the mix being extremely dry or stiff without exhibiting any slump due to low water/cement ratios, the maximum size is generally restricted to the range of 10 to 40 mm. Table 7.2 recommends the size of the aggregate based on the sectional dimensions and the distribution of reinforcements in the section.

(b) The target mean strength is determined from Table 7.3 depending upon the quality control measures and the standard deviation.

(c) The water/cement ratio required is determined from the dual criterion of strength (Table 7.5) and durability (Table 7.6).

Table 7.5: Water/cementing material ratio and compressive strength relationship

28-day compressive strength# (N/mm²)	Water/cementing material ratio by mass*	
	Non-air-entrained concrete	Air-entrained concrete
45	0.38	0.30
40	0.42	0.34
35	0.47	0.39
30	0.54	0.45
25	0.61	0.52
20	0.69	0.60
15	0.97	0.70

Note: * Maximum nominal size of aggregate to be about 20 to 28 mm
Strength is based on moist-cured cylinders

Table 7.6: Maximum permissible water/cement ratio in severe exposure conditions

Type of structure	Continuously wet structure exposed to frequent freezing and thawing	Structures exposed to sea water or sulphates
Thin sections (railings, curbs, sills, ledges, ornamental work) and sections with less than 25 mm cover over steel	0.45	0.40
All other structures	0.50	0.45

(d) The approximate mixing water required is selected from Table 7.8 to suit the maximum size of the aggregate and the required consistency.

(e) The quantity of cement required for cubic metre of concrete is determined from (c) and (d).

(f) The coarse aggregate content is estimated from Table 7.7 for the maximum size of aggregate and the fineness modulus of sand.

(g) For consistencies other than plastic, a multiplying factor is applied to the volume of coarse aggregate determined from (f) using the Table 7.9.

(h) The fine aggregate content is determined by subtracting the sum of the volumes of coarse aggregate, cement, water and air content from the unit volume of concrete.

(i) Suitable adjustments are made to the dry mix proportions to account for the moisture contained in the aggregates to finalize the field mix quantities.

Table 7.7: Bulk volume of coarse aggregate per unit volume of concrete for different fineness modulus of fine aggregate

Nominal maximum size of coarse aggregate (mm)	Bulk volume of oven-dry-rodded coarse aggregate (m³) fineness modulus of fine aggregate			
	2.40	2.60	2.80	3.00
10	0.50	0.48	0.46	0.44
14	0.59	0.57	0.55	0.53
20	0.66	0.64	0.62	0.60
28	0.71	0.69	0.67	0.65
40	0.75	0.73	0.71	0.69
56	0.78	0.76	0.74	0.72
80	0.82	0.80	0.78	0.76
150	0.87	0.85	0.83	0.81

7.4 LIMITATIONS OF THE ACI METHOD

The ACI standard recommendations for selecting proportions of concrete having different consistencies are based on experimental investigations using well-shaped aggregates within the range of generally accepted specifications. If the aggregates available at site depart from the standard gradings and have less favourable shape with an increased angularity number, suitable precautions are necessary to maintain the consistency of the mix by increasing the cement content. The water/cement ratio and strength relations of Table 7.4 are based on the use of ordinary Portland cement (Type–I)

Table 7.8: Approximate mixing water requirements for different consistencies and maximum sizes of aggregates

Description	Slump (mm)	Vebe (sec)	Compacting factor	Relative water content	Water (kg/m^3) for indicated maximum size of coarse aggregate				
					10 mm	12 mm	20 mm	25 mm	40 mm
				Non-air-entrained concrete					
Extremely dry	–	32–18	–	78	177	168	158	148	137
Very stiff	–	18–10	0.70	83	188	182	168	158	148
Stiff	0–25	10–5	0.75	88	196	192	177	168	158
Stiff plastic	25–50	5–3	0.85	92	206	196	182	177	162
Plastic	75–100	3–0	0.91	100	226	217	203	192	177
Flowing	150–175	–	0.95	106	240	226	212	203	188
Approximate amount of entrapped air in non-air-entrained concrete (per cent)					3	2.5	2	1.5	1

(Contd...)

(Contd....)

				Air-entrained concrete					
Extremely dry	—	32–18	—	78	158	148	137	133	123
Very stiff	—	18–10	0.70	83	168	158	148	137	133
Stiff	0–25	10–5	0.75	88	177	168	158	148	137
Stiff plastic	25–50	5–3	0.5	92	182	177	162	152	143
Plastic	75–100	3–0	0.91	100	203	192	177	168	158
Flowing	150–175	—	0.95	106	212	203	188	177	168
Recommended average total air content (per cent)**					8	7	6	5	4.5

* These quantities of mixing water are for use in computing cement factors for trial batches. They are for reasonably well-shaped angular coarse aggregates graded within limits of accepted specifications. If more mater is required than shown, the cement factor, estimated from these quantities, should be increased to maintain desired water/cement ratio, except as otherwise indicated by laboratory tests for strength. If less water is required than shown, the cement factor, estimated from these quantities, should not be decreased except as indicated by laboratory tests for strength.

** For consistencies below 25 mm slump, the volume of air entrained by wither an air-entraining cement or the amount of air-entraining admixture used for more plastic mixtures may be significantly lower than those shown. For these mixtures, it is recommended that the air content resulting from the use of air-entraining cement or the usual amount of air-entraining admixture per unit of cement for more plastic mixtures be accepted as adequate for ensuring durability. In the absence of such information for a particular air-entraining admixture, the amount to use per unit of cement can be determined on a trial mix having a slump in the 75 to 100 mm range, or by determining the amount needed to obtain 19 ± 3 per cent air in mortar prepared according to ASTM C–185.

in the mix. If rapid hardening Portland cement (Type–III) is used, the corresponding strength at 28 days will be higher for the same water/cement ratios. The strength, and water/cement ratios curves of Fig. 5.1 are useful in this regard.

Table 7.9: Factors to be applied to the volume of coarse aggregate calculated on the basis of Table 7.8, for mixes of consistencies other than plastic

| Consistence | *Factors of maximum size aggregate* | | | | |
	10 mm	*12.5 mm*	*20 mm*	*25 mm*	*40 mm*
Extremely dry	1.90	1.70	1.45	1.40	1.30
Very stiff	1.60	1.45	1.30	1.25	1.25
Stiff	1.35	1.30	1.15	1.15	1.20
Stiff plastic	1.08	1.06	1.04	1.06	1.09
Plastic (Reference)	1.00	1.00	1.00	1.00	1.00
Fluid	0.97	0.98	1.00	1.00	1.00

It is important to note that the mix design tables serve as a guide in selecting proportions and suitable minor adjustments should be effected in the field for any departures in the quality of aggregates and type of cement used. The volume of dry rodded coarse aggregate recommended in Table 7.7 is only applicable to the aggregates of the given specific gravity ρ which in this case is 2.68. If the aggregate used has a specific gravity ρ_1, the volume of coarse aggregates specified in the tables should be multiplied by the ratio (ρ_1/ρ) to account for the gross apparent volume of the solid particles.

Any desirable grading in their finer and coarser fractions of the aggregates may be corrected to a desirable particle size distribution by separation of the material into more size fractions and recombining them in suitable proportions and also be supplementing deficient size from other sources. The analytical and graphical methods mentioned in section 6.3 are useful in combining the aggregates to conform to a desirable standard grading.

7.5 MIX DESIGN EXAMPLES

7.5.1 Design a Concrete Mix and Estimate the Batch Quantities of Ingredients for Cubic Metre of Concrete to Suit the Following Requirements

Concrete is required for casting the interior columns in the ground floor of a multistoried building. Structural considerations specify a target mean cylinder compressive strength of 20 N/mm^2 at 28 days. The coarse aggregate locally available is well graded having a maximum size of 20 mm and dry rodded weight of 1600 kg/m^3. Fineness modulus of sand available at site is 2.8. Slump specified for the job is 75 to 100 mm. Specific gravities of cement, coarse and fine aggregates are 3.15, 2.68 and 2.68 and 2.64, respectively. Type of cement available at site is ordinary Portland (Type–I).

Design of Mix

Non-air-entrained concrete is selected since the structure is not exposed to adverse weather conditions.

Target mean strength of concrete = 25 N/mm^2.

Water/cement ratio (from Table 7.5) = 0.69.

Approximate mixing water for a slump of 75 to 100 mm (from Table 7.4) = 205 kg/m^3.

Volume of coarse aggregate per m^3 of concrete (from Table 7.7) = 0.62 m^3.

Weight of coarse aggregate required = (0.62 × 1600) = 992 kg.

Weight of cement required = $\dfrac{205}{0.69}$ = 297 kg.

Entrapped air for 20 mm aggregates (from Table 7.4) = 3.5 per cent.

Knowing the amount of cement, water, coarse aggregate and entrapped air in unit volume of concrete, the quantity of fine aggregate (sand) is determined by the absolute volume method outlined below:

Item no	Ingredient	Weight (kg)	Solid volume (cm)^3	
1	Cement	297	$\dfrac{297 \times 10^3}{3.15}$	= (94.2×10^3)
2	Water	205	–	= (205×10^3)
3	Coarse aggregate	992	$\dfrac{992 \times 10^3}{2.68}$	= (370×10^3)
4	Entrapped air (3.5%)	–	–	= (35×10^3)
5	Total volume of ingredients except sand			= (704.2×10^3)

Solid volume of dry sand required = (1000–704.2) = (295.8×10^3) cm^3

Weight of dry sand required = [295.8 × 2.64] = 781 kg

Estimated batch quantities per m^3 of concrete are compiled as

Cement   = 297 kg

Water	= 205 kg
Coarse aggregate	= 992 kg
Fine aggregate	= 781 kg
Density of fresh concrete	= 2275 kg/m^3

Mix proportions by weight of dry materials are expressed as:

Cement	Sand	Coarse aggregate	Water
1	2.62	3.34	0.69

Adjustments for field mix proportions

If the coarse and fine aggregates are found to contain 1 and 5 per cent of moisture by weight respectively, estimate the field mix proportions adjusting for water content in the aggregates.

Water contained in coarse aggregate	= $(0.01 \times 992) = 9.92$ kg
Water contained in fine aggregate	= $(0.05 \times 781) = 39.05$ kg
Total water contained in the aggregates	= 48.97 kg $\approx$ 49 kg
Free water to be added to the mix	= [205–49] = 156 kg

Field mix quantities:

Cement	= 297 kg
Water	= 156 kg
Sand	= $(781 + 39) = 820$ kg
Coarse aggregate	= $(992 + 10) = 1002$ kg

7.5.2 Design a Concrete Mix to Suit the Following Data

Concrete is required to cast the abutments of a bridge which is subjected to severe exposure conditions. The target mean cylinder strength required is 25 N/mm^2 at 28 days. Ordinary Portland cement (Type–I) and graded coarse aggregate of maximum particle size 40 mm are available at site. Dry rodded weight of coarse aggregate is 1760 kg/m^3. Fineness modulus of sand is 2.8. Desirable slump for the work is 30 to 50 mm. Specific gravity of cement is 3.15 and that of coarse and fine aggregates are 2.68 and 2.64, respectively.

Design of Mix

Air-entrained concrete will be used since the structure is exposed to severe atmospheric conditions.

Target mean strength = 25 N/mm^2

Water/cement ratio

(a) From strength considerations (Table 7.5) = 0.52

(b) From durability considerations (Table 7.6) = 0.50

Minimum water/cement ratio of 0.50 is adopted.

Approximate water content for the desired slump of 30 to 50 mm and maximum aggregate size of 40 mm is read out from Table 7.4 as 150 kg/m^3.

Desirable total air content for 40 mm size aggregate (severe exposure) = 5.5 per cent.

$$\text{Weight of cement required} = \frac{150}{0.50} = 300 \text{ kg/m}^3$$

Volume of coarse aggregate required for 40 mm aggregate and fineness modulus of sand of 2.80 are read out from Table 7.7 as 0.71 m^3.

Weight of coarse aggregate = (0.71 × 1760) = 1250 kg.

Item no	Ingredient	Weight (kg)	Solid volume (cm)^3
1	Cement	300	$\frac{300 \times 10^3}{3.15} = (95.2 \times 10^3)$
2	Water	150	$-$ $= (150 \times 10^3)$
3	Coarse aggregate	1250	$\frac{1250 \times 10^3}{2.68} = (466 \times 10^3)$
4	Entrapped air (5.5%)	$-$	$-$ $= (55 \times 10^3)$
5	Total volume of ingredients except sand		$= (766 \times 10^3) \text{ cm}^3$

Total volume of ingredients except sand = $(766 \times 10^3) \text{ cm}^3$

Solid volume of sand required = $(1000–766) = (234 \times 10^3) \text{ cm}^3$

Weight of sand required = $(2.64 \times 234) = 617 \text{ kg}$

Estimated batch quantities per m^3 of concrete is given as:

Cement = 300 kg

Water = 150 kg

Fine aggregate = 617 kg

Coarse aggregate = 1250 kg

Density of concrete = 2317 kg/m^3

7.5.3 Design a Concrete Mix for the Construction of a Heavy Bridge Pier Exposed to Fresh Water in a Severe Climate to Suit the Following Data

Target compressive cube strength = 26 N/mm^2

Type of cement = Ordinary Portland (Type–I)

Placement conditions require a slump of 30 to 50 mm

Graded coarse aggregate of maximum size	= 28 mm
Dry rodded weight of coarse aggregate	= 1520 kg/m³
Fineness modulus of fine aggregate	= 2.8
Specific gravity of cement, FA and CA	= 3.15, 2.64 and 2.68

Design of Mix

Air-entrained concrete is selected due to severe exposure conditions. Corresponding cylinder strength = (0.8 × cube strength) = (0.8 × 26) = 21 N/mm².

Water/cement ratio:

(a) From strength considerations (Table 7.5) = 0.58

(b) From durability considerations (Table 7.6) = 0.50

Minimum water/cement ratio of 0.50 is adopted

Approximate mixing water required (Table 7.4) = 160 kg/m³

Desirable air content = 6 per cent

Weight of cement required = $\dfrac{160}{0.5}$ = 320 kg

Volume of coarse aggregate required (Table 7.7) = 0.67 m³

Weight of coarse aggregate required = (0.67 × 1520) = 1018 kg

Item no	Ingredient	Weight (kg)	Solid volume (cm)³	
1	Cement	320	$\dfrac{320 \times 10^3}{3.15}$	= (102 × 10³)
2	Water	150	–	= (160 × 10³)
3	Coarse aggregate	1018	$\dfrac{1018 \times 10^3}{2.68}$	= (380 × 10³)
4	Entrapped air (6%)	–	–	= (60 × 10³)
5	Total volume of ingredients except sand			= (702 × 10³) cm³

Total volume of ingredients except sand	= (702 × 10³) cm³
Solid volume of sand required	= (1000–702)
	= (298 × 10³) cm³
Weight of sand required	= (2.64 × 298) = 786 kg

Estimated batch quantities per m³ of concrete is given as:

Cement	= 320 kg
Water	= 160 kg

Fine aggregate	= 786 kg
Coarse aggregate	= 1018 kg
Density of concrete	= 2284 kg/m³

7.5.4 Design the Batch Quantities per Cubic Metre of Concrete which is Required for Casting Prestressed Concrete Girders of a Bridge Exposed to Severe Weather with Frequent Alterations of Freezing and Thawing. Design a Suitable Mix to Suit the Following Data

Target mean cylinder compressive strength = 35 N/mm²

Workability requirements: Very heavy internal and external vibration equipment is available for good compaction. Very stiff consistency of mix without slump is recommended.

Graded coarse aggregates of maximum size 20 mm are available at site.

Dry rodded weight of coarse aggregate	= 1600 kg/m³
Fineness modulus of sand available at site	= 2.8
Specific gravity of cement, FA and CA	= 3.15, 2.64 and 2.68

Estimate the batch quantities on dry basis and the field mix quantities assuming coarse and fine aggregates to contain 1 and 5 per cent moisture respectively.

Design of mix

Air-entrained concrete mix is to be used for severe exposure conditions.

Water/cement ratio:
  (a) From strength consideration (Table 7.5) = 0.39
  (b) From durability considerations (Table 7.6) = 0.50

Minimum water/cement ratio of 0.39 is adopted

Approximate mixing water required for very stiff consistency (Table 7.8) = 148 kg/m³

Desirable air content = 6 per cent

Weight of cement required = $\dfrac{148}{0.39}$ = 379 kg

Volume of coarse aggregate required (Table 7.7) = 0.62 m³

Volume of CA after applying the correction factor (Table 7.9) = (1.3 × 0.62) = 0.806 m³

Weight of coarse aggregate required = $(0.806 \times 1600) = 1290$ kg

Item no	Ingredient	Weight (kg)	Solid volume (cm)^3
1	Cement	379	$\dfrac{379 \times 10^3}{3.15} = (120 \times 10^3)$
2	Water	148	$- \quad = (148 \times 10^3)$
3	Coarse aggregate	1290	$\dfrac{1290 \times 10^3}{2.68} = (481 \times 10^3)$
4	Entrapped air (6%)	–	$- \quad = (60 \times 10^3)$
5	Total volume of ingredients except sand		$= (809 \times 10^3)$ cm^3

Total volume of ingredients except sand $= (809 \times 10^3)$ cm^3

Solid volume of sand required $= (1000–809) = (191 \times 10^3)$ cm^3

Weight of sand required $= (2.64 \times 191) = 504$ kg

Estimated batch quantities per m^3 of concrete are given as:

Cement $= 379$ kg

Water $= 148$ kg

Fine aggregate $= 504$ kg

Coarse aggregate $= 1290$ kg

Density of concrete $= 2321$ kg/m^3

Adjustments for water contained in aggregates:

Weight of wet sand required $= [504 + (0.05 \times 504)] = 529$ kg

Weight of moist coarse aggregate required $= [1290 + (0.01 \times 1290)] = 1303$ kg

Weight of free water in sand $= (0.05 \times 504) = 25$ kg

Weight of free water in coarse aggregate $= (0.01 \times 1290) = 13$ kg

Total weight of free water in combined aggregates $= (25 + 13) = 38$ kg

Weight of water required for the field mix $= (148–38) = 110$ kg

Estimated batch quantities for the field mix are given as:

Cement $= 379$ kg

Water $= 110$ kg

Fine aggregate $= 529$ kg

Coarse aggregate $= 1303$ kg

Density of concrete $= 2321$ kg/m^3

Design of Concrete Mixes Based on Surface and Angularity Index of Aggregate

8.1 SURFACE AND ANGULARITY INDEX

In the design of concrete mixes, the grading of the coarse and fine aggregates is an important factor since it affects the resulting workability. The quantity of water required to produce a given workability depends to a large extent on the surface area of the aggregates. The surface area per unit weight of the material is termed as "specific surface" and this is an indirect measure of the aggregate grading[81]. The specific surface increases with a reduction in the size of material so that fine sand contributes very much more to the surface area than does the coarse aggregate. The workability of a mix is, therefore, influenced more by the finer fractions than the coarser particles of the aggregate.

Specific surface gives a somewhat misleading picture of the workability to be expected and to overcome this difficulty, Murdock[13] has suggested the use of surface index, which is an empirical number, related to the specific surface of the particles with more weightage assigned to the coarser material. The empirical numbers representing the surface index of aggregate particles within a set of sieve sizes are given in Table 8.1.

The total surface index (f_s) of a mixture of aggregates is calculated by multiplying the percentage weight of material retained on each sieve and the corresponding surface indices and to their sum is added a constant of 330 and the result is divided by 1000. The method of computing the total surface index for any given grading is shown in Table 8.2., in which the standard grading curve no. 1 of Road Note No. 4 (refer Fig. 2.2) is used. The values of total surface

Table 8.1: Surface index of aggregate particles

Sieve size within which particles lie	Surface index for particles within sieve sizes indicated
80–40 mm	–2.5
40–20 mm	–2
20–10 mm	–1
10–4.75 mm	1
475–2.36 mm	4
2.36–1.18 mm	7
1.18–600 micron	9
600–300 micron	9
300–150 micron	7
Smaller than 150 micron	2

Table 8.2: Surface index of combined grading

Sieve sizes within which particles lie	Percentage of particles within sieve size	Surface index for particles within sieve size	Surface index (f_s)
20–10 mm	55	–1	–55
10–4.75 mm	15	1	15
4.75–2.36 mm	7	4	28
2.36–1.18 mm	7	7	49
1.18–600 micron	7	9	63
600–300 micron	7	9	63
300–150 micron	2	7	14
			177
		Add constant =	+ 330
			507

Surface index $(f_s) = \dfrac{507}{1000} = 0.507$

index vary between the narrow limits of 0.507 and 0.691, for standard grading curves from 1 to 4, when 20 mm maximum size aggregates are used.

The specific surface varies with different types of aggregates due to variations in the angularity. The angularity index (f_a) depends upon the grading of coarse and fine aggregates, angularity number[82] and the relative proportion of coarse and fine aggregates in the final mix. For single sized aggregates, the angularity index (f_a) can be expressed in the form,

$$\left[f_a = 11.05 - \frac{15\ W}{VG} \right]$$

where,

W = weight of single size aggregates in grams, compacted in a cylinder of known volume

G = specific gravity of the aggregate particles

V = volume of the cylinder (ml)

The angularity index of combined aggregate is determined by combining the angularity index for each single size in proportion to the amount present in the mixture. Typical values of angularity index for different sizes and shapes of aggregates are given in the Table 8.3[13]. Crushed stones have a higher angularity index than the irregular and rounded gravels.

Table 8.3: Angularity index of aggregates

Type of aggregate	Size	Angularity index (f_a)
Rounded flint gravel	20–10 mm	1.05
Slightly rounded flint gravel	20–10 mm	1.44
	10–4.75 mm	1.87
Irregular flint gravel	20–10 mm	2.08
	10–4.75 mm	2.47
Angular granite	20–10 mm	2.62
Angular crystalline lime stone	20–10 mm	2.53
Quartz sand	4.75–2.36 mm	2.47
	2.36–1.13 mm	2.47
	1.18–600 micron	2.25
	600–300 micron	1.62
	300–150 micron	1.84

8.2 MIX DESIGN CHARTS

Based on experimental investigations, Murdock and Black-ledge[83] have presented design curves relating compressive strength, water/cement ratio and surface index, shown in Fig. 8.1. On the basis of this study, an optimum surface index of 0.6 has been suggested for use

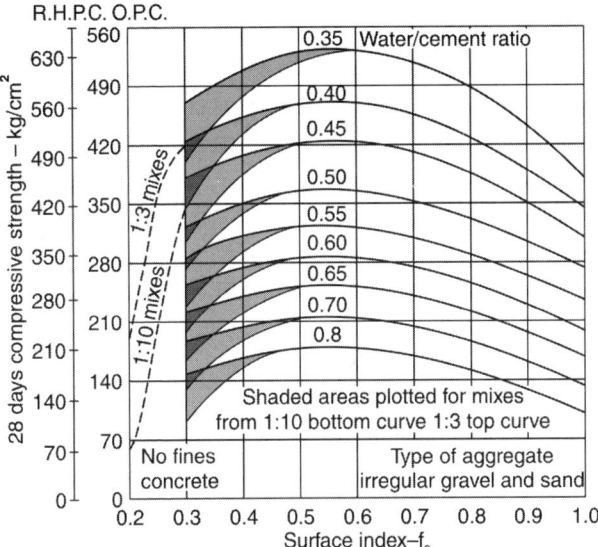

Fig. 8.1: Relation between compressive strength, water/cement ratio and surface index

in design for 20 mm maximum size aggregates. The corresponding values for 10 and 40 mm aggregates being 0.70 and 0.55 respectively.

Figures 8.2, 8.3 and 8.4 show the relationship between aggregate/ cement ratio, water/cement ratio compacting factor and cement content for different types of aggregate. The charts have been prepared for known values of the surface and angularity index and for two different types of cement commonly used for a majority of constructional works.

8.3 MIX DESIGN PROCEDURE

The mix design charts can be conveniently used for the design of concrete mixes of desirable strength and workability. The procedure to be followed is outlined below:

(a) The target or design strength is obtained by applying suitable control factors to the specified minimum compressive strength.

(b) The water/cement ratio required to achieve the design strength is obtained from Fig. 8.1.

(c) The cement content and the aggregate/cement ratio are determined from Figs 8.2, 8.3 or 8.4 for the desired workability expressed as compacting factor and the type of aggregate used.

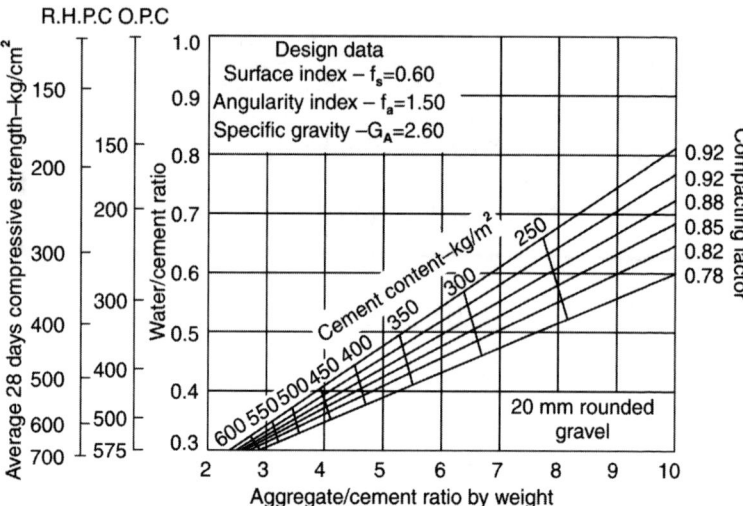

Fig. 8.2: Relationship between A/C ratio, water/cement ratio and cement content for concrete mix design for estimating purposes

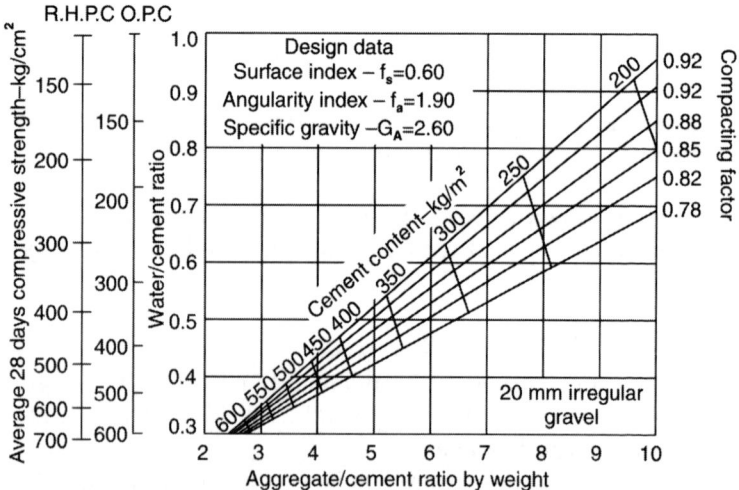

Fig. 8.3: Relationship between A/C ratio, W/C ratio and cement content for concrete mix design for estimating purposes

(d) Knowing the grading of fine and coarse aggregates available at site, the surface indices of the aggregates are evaluated using Table 8.1.

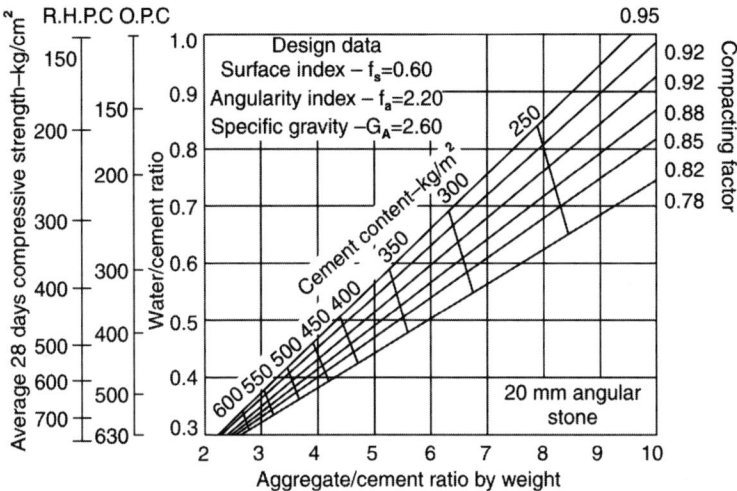

Fig. 8.4: Relationship between A/C ratio, W/C ratio and cement content for concrete mix design for estimating purposes

(e) The fine and coarse aggregates of known surface index are combined to yield an aggregate mixture having a specified or desirable surface index.

Let x = surface index of fine aggregate
y = surface index of coarse aggregate
z = surface index of combined aggregate
If a = proportion of fine to coarse aggregate

Then $a = \dfrac{(z-y)}{(x-z)}$

The proportion of cement, water, fine and coarse aggregates is determined, having a knowledge of the water/cement and aggregate/cement ratio, the cement content and the ratio of fine to coarse aggregate.

(g) The quantities of ingredients required for a unit volume of concrete are evaluated from the known proportions.

8.4 MIX DESIGN EXAMPLES

8.4.1 Design the Concrete Mix to Suit the Following Data

Specified minimum 28 days cube strength = 210 kg/cm²
Control factor = 0.75

Concrete is required for normal reinforced concrete work with "Medium" workability.

Corresponding compacting factor = 0.95
Type of cement = Ordinary Portland
Type of sand = Irregular gravel 20 mm maximum size
Angularity index = 1.90

The particle size distribution is as given below:

Sieve size within which particles lie	Percentage of particles within sieve sizes
Coarse aggregate	
20–10 mm	65
10–4.75 mm	35
Fine aggregate	
4.75–2.36 mm	10
2.36–1.18 mm	20
1.18–600 micron	20
600–300 micron	30
300–150 micron	15

$$\text{Mean strength} = \frac{210}{0.75} = 280 \text{ kg/cm}^2$$

Water cement ratio required (Fig. 8.1) = 0.61
For W/C = 0.61 and CF = 0.95
Using 20 mm maximum size irregular gravel
Cement content (Fig. 8.3) = 320 kg/m^3
Aggregate cement ratio = 5.9 by weight.

Surface Index of Aggregates

Sieve size within which particles lie	Percentage of particles within sieve size	Surface index for particles within sieve size	Surface index (f_s)
Coarse aggregate			
20–10 mm	65	–1	–65
10–4.75 mm	35	1	35
			–30
		Add constant	+330
			300

$$\text{Surface index of coarse aggregate } = \frac{300}{1000} = 0.30$$

Fine aggregate			
4.75–2.36 mm	10	4	40
2.36–1.18 mm	20	7	140
1.18–600 micron	20	9	180
600–300 micron	30	9	270
300–150 micron	15	7	105
			735
		Add constant	330
			1065

$$\text{Surface index of fine aggregate } = \frac{1065}{1000} \qquad = 1.065$$

Surface index of combined aggregate required = 0.60

Let x = surface index of FA = 1.065
    y = surface index of CA = 0.30
    z = surface index of combined aggregate = 0.60
If a = proportion of fine to coarse aggregate

$$a = \frac{(z-y)}{(x-z)} = \frac{(0.60-0.30)}{(1.065-0.60)} = \frac{1}{1.55}$$

∴   FA : CA
      1 : 1.55

Weight of materials for 1 m³ of concrete:

Cement   = 320 kg

Weight of combined aggregate = (5.9 × 320) = 1900 kg

Weight of fine aggregate   = $\dfrac{1}{2.55}$×1900 = 750 kg

Weight of coarse aggregate   = $\dfrac{1.55}{2.55}$×1900 = 1150 kg

Weight of water   = (0.61 × 320) = 195 kg

Density of fresh concrete   = 2415 kg/m³

8.4.2 Design a Concrete Mix for Producing Prestressed Precast Elements to Suit the Following Data

Specified minimum cube strength at 28 days = 350 kg/cm²

Control factor   = 0.80
Concrete to be compacted by vibration and required workability
"Low" and compacting factor   = 0.83
Type of cement   = Rapid hardening Portland
Type of coarse aggregate   = 28 mm maximum size
                                 irregular gravel
Type of sand: Natural sand
Angularity index of aggregates = 1.90
The grading of the aggregates available at the site is as given below:

| I.S. sieve size | Percentage passing | |
	Coarse aggregate	Fine aggregate
20 mm	100	–
10 mm	45	–
4.75 mm	–	100
2.36 mm	–	77
1.18 mm	–	53
600 micron	–	30
300 micron	–	10
150 micron	–	–

Mix design

Mean strength $= \dfrac{350}{0.80}$ $= 440 \text{ kg/cm}^2$
Water/cement ratio (Fig. 8.1)   = 0.50
For W/C   = 0.50 and CF   = 0.83
Cement content (Fig. 8.3)   = 340 kg/m^3
Aggregate/cement ratio   = 5.7 by weight

Surface index of aggregates

Sieve size within which particles lie	Percentage of particles within sieve size	Surface index for particles within sieve size	Surface index (f_s)
Coarse aggregate			
20–10 mm	55	–4	–55
10–4.75 mm	45	1	45
		Add constant	+ 330
			320

$$\text{Surface index of coarse aggregate} = \frac{320}{1000} = 0.32$$

Fine aggregate

4.75–2.36 mm	23	4	92
2.36–1.18 mm	24	7	168
1.18–6000 micron	23	9	207
600–300 micron	20	9	180
300–150 micron	10	7	70
			717
		Add constant	330
			1047

$$\text{Surface index of fine aggregate} = \frac{1047}{1000} \qquad = 1.05$$

Let x = surface index of FA $\qquad = 1.05$

$\quad y$ = surface index of CA $\qquad = 0.32$

$\quad z$ = surface index of combined aggregate $= 0.60$

If $\quad a$ = proportion of fine to coarse aggregate,

$$a = \frac{(z-y)}{(x-z)} = \frac{(0.60-0.32)}{(1.05-0.60)} \qquad = \frac{1}{1.61}$$

$\therefore \quad$ FA : CA

$\qquad$ 1 : 1.55

Batch quantities per m^3 of concrete:

Cement $\qquad\qquad\qquad\qquad$ = 340 kg

Weight of combined aggregate = (340 × 5.7) = 1940 kg

Weight of fine aggregate $\qquad = \dfrac{1}{2.61} \times 1940 = 740$ kg

Weight of coarse aggregate $\qquad = \dfrac{1.61}{2.61} \times 1940 = 1200$ kg

Weight of water $\qquad\qquad\quad$ = (0.50 × 340) = 170 kg

Density of fresh concrete $\qquad$ = 2450 kg/m^3

9

Design of High-Strength Concrete Mixes

9.1 HIGH-STRENGTH CONCRETE

In the case of low and medium strength concretes, the strength is mainly influenced by the water/cement ratio, and is almost independent of the other parameters. The properties of high strength concretes with a compressive strength above 400 kg/cm², are influenced by the properties of the aggregate in addition to that of the water/cement ratio. To obtain high strengths, it is necessary to use the lowest possible water/cement ratio which affects the workability of the mix and necessitates the use of normal vibration techniques for proper compaction. In this context, it is to be noted that high strengths have to be achieved by suitable proportioning and not be stream curing[84] or application of pressure. In the present state of the art, concrete which has a desired 28 days compressive strength up to 700 kg/cm², can be made by suitable proportioning of the ingredients using normal vibrational methods for compacting the mixes. The highest strengths reported by Collins[85] are in the vicinity of 1100 kg/cm².

Recent report by Parott[86,87] outlines the experimental investigations on the production and properties of high-strength concrete, pursued at the Cement and Concrete Association, London.

9.2 ERNTROY AND SHACKLOCK'S EMPIRICAL GRAPHS

Extensive experimental investigations by Erntroy and Shacklock[88] have indicated that in high-strength concrete mixes, workability, type and maximum size of aggregate and the strength requirement, influence the selection of the water/cement ratio. Crushed rock aggregates, being angular, produce stronger concretes at the same age in comparison with gravel aggregates. In the case of high-strength mixes, in addition to the type of coarse aggregate, either the aggregate/cement ratio or the workability has to be known in order to choose the water/cement ratio for a required strength. For purposes of design, Erntroy and Shacklock have suggested empirical graphs relating compressive strength to an arbitrary 'Reference Number' for concretes made with crushed granite coarse aggregates and irregular gravel. These graphs are reproduced in Figs 9.1 and 9.2 for mixes with ordinary Portland cement and in Figs 9.3 and 9.4 for mixes with rapid hardening Portland cement.

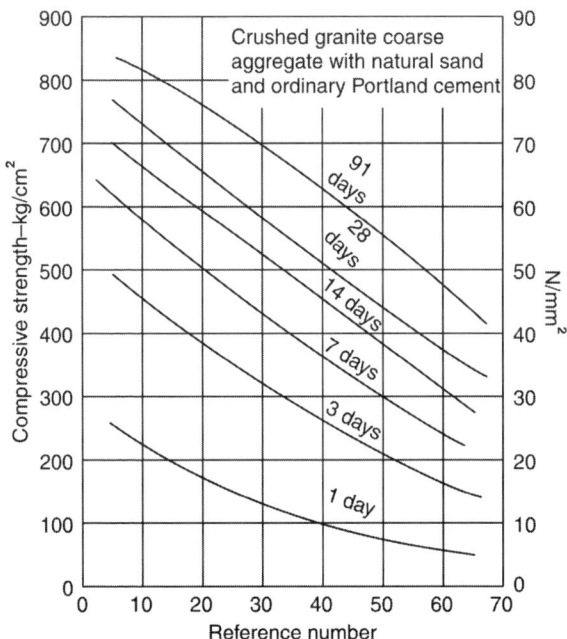

Fig. 9.1: Relation between compressive strength and reference number for mixes containing crushed granite coarse aggregate, natural sand and ordinary Portland cement

The relation between water/cement ratio and reference number for 20 and 10 mm maximum size aggregates is shown in Fig. 9.5 in which four different degrees of workability are considered. The range of the degrees of workability varying from 'Extremely Low' to 'High' corresponds to the compacting factor values of 0.65 and 0.95,

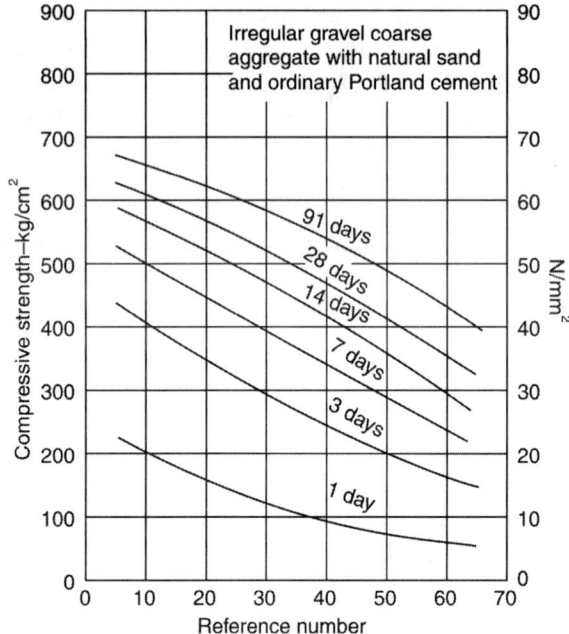

Fig. 9.2: Relation between compressive strength and reference number for mixes containing irregular gravel coarse aggregate, natural sand and ordinary Portland cement

respectively. The relations between the aggregate/cement and water/cement ratios to achieve the desired workability with a given type and maximum size of aggregate are compiled in Tables 9.1 and 9.2 for two different types of cement. The limitations of these design tables being that they were obtained with aggregates containing 30 per cent of material passing the 4.75 mm IS sieve and if other gradings are used, suitable adjustments have to be made. The aggregates available at site may be suitably combined to obtain the type grading curve no. 1 of Fig. 2.2. In view of the considerable variations in the properties of aggregates, it is generally recommended that trial mixes be first made and suitable adjustments in grading and mix proportions affected to achieve the desired results.

9.3 MIX DESIGN PROCEDURE

(a) The mean design strength obtained by applying suitable control factors to the specified minimum strength.

(b) For a given type of cement and aggregate used, the reference number corresponding to the design strength at a particular age is interpolated from Figs 9.1 to 9.4.

(c) The water/cement ratio to achieve the required workability is obtained from Fig. 9.5 for aggregates with maximum size of 20 and 10 mm.

(d) The aggregate/cement ratio to give the desired workability with the known water/cement ratio is obtained from Tables 9.1 and 9.2.

(e) Knowing the water/cement and aggregate/cement ratio and the specific gravities of the ingredients of the mix, the cement content is obtained by the absolute volume method outlined in Sec. 6.4.

(f) Batch quantities are worked out after adjustments for moisture content in the aggregates.

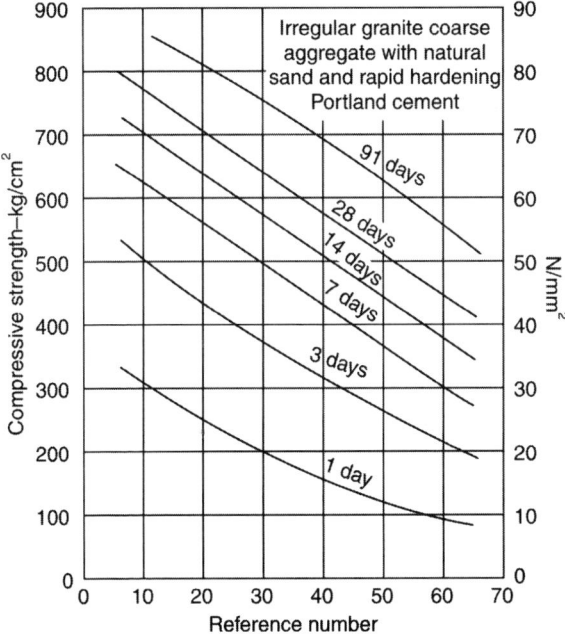

Fig. 9.3: Relation between compressive strength and reference number for mixes containing crushed granite coarse aggregate, natural sand and rapid hardening Portland cement

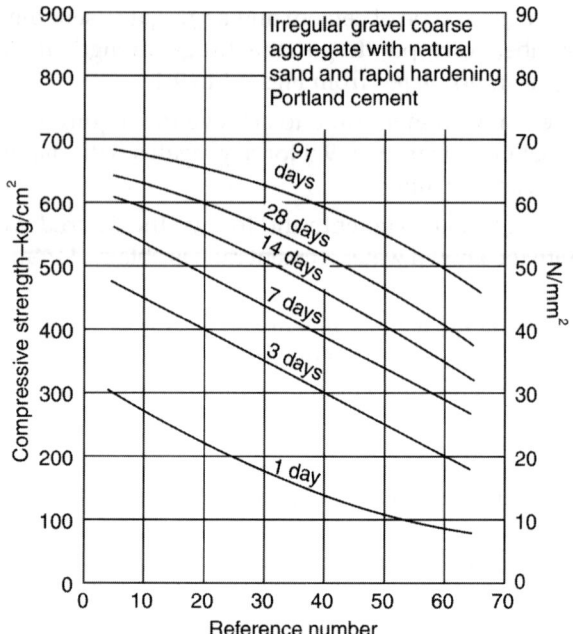

Fig. 9.4: Relation between compressive strength and reference number for mixes containing irregular gravel coarse aggregate, natural sand and rapid hardening Portland cement

9.4 MIX DESIGN EXAMPLES

9.4.1 Design a High-strength Concrete Mix for Use in the Production of Precast Prestressed Concrete Elements, to Suit the Following Requirements

Specified 28 days works cube strength   = 500 kg/cm^2

Very good degree of control; control factor = 0.80

Degree of workability: Very low

Type of cement: Ordinary Portland

Type of coarse aggregate: Crushed granite (Angular)

Maximum size: 10 mm

Type of fine aggregate: Natural sand

Specific gravity of cement   = 3.15

Sand   = 2.60

Coarse aggregate   = 2.50

Table 9.1: Aggregate/cement ratio (by weight) required to give four degrees of workability with different water/cement ratios using ordinary Portland cement

Type of coarse aggregate*	Irregular gravel								Crushed granite							
Maximum size of aggregate	20 mm				10 mm				20 mm				10 mm			
Degree of workability+	EL	VL	L	M	EL	VL	L	M	EL	VL	L	M	EL	VL	L	M
Water/cement ratio 0.30 by weight	3.0	–	–	–	2.4	–	–	–	3.3	–	–	–	2.9	–	–	–
0.32	3.8	2.5	–	–	3.2	–	–	–	4.0	2.6	–	–	3.6	2.3	–	–
0.34	4.5	3.0	2.5	–	3.9	2.6	–	–	4.6	3.2	2.6	–	4.2	2.8	2.3	–
0.36	5.2	3.5	3.0	2.5	4.6	3.1	2.6	–	5.2	3.6	3.1	2.6	4.7	3.2	2.7	2.3
0.38	–	4.0	3.4	2.9	5.2	3.5	3.0	2.5	–	4.1	3.5	2.9	5.2	3.6	3.0	2.6
0.40	–	4.4	3.8	3.2	–	3.9	3.3	2.7	–	4.5	3.8	3.2	–	4.0	3.3	2.9
0.42	–	4.9	4.1	3.5	–	4.3	3.6	3.0	–	4.9	4.2	3.5	–	4.4	3.6	3.1
0.44	–	5.3	4.5	3.8	–	4.7	3.9	3.3	–	5.3	4.5	3.7	–	4.8	3.9	3.3
0.46	–	–	4.8	4.1	–	5.1	4.2	3.6	–	–	4.8	4.0	–	5.1	4.2	3.6
0.48	–	–	5.2	4.4	–	5.4	4.5	3.8	–	–	5.1	4.2	–	5.5	4.5	3.8
0.50	–	–	5.5	4.7	–	–	4.8	4.0	–	–	5.4	4.5	–	–	4.7	4.0

* Natural sand used in combination with both types of coarse aggregate

+ EL = 'Extremely low'
  VL = 'Very low'
  L = 'Low'
  M = 'Medium'

Table 9.2: Aggregate/cement ratio (by weight) required to give four degrees of workability with different water/cement ratios using hardening Portland cement

Type of coarse aggregate*	Irregular gravel								Crushed granite							
Maximum size of aggregate	20 mm				10 mm				20 mm				10 mm			
Degree of workability(†)	EL	VL	L	M	EL	VL	L	M	EL	VL	L	M	EL	VL	L	M
Water/cement ratio by weight 0.32	2.6	–	–	–	–	–	–	–	2.9	–	–	–	2.5	–	–	–
0.34	3.4	2.2	–	–	2.8	–	–	–	3.6	2.4	–	–	3.2	–	–	–
0.36	4.1	2.7	2.3	–	3.5	2.4	–	–	4.3	2.9	2.4	–	3.9	2.5	–	–
0.38	4.8	3.2	2.8	2.3	4.2	2.9	2.4	–	4.9	3.4	2.9	2.4	4.5	3.0	2.5	–
0.40	5.5	3.7	3.2	2.7	4.9	3.3	2.8	2.3	5.5	3.9	3.3	2.7	5.5	3.4	2.9	2.4
0.42	–	4.2	3.6	3.0	–	3.7	3.0	2.6	–	4.2	3.6	3.0	–	3.8	3.2	2.7
0.44	–	4.6	4.0	3.4	–	4.1	3.5	2.9	–	4.7	4.0	3.3	–	4.2	3.5	3.0
0.46	–	5.0	4.3	3.7	–	4.5	3.8	3.2	–	5.1	4.3	3.6	–	5.0	4.1	3.4
0.48	–	5.5	4.7	4.0	–	4.9	4.1	3.5	–	5.5	4.6	3.9	–	5.0	4.1	3.4
0.50	–	–	5.0	4.3	–	5.2	4.4	3.7	–	–	4.9	4.1	–	5.3	4.4	3.7

(*) Natural sand used in combination with both types of coarse aggregate

(†) EL = 'Extremely low'

    VL = 'Very low'

    L = 'Low'

    M = 'Medium'

The fine and coarse aggregates contain 5 and 1 per cent moisture respectively and they have a grading characteristic which is detailed above.

IS sieve size	Percentage passing	
	Coarse aggregate	Fine aggregate
20 mm	100	–
10 mm	96	100
4.75 mm	8	98
2.36 mm	–	80
1.18 mm	–	65
600 micron	–	50
300 micron	–	10
150 micron	–	0

Design of mix

$$\text{Average strength} = \frac{500}{0.80} \quad = \ 630 \ \text{kg/cm}^2$$

Reference number (Fig. 9.1) $\quad = \ 25$

Water/cement ratio (Fig. 9.5) $\quad = \ 0.35$

For 10 mm maximum size aggregate and 'Very low' workability Aggregate/cement ratio for the desired workability (Table 9.1) = 3.2. The aggregates are combined by the graphical method shown in Fig. 9.6, so that 30 per cent of the material passes through the 4.75 mm IS sieve.

Ratio of fine to total aggregate $\ = \ 2.5\%$

Required proportions by weight of dry materials.

Cement	:	FA	:	CA	:	Water
1	:	$\left(\dfrac{25}{100}{\times}32\right)$	:	$\left(\dfrac{75}{100}{\times}3.2\right)$	:	0.35
1	:	0.8	:	2.4	:	0.35

If C = weight of cement required per m^3 of concrete

Then

$$\frac{C}{3.15} + \frac{0.8C}{2.6} + \frac{2.4C}{2.5} + \frac{0.35C}{1} = 100$$

$\therefore C = 520$ kg

Batch quantities per m^3 of concrete,

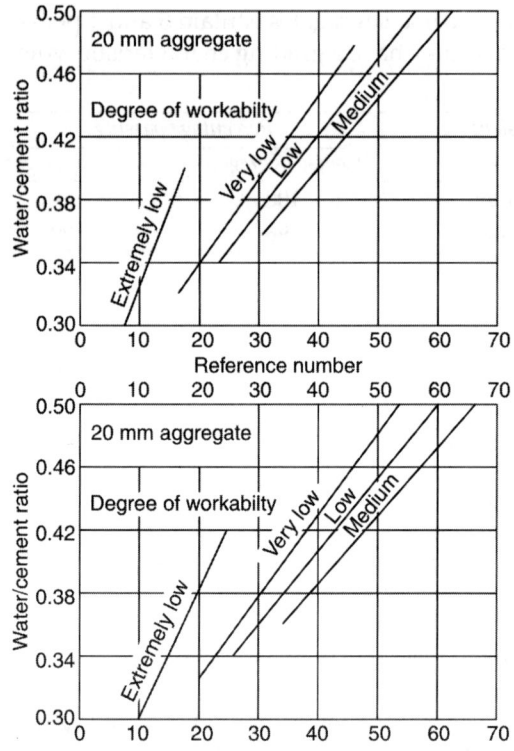

Fig. 9.5: Relation between water/cement ratio and reference number for 10 and 20 mm maximum size aggregates

Ingredient	kg	
	Dry aggregates	Moist aggregates
Cement	520	520
Water	182	148
Fine aggregates	416	437
Coarse aggregates	1250	1263

9.4.2 Design the Concrete Mix and Calculate the Batch Quantities Required for a Cubic Metre of Concrete to Suit the following Data

Average 7 days cylinder strength = 350 kg/cm²

Degree of workability = Low

    Type of cement: Rapid hardening Portland

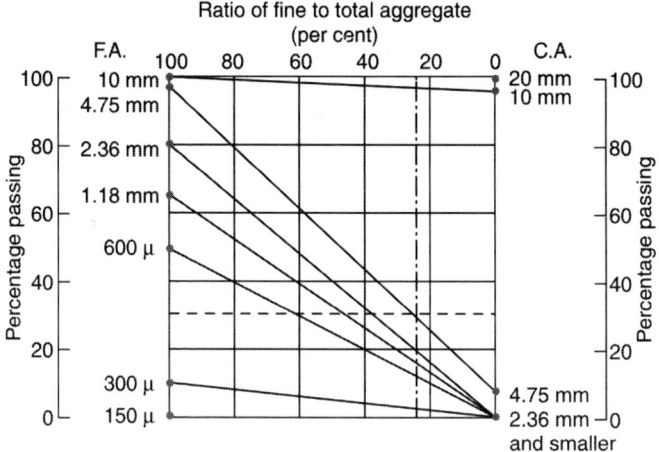

Fig. 9.6: Combining of coarse and fine aggregates

Type of coarse aggregate: Irregular gravel (20 mm maximum size)

Specific gravities of cement, sand and gravel are 3.15, 2.60 and 2.50, respectively.

The fine and coarse aggregates are stockpiles separately at site and having the following gradings:

I.S. sieve size	Percentage passing	
	Coarse aggregate	Fine aggregate
20 mm	100	–
10 mm	45	–
4.75 mm	–	100
2.36 mm	–	77
1.18 mm	–	53
600 micron	–	30
300 micron	–	10
150 micron	–	–

Design of mix

Average design cube strength,

Corresponding to cylinder strength (Fig. 5.2) = 400 kg/cm²

Reference number (Fig. 9.4)   = 40

Water/cement ratio (Fig. 9.5)   = 42

(For 20 mm maximum size aggregate with 'Low' workability)

Aggregate/cement ratio for the desired workability (Table 9.2) = 3.6

The aggregates are combined by the graphical method so that 30 per cent of the combined aggregate passes through the 4.75 mm IS sieve. This results in a ratio of fine to total aggregate = 30%.

Required proportions by weight of dry materials.

Cement	:	FA	:	CA	:	Water
1	:	$\left(\dfrac{30}{100}\times3.6\right)$	:	$\left(\dfrac{70}{100}\times3.6\right)$	:	0.42
1	:	1.08	:	2.52	:	0.42

If C = weight of cement required per m³ of concrete,

Then

$$\frac{C}{3.15}+\frac{1.08C}{2.6}+\frac{2.52C}{2.5}+\frac{0.42C}{1}=1000$$

$\therefore C = 465$ kg

Batch quantities per m³ of concrete,

Cement = 465 kg

Water = (0.42 × 465) = 196 kg

F.A. = (1.08 × 465) = 504 kg

C.A. = (2.52 × 465) = 1180 kg

9.5 EXPERIMENTAL REVIEW OF HIGH-STRENGTH CONCRETE MIXES WITH SUPERPLASTICIZERS

9.5.1 General Features

High-strength concrete mixes were designed using C-33 and C-53 grade ordinary Portland cements conforming to IS: 269–1967[14] and IS: 10262–1982[179] codes using superplasticizers. Several well-established empirical methods are available for designing the mix proportions of high-strength concrete of grades exceeding M-60. In the present investigation, the following methods were used to estimate the tentative mix proportions.

    (i) Erntroy and Shacklock's method[88]

    (ii) Teychenne, Franklin and Erntroy's method[182]

Conventional crushed granite coarse aggregate of maximum size 20 and 10 mm and river sand conforming to the grading requirements of IS: 380–1963 were used in the mixes. High-strength concrete mixes generally comprise very low water/cement ratios. Hence to improve workability, superplasticizers are invariably used to facilitate proper compaction resulting in dense concrete. The experimental investigations were conducted at the concrete laboratories of M S Ramaiah Institute of Technology, Bangalore as a part of the post-graduate research programme.

9.5.2 High-strength Concrete using C-33 Grade Portland Cement

(a) *Materials and mix proportions*: Ordinary Portland cement (ACC) of C-33 grade conforming to IS: 269–1967 code specifications. Crushed granite coarse aggregate of 20 and 10 mm maximum sizes conforming to the grading requirements of IS: 383–1963. Superplasticizers: Conplast-337 supplied by Fosroc Chemicals Ltd.

Degree of workability assumed: Extremely low since vibration were available in the laboratory.

Degree of control: Very good

Value of standard deviation specified in IS: 10262 are $s = 6.8$ and $t = 1.65$

Characteristic compressive strength of concrete $= f_{ck} = 60 \text{ N/mm}^2$

$$
\begin{aligned}
\text{Target mean strength} &= F_{ck} = (f_{ck} + ts) \\
&= (60 + 1.65 \times 6.8) \\
&= 72 \text{ N/mm}^2
\end{aligned}
$$

The design mix proportions obtained by the two methods are compiled in Table 9.3.

Table 9.3: Design mix proportions

Design method	Maximum size of C.A.	Mix proportions (by weight)			
		Cement	F.A.	C.A.	W/C ratio
Erntroy and Shacklock	20 mm	1	0.8	2.40	0.3
	10 mm	1	0.7	1.45	0.3
Teychenne, Franklin	20 mm	1	0.82	2.48	0.3
and Erntroy	10 mm	1	0.99	2.32	0.3

Based on the results of mix design, it was proposed to modify the design mix and adopt mix proportions which W/C ratios varying from 0.28 to 0.3 and the dosage of superplasticizer varying from 0 to 1.5 per cent by weight of cement.

The modified mix proportions used in the experimental investigation are compiled in Table 9.4.

(b) *Mixing of concrete and test specimens*: The ingredients were thoroughly mixed dry and then water with superplasticizer was added to the mix. The concrete was mixed in a laboratory

Table 9.4: Experimental mix proportions

Mix group	Maximum C.A. size	Superplasticizer conplast-337 (per cent by wt. of cement)	Mix proportions (by weight)			
			C	F.A.	C.A.	W/C
G-1	20 mm	0.0	1	1.0	2.0	0.30
		0.0	1	1.0	2.0	0.28
		0.5	1	1.0	2.0	0.28
		1.0	1	1.0	2.0	0.28
		1.5	1	1.0	2.0	0.28
G-2	10 mm	0.0	1	0.5	1.5	0.30
		0.0	1	0.5	1.5	0.28
		0.5	1	0.5	1.5	0.28
		1.0	1	0.5	1.5	0.28
		1.5	1	0.5	1.5	0.28
G-3	10 mm	0.0	1	0.2	1.8	0.30
		0.0	1	0.2	1.8	0.28
		0.5	1	0.2	1.8	0.28
		1.0	1	0.2	1.8	0.28
		1.5	1	0.2	1.8	0.28

type drum mixer. The thoroughly mixed concrete was placed in 3 to 4 layers in the 150 mm standard cube moulds and compacted using a platform vibrator. 8 cubes were cast for each mix with varying percentages of superplasticizer.

(c) *Workability of fresh concrete*: The workability tests comprising the slump, compaction factor and Vebe time were conducted according to the procedure recommended in the Indian Standard IS: 1199[77]. A comparative analysis of the workability characteristics of fresh concrete of the various mixes comprising the groups G-1, G-2 and G-3 is presented in Figs 9.7 to 9.9. The general trend of increasing workability with the percentage increase in the superplasticizer admixture is clearly evident in the test results.

In general, the test results indicate that an increase of admixture content from 0 to 1.5 per cent resulted in the increase of compaction factor from 0.7 to 0.9 while the Vebe time decreased from 35 to 20 seconds. The slump test was of little use with stiff concrete mixes since the variation of

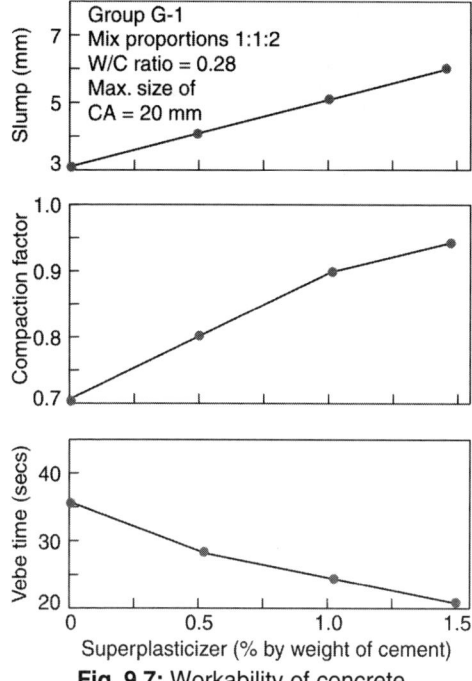

Fig. 9.7: Workability of concrete

slump with the addition of super plasticizer was in the narrow range of 0 to 5 mm.

(d) *Compressive strength of concrete*: The standard cube specimens were tested for 28 days compressive strength in a 2000 kN capacity hydraulic compression testing machine in accordance with the procedure specified in the Indian Standard IS: 516[72]. The results of compressive strength with varying percentages of admixture are shown in Figs 9.10 to 9.12 for the mix groups G-1, G-2 and G-3. Significant increase in compressive strength have been recorded when the super-plasticizer content varied from 0 to 1.5 per cent. Maximum compressive strength of 66 N/mm² was recorded for the G-3 mix having mix proportions of 1 : 0.2 : 1.8 with a water/cement ratio of 0.28 and an aggregate/cement ratio of 2.

In general, the compressive strength increased at a faster rate for small percentages of admixture (up to 0.5 per cent) and the rate of increase decreased for the higher percentages. The

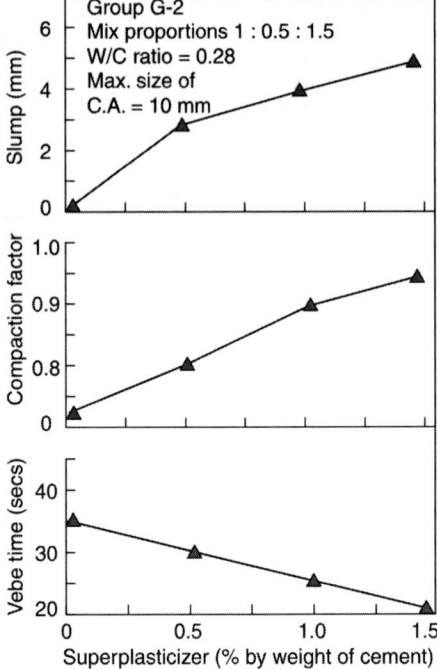

Fig. 9.8: Workability of concrete

percentage increase in compressive strength was observed to be 35, 45 and 50 per cent for admixture contents of 0.5, 1.0 and 1.5 per cent, respectively.

9.5.3 High-strength Concrete using C-53 Grade Portland Cement

(a) *Material and mix proportions*: Type of cement—ordinary Portland cement C-53 grade conforming to IS: 12269–1987 code specifications.

Coarse aggregate: Crushed granite aggregate of maximum size 20 mm with a specific gravity of 2.60.

Fine aggregate: River sand with a specific gravity of 2.59.

Superplasticizer: Conplast SP-430 marketed by Fosroc Chemicals Limited.

The aggregates conformed to the grading requirements of IS: 383 code specifications.

Characteristic compressive strength $= f_{ck} = 70$ N/mm^2

Degree of workability: Extremely low

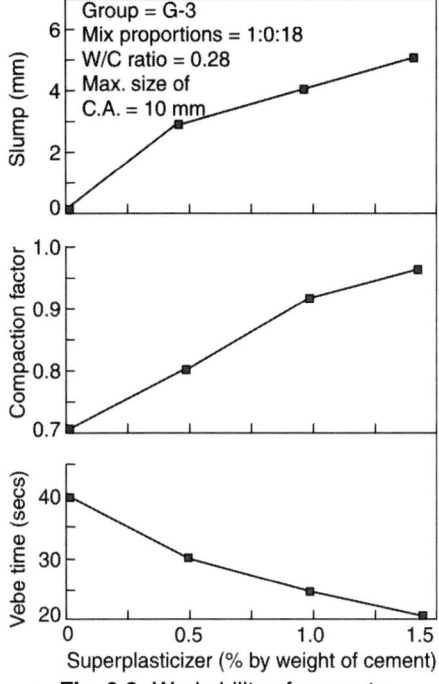

Fig. 9.9: Workability of concrete

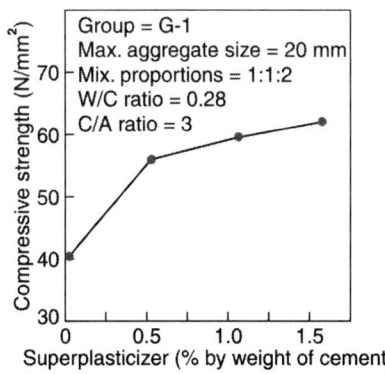

Fig. 9.10: Variation of compressive strength with superplasticizer

Degree of control: Very good

$$\text{Target mean strength} = F_{ck} = (f_{ck} + t.s)$$
$$= (70 + 1.65 \times 6.8)$$
$$= 81.22 \text{ N/mm}^2$$

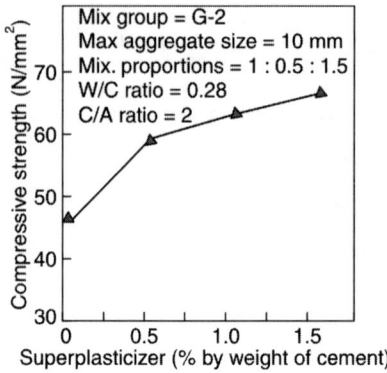

Fig. 9.11: Variation of compressive strength with superplasticizer

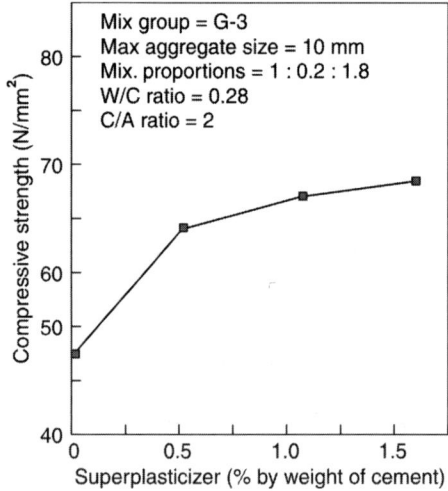

Fig. 9.12: Variation of compressive strength with superplasticizer

The design mix proportions obtained by the two methods used are compiled in Table 9.5.

Table 9.5: Design mix proportions

Design method	Mix proportions (by weight)			
	Cement	FA	CA	W/C ratio
Erntroy and Shacklock	1	0.75	2.25	0.3
Teychenne, Franklin and Erntroy	1	0.82	2.44	0.3

The designed mix proportions were modified in conjunction with the use of superplasticizer Conplast SP-430 marketed by Fosroc Chemical Limited.

The details of mix proportions adopted for the experimental investigations are compiled in Table 9.6.

Table 9.6: Experimental mix proportions

Mix group	Super-plasticizer	Conplast SP-430 (per cent by weight) mix proportions (by weight)			
		C	FA	CA	W/C ration
S-1	0.0	1	1	2	0.28
	0.5	1	1	2	0.28
	1.0	1	1	2	0.28
	1.5	1	1	2	0.28
S-2	0.0	1	1	2	0.30
	0.5	1	1	2	0.30
	1.0	1	1	2	0.30
	1.5	1	1	2	0.30

(b) *Mixing of concrete and test specimens*: The mix ingredients were thoroughly mixed dry and then water mixed with super-plasticizer was added and the concrete was placed and compacted in 150 mm size standard cube moulds in several layers using a platform vibrator. Six cubes were cast for each mix with varying percentages of superplasticizer for tests at 7 and 28 days.

(c) *Workability of fresh concrete*: The workability tests comprising the slump, compaction factor and Vebe time were conducted according to the procedure outlined in IS: 1199[70]. A comparative analysis of the workability characteristics of fresh concrete of the various mixes comprising the groups S-1 and S-2 is presented in Figs 9.13 and 9.14.

In the S-1 group with a W/C ratio of 0.28, the mix did not exhibit any slump while the compaction factor increased with the dosage of superplasticizer and the vebe time correspondingly decreased showing significant improvement in the workability characteristics. The same trend was observed in the S-2 group. In this group, a small increase of slump (5 mm) was recorded for admixture percentage of 0.5 to 1.5 per cent.

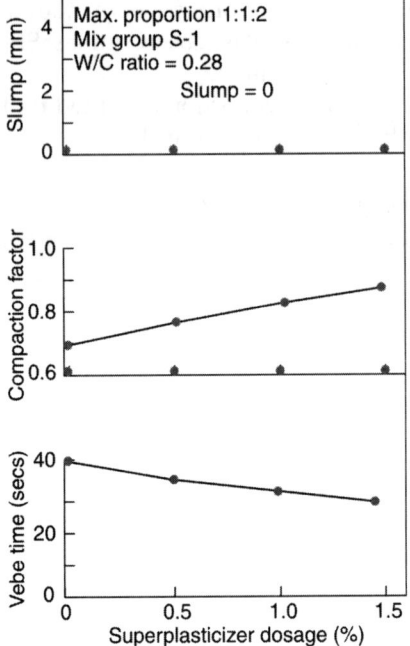

Fig. 9.13: Workability of concrete

The S-2 group mix indicated improved workability characteristics since the W/C ratio was 0.3 in comparison with the S-1 group which had a W/C ratio of 0.28.

The workability tests indicate that the use of superplasticizer in stiff mixes with low water/cement ratios significantly improves the workability leading to better compaction and higher strength of hardened concrete.

(d) *Compressive strength of concrete*: The standard 150 mm size cube specimens were tested in a 200 kN compression testing machine for compressive strength at 7 and 28 days according to the procedure prescribed in IS: 516[72].

The results of average compressive strength of various cube specimens of group S-1 and S-2 at 7 and 28 days are shown in Figs 9.15 and 9.16. The compressive strength increased significantly with the increase in dosage of superplasticizer from 0 to 1.5 per cent.

The percentage increase in compressive strength at 7 days was observed to be 11, 28 and 35 per cent for water/cement ratio

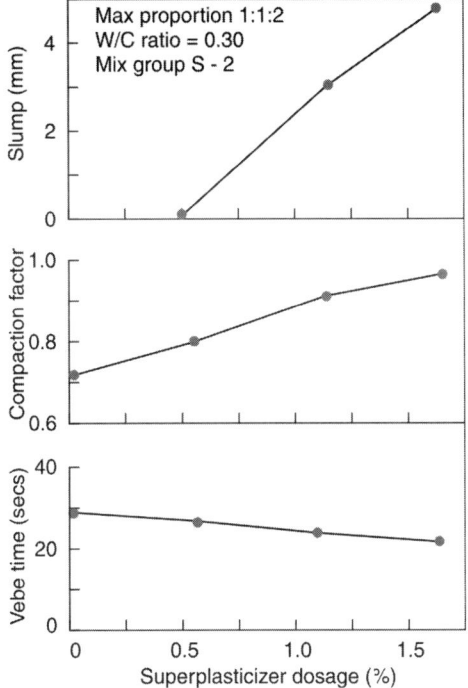

Fig. 9.14: Workability of concrete

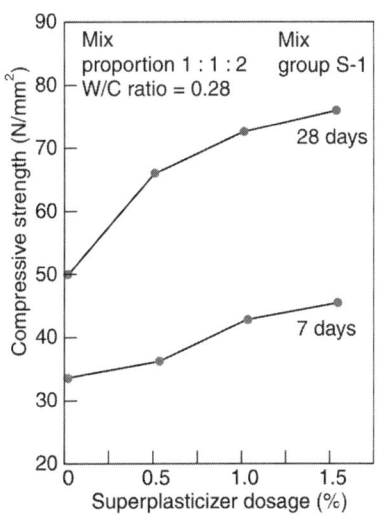

Fig. 9.15: Variation of compressive strength with superplasticizer content

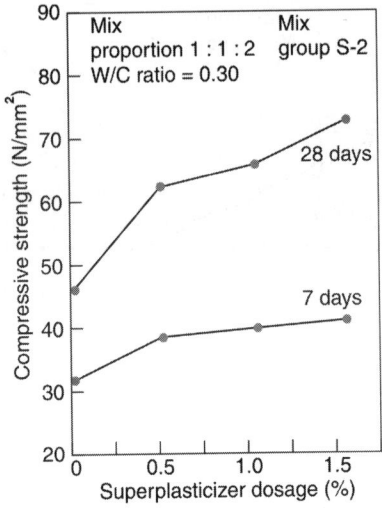

Fig. 9.16: Variation of compressive strength with superplasticizer content

of 0.3 for admixture dosage of 0.5, 1.0 and 1.5 per cent, respectively. The percentage increase in compressive strength at 28 days was observed to be 32, 44 and 50 per cent for water/cement ratio of 0.3 for admixture dosage of 0.5, 1.0 and 1.5 per cent, respectively.

The maximum compressive strength of 75 N/mm² at 28 days was recorded for 1 : 1 : 2 concrete mix proportions with a superplasticizer content of 1.5 per cent.

Design of High Alumina Cement Concrete Mixes

10.1 STRENGTH DEVELOPMENT OF ALUMINOUS CEMENT CONCRETE

High alumina cement concrete has a very high rate of strength development and it is found that as much as 80 per cent of its ultimate compressive strength is achieved at the age of 24 hours, and even at 6 to 8 hours, the concrete is strong enough for the side form work to be struck and for the preparation for further concreting to take place. Neville[89] has reported the strength development of aluminous cement concretes with water/cement ratios in the range from 0.5 to 0.90, at different ages. With a water/cement ratio of 0.35, strength of the order of 700 kg/cm^2 may be reached at the age of two weeks.

Aluminous cement is slow setting, but the final set follows the initial set more rapidly in comparison with the Portland cement. According to the British Standard[90], the initial setting time is not to exceed 6 hours after mixing, while the final set should take place in not more than two hours after the initial set.

The compressive strength attainable with high alumina cement and its relation with water/cement ratio is given in Fig. 10.1, as reported by Newman[91]. Aluminous concretes gain strength even at very low temperatures and in this respect it is assisted by the high rate of heat evolution. A concrete at different temperatures has been reported by Hussey and Robson[92]. Gottlieb[93] has reported the excellent results obtained with high alumina cement concretes when used at temperatures of the order of 0°C.

The strength of high alumina concrete is adversely affected by a rise in temperature, when the concrete is stored under moist conditions. The loss of strength is attributed to the conversion of unstable aluminate hydrates at temperatures exceeding 80° F[94]. Although the rate of conversion is found to be very slow at ordinary temperatures, over a long period, the loss in strength being cumulative, may be as much as 50 to 70 per cent of the maximum strength after 21 years, according to date of Neville[95]. In view of these adverse effects, the European Committee on concrete has recommended that the use of high alumina cement be subject to special justification.

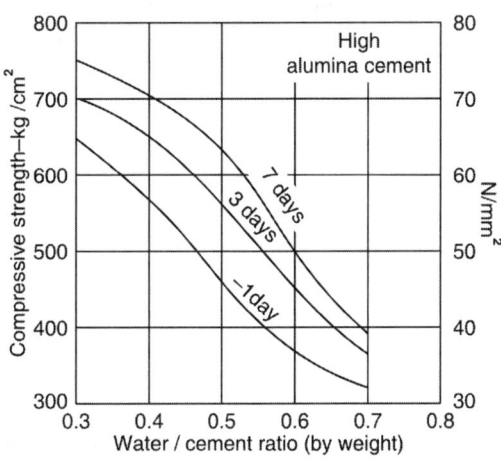

Fig. 10.1: Relationship of compressive strength and water/cement ratio for high alumina cement

10.2 MIX DESIGN USING EMPIRICAL TABLES

For many years, the design of high alumina cement concrete mixes has been based on nominal mix proportions obtained by trial mixes. The experimental investigations by Newman[91] has resulted in empirical design tables which are similar to those recommended by the Road Note No. 4[12]. The workability, aggregate/cement ratio and the effective water/cement ratio for the three type gradings (Refer Figs 2.1 to 2.3) are given in Tables 10.1 and 10.2 for 20 mm maximum size irregular gravel and angular aggregate, respectively. Type grading no. 1 is not included since the mixes with the high proportion of sand are liable to segregation.

The method of proportioning of a mix is the same as that recommended by Road Note No. 4 for Portland cement concrete mixes in chapter 9. However, the method is limited to the type of

Table 10.1: Aggregate/cement ratio required to four degrees of workability with different effective water/cement ratios and gradings for 20 mm angular aggregate

Degree of workability		Very low			Low			Medium			High		
Grading no.		2	3	4	2	3	4	2	3	4	2	3	4
Effective water/cement	0.35	4.2	3.8	3.6	3.0	2.9	2.7	–	–	–	–	–	–
ratio by weight	0.40	5.2	4.4	4.2	3.3	3.8	3.6	3.6	3.2	3.0	3.0	2.9	2.7
	0.45	5.8	5.1	4.8	4.8	4.3	4.1	4.0	3.7	3.5	3.7	3.4	3.3
	0.50	6.4	5.7	5.4	5.3	4.9	4.6	4.4	4.2	3.9	4.0	3.9	3.6
	0.55	7.0	6.2	5.8	5.8	5.4	5.1	4.8	4.7	4.4	4.3	4.4	4.0
	0.60	7.6	6.8	6.4	6.3	5.9	5.6	5.4	5.2	4.9	x	4.8	4.6
	0.65	8.0	7.3	7.0	6.9	6.3	6.0	5.7	5.5	5.3	x	5.2	4.9
	0.70	–	7.7	7.5	7.3	6.8	6.5	6.1	5.9	5.7	x	5.5	5.3
	0.75	–	–	–	7.6	7.1	6.9	x	6.2	6.1	x	5.8	5.7

– Indicates that the mix was outside the range tested

x Indicates that the mix might segregate

Table 10.2: Aggregate/cement ratio required to four degrees of workability with different effective water/cement ratios and grading for 20 mm irregular gravel aggregate

Degree of workability	Very low			Low			Medium			High		
Grading no.	2	3	4	2	3	4	2	3	4	2	3	4
Effective water/cement ratio by weight												
0.30	3.5	3.5	3.0	2.8	2.5	2.5	–	–	–	–	–	–
0.35	4.5	4.3	4.0	3.7	3.6	3.3	3.0	3.0	2.7	2.5	2.6	2.4
0.40	5.7	5.2	4.9	4.6	4.4	4.2	3.7	3.7	3.5	3.3	3.3	3.1
0.45	7.1	6.2	5.7	5.6	5.3	4.9	4.3	4.4	4.2	3.9	3.9	3.9
0.50	8.0	7.2	6.5	6.6	6.0	5.4	5.1	5.0	4.8	4.6	4.6	4.5
0.55	–	8.1	7.2	7.2	6.8	6.0	5.7	5.5	5.2	5.0	5.0	4.9
0.60	–	–	7.7	7.7	7.4	6.6	6.1	6.1	5.7	5.5	5.4	5.3
0.65	–	–	8.2	8.0	7.8	7.0	x	6.5	6.1	x	5.9	5.7
0.70	–	–	–	–	–	7.3	x	x	6.5	x	6.2	6.1

– Indicates that the mix was outside the range tested

x Indicates that the mix might segregate

aggregates and the range of parameters used to formulated the empirical design tables.

The water/cement ratio required to produce the desired mean strength is obtained from Fig. 10.1 and the aggregate/cement ratio to give the desired workability is directly obtained from Tables 10.1 and 10.2. If the aggregates available at site do not conform to the type gradings, they should be suitably combined by the analytical or graphical methods to conform to a standard types grading.

10.3 MIX DESIGN EXAMPLES

10.3.1 Design a Concrete Mix Suitable for Use in Prestressed Concrete Work using High Alumina Cement

The strength and other requirements are as given below:

Specified minimum 7 days cube strength = 500 kg/cm^2
Control factor = 0.75
Type of coarse aggregate: Irregular gravel of 20 mm maximum size
Type of fine aggregate: Natural sand
Workability: Low
The grading of aggregates is such that the ratio of fine to total aggregates should be 28 per cent to obtain the type grading curve no. 2.
Specific gravity of cement = 3.10
Specific gravity of aggregate = 2.60

Design

Seven days average strength = $\dfrac{500}{0.75}$ = 670 kg/cm^2

Water/cement ratio (Fig. 10.1) = 0.45
Aggregate/cement ratio (Table 10.1) = 5.6
Proportion of ingredients:

Cement	:	F.A.	:	C.A.	:	Water
1	:	$\left(\dfrac{28}{100}\times5.6\right)$	:	$\left(\dfrac{72}{100}\times5.6\right)$	:	0.45
1	:	1.57	:	4.05	:	0.45

If C = Cement required per m^3 of concrete,
Then

$$\frac{C}{3.1}+\frac{1.57C}{2.6}+\frac{4.05C}{2.6}+\frac{0.45C}{1} = 1000$$

$\therefore C \quad = 345 \text{ kg}$
Batch quantities:
Cement = 345 kg
F.A. = 540 kg
C.A. = 1400 kg
Water = 156 kg

10.3.2 Design a Concrete Mix to Suit the Following Conditions

Specified minimum compressive strength:

At 7 days	$= 400 \text{ kg/cm}^2$
Control factor	$= 0.8$
Workability required	= Medium
Type of cement	= High alumina
Type of coarse aggregate	= Crushed granite
	20 mm maximum size
Type of fine aggregate	= Natural sand

The aggregates available at work site have the following grading.

Specific gravity of cement = 3.10
F.A. = 2.60
C.A. = 2.70

I.S. sieve size	Percentage passing	
	C.A.	F.A.
20 mm	100	–
10 mm	55	–
4.75 mm	–	100
2.36 mm	–	80
1.18 mm	–	60
600 micron	–	40
300 micron	–	10
150 micron	–	–

Seven days average strength $= \dfrac{400}{0.8} = 500 \text{ kg/cm}^2$

Water/cement ratio (Fig. 10.1) = 0.60
Aggregate/cement ratio (Table 10.2) = 5.4

The aggregates are combined by graphical method to correspond to type grading curve no. 2, which gives the ratio of fine to total aggregate at 35%.

Proportions of ingredients:

Cement	:	F.A.	:	C.A.	:	Water
1	:	$\left(\dfrac{35}{100}\times 5.4\right)$	:	$\left(\dfrac{65}{100}\times 5.4\right)$	:	0.60
1	:	1.9	:	3.5	:	0.60

If C = Weight of cement per m^3 of concrete

Then

$$\frac{C}{3.1}+\frac{1.9C}{2.7}+\frac{3.5C}{2.7}+\frac{0.60C}{1} \;=\; 1000$$

$$\therefore C \qquad\qquad\qquad = 340 \text{ kg}$$

Weights of materials required per m^3 of concrete

Cement = 340 kg

Water = 204 kg

F.A. = 650 kg

C.A. = 1200 kg

11

Concrete Mix Design for Strength in Flexure

11.1 CRITERION OF FLEXURAL STRENGTH

Generally, concrete used in the construction industry should invariably conform to a specification of minimum compressive strength, which is considered as an overall measure of the quality of concrete. However, in specific applications, like roads and airport runways, the flexural strength of concrete is equally important as the compressive strength and in many cases, specifications[96] for concrete used in runways require a minimum flexural strength rather than a minimum compressive strength.

The American Concrete Institute Standard ACI 617–58[97], which covers the application of concrete pavements for both highways and airports, specifies an average 28 days compressive and flexural strength of concrete of 280 and 45 kg/cm^2 respectively as the minimum criterion of acceptance. These values indicate the importance of flexural strength as the basis of mix design in contrast to that of compressive strength.

The effects of shape and texture are particularly significant in influencing the flexural strength in the case of high-strength concrete according to the data of Kaplan[98]. The flexural strength is more sensitive to inadequate curing[99] since the effects of non-uniform shrinkage adversely affect the flexural strength more than the compressive strength. However, in the case of rich and strong mixes, air-entrainment lowers the compressive strength more than the flexural strength. Kaplan's[100] data indicate that the effect of incomplete compaction on strength is similar to that of entrained air.

11.2 RELATION BETWEEN WATER/CEMENT RATIOS AND FLEXURAL STRENGTH

When concrete is fully compacted, the water/cement ratio law generally applicable to the compressive strength is also qualitatively true in the case of flexural strength. However, in contrast to the compressive strength which mainly depends on the water/cement ratio, almost independent of the other parameters, the flexural strength is also sensitive to the type of aggregates used in the mix. Fig. 11.1 shows the relation between water/cement ratio and flexural strength of concrete according to the data reported by Wright[101].

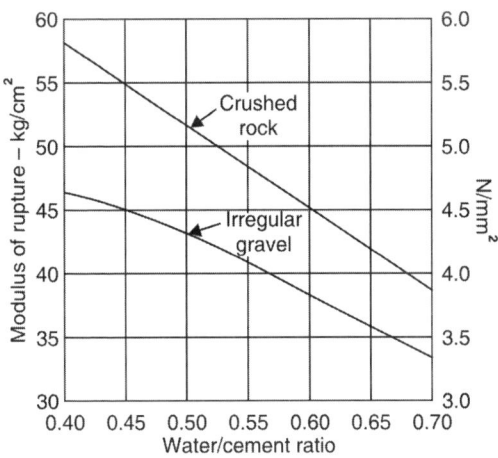

Fig. 11.1: Relation between modulus of rupture at 28 days and water/cement ratio

The modulus of rupture is generally determined by standard tests[72,73] on beam specimens under third point loading. The use of angular aggregates obtained from crushed rock result in stronger concretes in comparison with gravel aggregates when the water/cement ratio is maintained at a constant value. The difference in strength between the two types is higher for low water/cement ratios corresponding to the high strength range.

11.3 MIX PROPORTIONING FOR FLEXURAL STRENGTH

In designing concrete mixes for flexural strength, it is important to note that the leanest mix is obtained, if crushed rock is used. This is in contrast to the proportioning of mixes for compressive strength where the leanest mix is achieved with rounded aggregates. In order to produce the same workability, the angular aggregate requires a richer mix than when rounded aggregate is used for the same water/cement ratio. However, due to the significant differences in the flexural strength developed with gravel and crushed rock aggregates, it is possible to use a higher water/cement ratio with the angular aggregate than that for gravel aggregate and yet produce the same degree of workability. The use of higher water/cement ratio, permit the use of higher aggregate/cement ratios at a constant workability, thus reducing the cement in the mix.

A mix design procedure based on Wright's data and the empirical tables of Road Note No. 4 is outlined below:

(a) The average design flexural strength is computed by applying a suitable control factor to the minimum strength.

(b) The water/cement ratio required to produce the desired average strength is interpolated from Fig. 11.1, for the known types of aggregate used.

(c) The aggregate/cement ratio to give the desired workability using the water/cement ratio is selected from Tables 6.2, 6.3 and 6.4, depending upon the type and size of aggregates to suit a particular grading.

(d) The coarse and fine aggregates available at site should be combined to achieve the desired grading resulting in the most economical mix.

(e) Suitable adjustments in the mix proportions may be necessary, if the aggregates used at site are moist.

The ACI-Standard (ACI: 617–58) recommends two methods of proportioning concrete required for pavements and bases. In the first method, the mix is designed for a specified minimum flexural strength, and a suitable consistency expressed as slump of 40 to 80 mm for unvibrated concrete or 12 to 40 mm for vibrated concrete. The flexural strengths at 7 and 28 days are to be established with four cement factors and these will serve as a basis for adjusting the mix proportions in the light of 7 days test results made during the progress of the work. In the second method, the proportions are based on fixed cement content. The suggested trial proportions for concrete of specified cement content and slump are compiled in Table 11.1 in which the water/cement ratio is restricted to a narrow range varying from 0.40 to 0.50, while the aggregate cement ratio varies between the limits of 5.0 and 5.8. The lowest water/cement ratios are to be used with rounded gravel while the higher ratios are associated with the crushed gravel or rock. It can be seen from Fig. 11.1 that a specified average flexural strength of 45 kg/cm^2 is assured with these water/cement ratios.

Table 11.1: Suggested trial proportions for concrete of specified cement and slump*

Type of concrete**	Type of coarse aggregate	Water litres	Fine†	Aggregate (kg) Coarse†	
				Small	Large
Plain	Round gravel	2.27	100	79	109
Plain	Crushed gravel or stone	25.0	109	74	109
Plain	Crushed gravel	27.2	120	58	85
Air-entrained	Round gravel	20.4	87	79	120
Air-entrained	Crushed gravel	22.7	98	74	109
Air-entrained	Crushed slag	25.0	106	57	85

* Proportions intended to produced concrete containing 333 kg of cement per cubic metre with slump of 40 to 80 mm, suitable for normal machine placement. When vibration is used, slump may be reduced to about 12 to 40 mm and batch quantities adjusted accordingly to maintain same yield and cement factor

** Air content assumed to be 1 per cent for plain mixes and 5.5 per cent for air-entrained mixes

*** Aggregate weights based on assumed bulk specific-gravity, saturated surface dry of 2.65 for sand gravel and stone and 2.25 for slag. For other specific gravities, aggregate weights should be adjusted in direct proportion to the specific gravity. Aggregate weights and quantity of added mixing water must be adjusted to allow for free moisture on aggregates

† Fine aggregates assumed to be well graded natural sand of average fineness (fineness modulus about 2.6 to 2.9)

† Coarse aggregates assumed to be well graded from 40 to 4.75 mm, or 50 to 4.75 mm, furnished in two sizes, separated on the 20 or 25 mm sieve, respectively and used in the proportion of approximately 40 per cent of the small and 60 per cent of the large size

11.4 MIX DESIGN EXAMPLES

11.4.1 Design a Concrete Mix for an Airport Pavement Using the Following Data

Minimum 28 days flexural strength = 40 kg/cm^2

Control factor = 0.75

Type of cement = Ordinary Portland

Type of coarse aggregate = Crushed granite of 20 mm maximum size

Type of fine aggregate = Natural sand

Good vibration equipment available at worksite and desired workability to be very low. The aggregates are graded to conform to type grading curve no. 1 of Fig. 2.2.

Specific gravity of cement = 3.15
F.A. = 2.60
C.A. = 2.50

Design

Average 28 days flexural strength = $\dfrac{40}{0.75}$ = 54 kg/cm^2

Water/cement ratio (Fig. 11.1) = 0.46

Aggregate/cement ratio for 'Very low' workability using 20 mm angular aggregate (Grading curve no. 1) (Table 6.3c) = 5.7.

Proportions of ingredients:

Cement	:	F.A.	:	C.A.	:	Water
1	:	$\left(\dfrac{30\times5.7}{100}\right)$	:	$\left(\dfrac{70\times5.7}{100}\right)$	:	0.46
1	:	1.71	:	3.99	:	0.46

If C = weight of cement required per m^3 of concrete. Then using the absolute volume method,

$$\frac{C}{3.15}+\frac{1.71C}{2.50}+\frac{3.90C}{2.50}+\frac{0.46C}{1} = 1000$$

$\therefore$ C = 345 kg

Batch quantities per m^3 of concrete:

Cement = 330 kg
Water = 152 kg
F.A. = 565 kg
C.A. = 1130 kg

11.4.2 A Concrete Mix is Required for a Highway Pavement to Suit the Following Data

Minimum 28 days flexural strength = 35 kg/cm^2
Control factor = 0.78
Type of cement = Ordinary Portland
Type of fine aggregate = Natural sand

Heavy vibration available and 'Very low' workability required for placement of concrete.

Both irregular gravel and crushed rock aggregates of 20 mm maximum size are available at worksite.

The specific gravities of cement, fine aggregate and coarse aggregate are 3.15, 2.60 and 2.50, respectively.

Design the mix proportions required using both the types of aggregates.

Design

Average 28 days flexural strength = $\dfrac{35}{0.78}$ = 45 kg/cm^2

Using Fig. 11.1, water/cement ratio required for

(a) Irregular gravel   = 0.45

(b) Crushed rock   = 0.60

If the aggregates are graded to correspond to the Type 1 grading curve of Road Note No. 4, the aggregate/cement ratio required to give 'Very low' workability is obtained from Tables.

A/C ratio for

(a) Irregular gravel   = 6.0

(b) Angular aggregate = 7.8

Proportions of ingredients:

(a) Using gravel

Cement	:	F.A.	:	C.A.	:	Water
1	:	$\left(\dfrac{30\times6.0}{100}\right)$	:	$\left(\dfrac{70\times6.0}{100}\right)$	:	0.45
1	:	1.8	:	4.2	:	0.45

(b) Using crushed rock

1	:	$\left(\dfrac{30\times7.8}{100}\right)$	:	$\left(\dfrac{70\times7.8}{100}\right)$	:	0.60
1	:	2.34	:	5.46	:	0.60

Batch quantities per m^3 of concrete

Ingredient	Gravel aggregate (kg)	Crushed rock aggregate (kg)
Cement	318	250
Water	144	150
F.A.	572	585
C.A.	1340	1370

12

Design of Lightweight Aggregate Concrete Mixes

12.1 TYPES OF LIGHTWEIGHT CONCRETE

Basically, all types of lightweight concrete are produced by including large quantities of air in the aggregate, or in the matrix, or between the aggregate particles. Correspondingly, we have three basic types[102] of lightweight concrete generally referred to as:

(a) Lightweight aggregate concrete
(b) Aerated concrete
(c) No-fines concrete

The four major types of processed lightweight aggregate widely used in United Kingdom, United States of America and Europe, belong to the category of:

(a) Expanded shales, clays and slates produced in a rotary kiln (Leca, Kermazite)
(b) Expanded shales or clay produced on a sintering grate (Aglite, Agloporite)
(c) Slags expanded mechanically or by water jet process (Foamed slag)
(d) Sintered pulverized fuel ash aggregate (Lytag)

Lightweight concrete is now a firmly established building material throughout the world. In the United States of America alone, the yearly production of structural lightweight concrete now is over three million cubic metres. The early applications of lightweight concrete were mainly in the block making industry[103] and for producing insulating screeds. The use of lightweight concrete in reinforced concrete construction was first proposed in 1936[104], and currently the material is widely used for a variety of constructions involving plain, reinforced and even prestressed concrete[105], where high strength is an important factor.

12.2 INFLUENCE OF LIGHTWEIGHT AGGREGATE PROPERTIES ON MIX DESIGN

Many of the characteristics of a chemically stable lightweight aggregate can influence the choice of mix proportions, ultimately affecting the properties of concrete. Some of the important properties include the bulk density, grading, particle shape and absorption characteristics of the aggregates.

The density of the concrete produced depends to a large extent on the bulk density of the aggregates. The unit weight of the aggregates varies depending upon the type of aggregates and the production plant. The coarse fraction, up to 20 mm size may have dry loose unit weights from 500 to 900 kg/m³ and the finer fractions from 700 to 1100 kg/m³. The grading characteristics of the finer fractions are known to affect the properties of lightweight concrete whether used in block type mixes[106] or in fully compacted concrete[107]. The grading limits of lightweight fine aggregate specified in various standards are compiled in Table 12.1. The British Standard, BS: 3797–1990[108] covers foamed slag while the standard BS EN 13055–1:2002(E)[109] deals with the other types of lightweight aggregates, like Aglite, Leca and Lytag. The corresponding American Standard ASTM C: 330–64T[110] embodies the tentative specifications for lightweight aggregates for structural concrete. The grading of most fine lightweight aggregates is harsher and the band of gradings is narrower than for fine natural aggregates.

Many of the lightweight aggregates have particle shapes ranging from rounded to very irregular. It has been found that aggregates which are very irregular, generally produce heavier concretes due to the higher apparent specific gravity. Table 12.2 shows the typical properties of some of the British lightweight aggregates[111] having relative densities varying from 0.40 to 0.77. The aggregates generally absorb from 5 to 20 per cent of moisture by weight of dry aggregate after immersion in water for 24 hours. Hobbs[112] has shown that lightweight aggregates will rapidly absorb about 9 per cent of its volume in water quite quickly and then go on to absorb a great deal more over a long period of time. However, the effect in practice is more complicated because the aggregates are subjected to histories of alternate wetting and drying in the stock piles and as Landgren, Hanson and Preifer[113] point out, 'Initially damp lightweight aggregates usually contain more total absorbed water after a short period of water immersion than is contained in similar, initially dry aggregates after immersion for the same period'.

Table 12.1: Grading limits of lightweight fine aggregates specified in various standards

| I.S. and B.S. test sieve | B.S. 877 | B.S. EN 13055 | | Requirements of A.S.T.M. standard C: 330–64T | |
		Grading zone-L1	Grading zone-L2	Sieve size	Percentage by weight passing sieve
4.76 mm	90–100	90–100	90–100	No. 4	85–100
2.38 mm	70–100	55–95	60–100	–	–
1.18 mm	45–90	35–70	40–80	No. 16	40–80
600	20–60	20–60	30–60	–	–
300	10–30	10–30	25–40	No. 50	10–35
150	5–20	5–19	20–35	No. 100	5–25

Percentage by weight passing I.S. or B.S. sieve size

Table 12.2: Typical properties of British lightweight aggregates

Property (of 6 to 10 mm fraction)	Aglite	Foamed slag	Leca	Lytag
Particle shape	Irregular	Very irregular	Rounded	Rounded
Relative density	0.58	0.62	0.40	0.77
Apparent specific gravity	1.6	2.6	0.9	1.7
24 hours absorption				
Per cent by weight	20	19	24	13
Per cent by volume	32	49	21	22

These considerations have an important bearing upon the methods of proportioning lightweight concrete mixes. The bulk density of the aggregates varies considerably depending upon the moisture content and unless the absorption capacity of the light-weight aggregate has been fully satisfied in the mixer, the aggregate will continue to absorb water from the matrix and there will be a disconcerting loss of workability between the mixer and the point of placing. With lightweight concrete mixes, it is important to allow generous mixing times or to take steps to prewet the aggregates[114].

In view of the complexity of lightweight aggregate absorptions and the wide variety of aggregates used, the American Concrete Institute recommends a method of trial mixes proportioned on the basis of cement content at the required consistency.

In the case of structural lightweight concrete, one of the essential requirements in selecting mix proportions is to produce a workable concrete which can be easily compacted on the site. The slump test does not seem to be suitable for measuring the workability of lightweight concrete since most of the mixes tested at the Building Research station had slumps between zero and 25 mm only and yet they were capable of being easily compacted[115]. Compacting factor and Vebe tests are being increasingly used for measuring the workability of lightweight concrete mixes. The range of compacting factors for structural lightweight concrete is reported to be smaller than for ordinary concrete and compacting factor values imply differing degrees of workability. The cohesiveness of lightweight concrete mixes can be improved by replacing the finer fractions of the lightweight aggregate by natural sand[116,117] or by the use of air entraining agents.

12.3 MIX PROPORTIONING USING EMPIRICAL GRAPHS

The empirical approach to the design of normal density concrete is well established and lightweight concrete is no exception to this. Based on the extensive experimental investigations at the Building Research Station, UK, Teychenne[118,119] has developed empirical graphs relating the important parameters in a lightweight concrete mix and these can be conveniently used to estimate the trial mix proportions of structural lightweight aggregate concrete.

The significant parameters influencing the mix proportions are:

(a) Type of aggregate
(b) Cement content
(c) Total water/cement ratio
(d) Workability
(e) Strength
(f) Relative density

The compressive strength depends primarily upon the total water/cement ratio and the type of lightweight aggregate used in the mix. The relationship between the compressive strength of water stored cubes and the total water/cement ratio is shown in Fig. 12.1 for four different types of most commonly used lightweight aggregates. Lytag generally produces the highest strength for a given water/cement ratio in comparison with other aggregates.

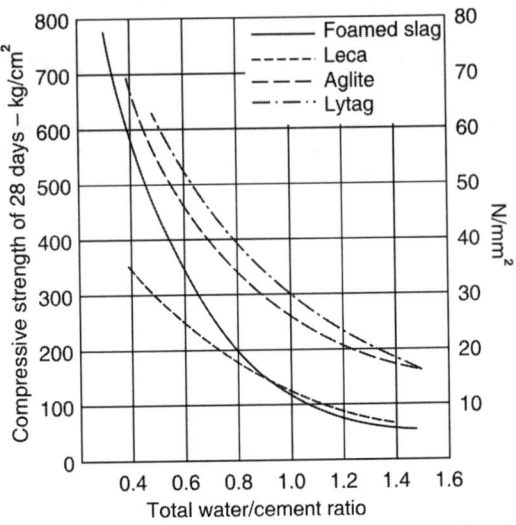

Fig. 12.1: Relationship between the compressive strength of water stored cubes and the total water/cement ratio

The investigations regarding workability showed that for a particular aggregate, the compacting factor is best related to the total water content in the mix. Figs 12.2 and 12.3 show the relationship between the total water/cement ratio and the cement content, the results for a particular aggregate being bounded by curves of constant water content, which also represent lines of approximately equal workability. A compacting factor of 0.8 and above, with a Vebe time of less than 12 seconds indicated good workability of the lightweight aggregate concrete mixes. The upper bound curves in the figures correspond to a range of workability from 'Medium to High' with compacting factor values exceeding 0.8.

The relative density depends primarily upon the type of aggregate, the cement and moisture content in the concrete. The relative densities shown in Figs 12.4 and 12.5 relate to freshly placed concrete and to concrete stored in air at 17°C and 65 per cent relative humidity for 28 days. Generally, the relative density increases more or less linearly with the cement content for fresh concrete, but it becomes still more pronounced as the concrete dries out since mixes with lower cement content lose a larger quantity of water.

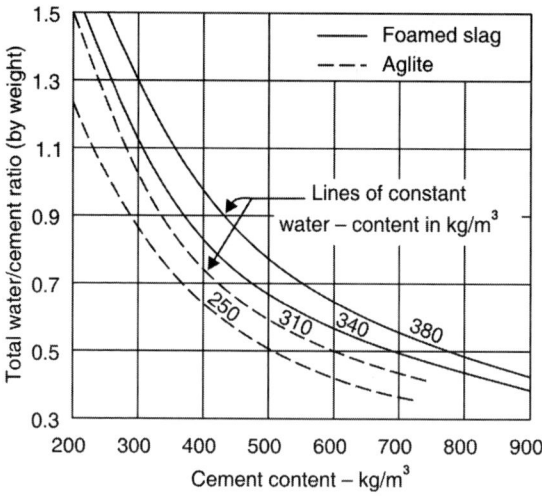

Fig. 12.2: Relationship between the total water/cement ratio and the cement content

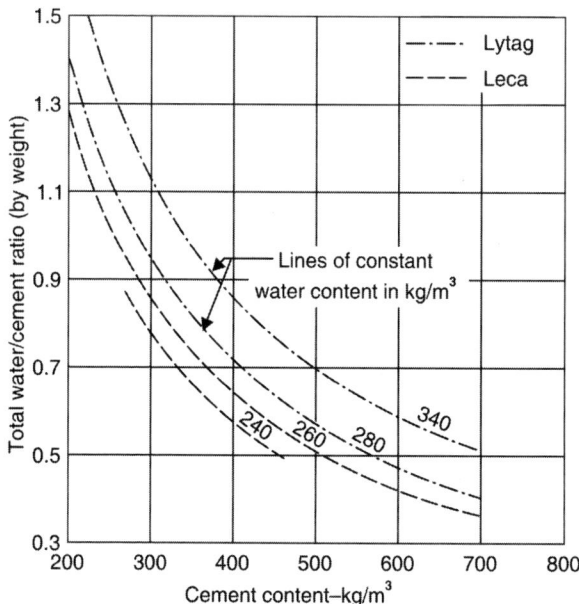

Fig. 12.3: Relationship between the total water/cement ratio and the cement content

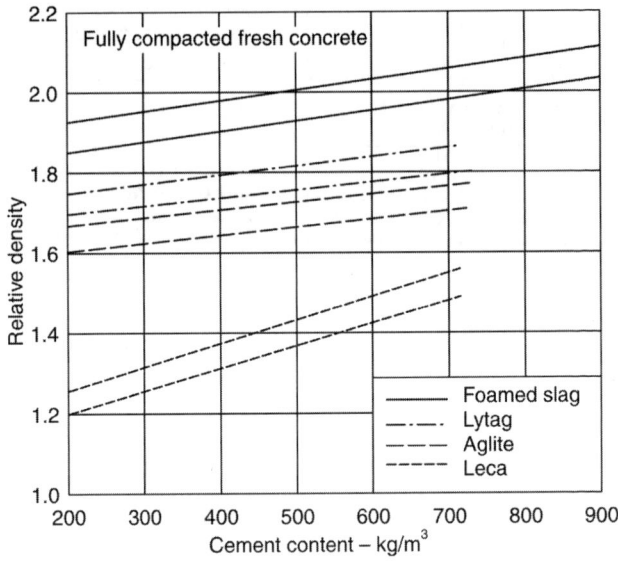

Fig. 12.4: Relationship between the relative density and the cement content for fully compacted fresh concrete

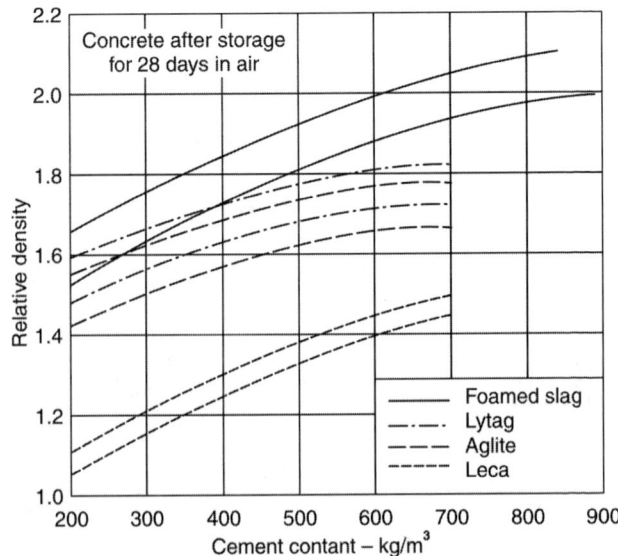

Fig. 12.5: Relationship between the relative density and the cement content for concrete stored in air for 28 days

A critical examination of the empirical graphs indicates that it is not possible to attain certain combinations of strength and density with all the types of aggregates. Very low relative densities of the order of 1.3 and less can be obtained only with a particular aggregate like Leca, using comparatively lower cement contents. Alternatively, high-strength concrete suitable for prestressed work can be produced using foamed slag, aglite or lytag with varying densities. The compressive strengths shown in the empirical graphs relate to the mean design strength which is to be computed by applying suitable control factors to the specified minimum strength.

The mix design procedure based on empirical graphs is summarized below:

(a) The total water/cement ratio required to achieve the mean design strength is selected from Fig. 12.1, depending upon the type of lightweight aggregate available at worksite.

(b) Using the water/cement ratio and the required workability, the cement content required for different types of aggregate is determined from Figs 12.2 and 12.3.

(c) The relative densities of concrete resulting from using the cement content are estimated from Figs 12.4 and 12.5 for different types of aggregates.

(d) The proportion of finer fractions in the combined aggregates is varied from 25 to 70 per cent by weight depending upon the cement content and the aggregate grading. For higher cement contents, lower proportion of fines may be used. Starting with a fines percentage of 50 for cement contents in the range of 300 to 500 kg/m^3, the mix proportions can be suitably adjusted from the results of trial mixes as for natural aggregates.

12.3.1 Example

Design a lightweight aggregate concrete mix to suit the following data:

Specified 28 days minimum cube strength	$= 120 \text{ kg/cm}^2$
Control factor	$= 0.75$
Type of aggregate available	$=$ Leca
Required workability	$=$ Medium to high
Maximum relative density (Air dry concrete)	$= 1.30$

Design

$$\text{Mean design strength} = \frac{120}{0.75} = 160 \text{ kg/cm}^2$$

Total water cement ratio (Fig. 12.1) = 0.87

Using the upper portion of the bands in Fig. 12.3, for the desired workability and water/cement ratio of 0.87, the cement content required = 310 kg/m³.

Using this cement content, the relative density of air dry concrete obtained from Fig. 12.5 is 1.16 to 1.21 using Leca.

The corresponding relative density of fully compacted fresh concrete = 1.25 to 1.30

Mean relative density of fresh concrete = 1.275

Batch quantities for one cubic metre of concrete:

$$\text{Total weight} = 1275 \text{ kg}$$
$$\text{Cement} = 310 \text{ kg}$$
$$\text{Water} = (0.87 \times 310) = 275 \text{ kg}$$
$$\text{Total weight of dry aggregates} = (1275 - 580) = 695 \text{ kg}$$

The fine and coarse aggregates are used in equal proportions for the first trial mix and suitable adjustments are made, based on the results of trial mixes.

12.3.2 Example

A lightweight concrete mix is required for structural concrete work. A minimum 28 days cube strength of 300 kg/cm² is required based on structural considerations. Control factor = 0.75. The relative density of the concrete, not to exceed a value of 1.75.

Workability required is medium to high.

Available aggregates are Foamed slag, Lytag, Aglite and Leca.

Design the most economical mix and set out the dry batch weights and also the field mix quantities per cubic metre of concrete, if the fine and coarse aggregates contain 5 and 3 per cent of moisture by dry weight, respectively.

Design

$$\text{Mean design strength} = \frac{300}{0.75} = 400 \text{ kg/cm}^2$$

From Fig. 12.1

Total water/cement ratio = 0.56 for foamed slag

= 0.69 for Aglite

= 0.77 for Lytag

The required strength cannot be achieved by using Leca.

For these water/cement ratios, from Figs 12.2 and 12.3 (using upper portions of bands)

The cement content required is 700 kg/m³ for Foamed slag

450 kg/m³ for Aglite

430 kg/m³ for Lytag

The corresponding relative densities of air dry concrete obtained from Figs 12.4 and 12.5 are, 1.96 to 2.03 for Foamed slag.

1.65 to 1.72 for Aglite

1.65 to 1.75 for Lytag

Foamed slag aggregates result in a concrete which exceeds the permissible limit of relative density and hence it cannot be used.

Both Aglite and Lytag produce a concrete satisfying the relative density limits and the choice between them depends upon their relative cost and availability.

The cement content required is the least, if Lytag is used and the relative density of fresh concrete is 1.75.

Dry batch weights for one cubic metre of concrete:

Cement	= 430 kg	
Water	= (0.77 × 430)	= 330 kg
Aggregates	= (1750 – 760)	= 990 kg

Equal proportions of fine and coarse aggregates are used for the trial mixes.

Field mix quantities after adjusting for moisture content in aggregates are as follows:

Cement	= 430 kg
Water	= 290 kg
Coarse aggregate	= 510 kg
Fine aggregate	= 520 kg

12.4 ACI METHOD OF PROPORTION LIGHTWEIGHT MIXES

The current ACI standard[120] provides a generally applicable method for selecting and adjusting mix proportions for structural lightweight concrete, using different types of lightweight and normal weight aggregates. According to the American standard, structural lightweight aggregate concrete is defined as concrete which is made with lightweight aggregates and having a 28-day compressive strength in excess of 175 kg/cm^2 and with an air dry weight not exceeding 1840 kg/m^3. In the case of lightweight concrete mixes, the net water/cement ratio required cannot be established with sufficient accuracy to be used as a basis for proportioning the mixes. This is due to the difficulty in determining the quantity of water absorbed by the aggregate and which is not available for reaction with cement while the concrete is in its plastic state. In view of this, lightweight concrete trial mixes are usually proportioned on a cement and air content basis at the required consistency.

Air entrainment is generally recommended in lightweight aggregate concrete mix since it enhances workability, improves resistance to freezing and thawing cycles[121], decreases bleeding and tends to obscure minor grading deficiencies[122]. An air content of 4 to 8 per cent when the maximum size of aggregate is 20 mm and 5 to 9 per cent with 10 mm maximum size aggregate, is generally preferred in lightweight concretes subjected to frost attack and deicer salts.

12.4.1 Estimation of Trial Mix Proportions

The ACI standard ACI 213R–03[123] overcomes the difficulties arising out of lightweight aggregate absorption and bulk specific gravity by incorporating the concept of 'Specific Gravity Factor' for the aggregates. The specific gravity factor is calculated as the relationship between the dry weight of each size of aggregate in the mix and the displaced volume it is assumed to occupy. This factor can be conveniently used in subsequent calculations as though it really was the apparent specific gravity.

The following sequence of operations is useful in estimating trial mix proportions.

(a) In well-proportioned mixes, the cement content-strength relationship is constant for given aggregates but may vary widely depending upon the type and source of aggregate[124,125]. An approximate range of cement content required to produce a concrete of given strength is shown in Table 12.3. When sufficient information is not available, the only alternative is to make a sufficient number of trial mixes with varying cement contents so as to achieve a range of compressive strengths including the one required.

(b) The dry bulk volume of aggregates required vary from 1.04 to 1.19 m^3 per cubic metre of concrete. The exact amount depends upon gradation, shape, size and surface texture of aggregate particles. With well graded and round particles, the amount may be as low as 1.04 m^3. Of this the finer fractions may vary from 40 to 60 per cent. To start with, trial mixes are recommended with equal proportions of coarse and fine aggregates with suitable adjustments in the subsequent mixes.

Table 12.3: Relationship between strength of lightweight aggregate concrete and cement content

Compressive strength (kg/cm^2)	Cement content (kg/m^3)
175	250 to 420
210	280 to 450
280	330 to 500
350	390 to 560

(c) Due to the very high absorption of moisture by lightweight aggregates, the quantity of water required to produce a unit volume of concrete depends on many factors and is usually fixed up by trials. Depending upon the type of aggregates, the water content may vary from 160 to 380 kg/m^3 of concrete. For the first trial, the cement and aggregate content in a unit volume of concrete is assumed and the water content as adjusted so as to obtain the required workability expressed as slump, compacting factor or Vebe time.

(d) Calculations are made for yield (the total batch weight divided by the plastic unit weight) and for actual weights of the ingredients per unit volume. Displaced volumes are calculated for the cement, air and total water. The remaining volume is then assigned to the coarse and fine aggregates, the volume occupied by each being assumed to be proportional to its dry loose unit weight. The specific gravity factor is calculated as the relationship between the dry weight of each size of aggregate in the mix and the displaced volume it is assumed to occupy.

(e) For the second and third trial mixes, using different cement content, the quantities of materials required can be quickly calculated using the specific gravity factor. Based on the results of tests on control specimens of trial mixes, the optimum cement factor which produces the desired strength is selected.

12.4.2 Example

The following example illustrates the method of selecting the trial mix proportions of lightweight aggregate.

Design a lightweight aggregate concrete mix for an average 28 days cylinder strength of 210 kg/cm². Placement conditions require a slump of 5 cm.

Dry loose densities of fine aggregates	=	900 kg/m³
Coarse aggregates	=	700 kg/m³

Bulk volume of fine and coarse aggregates in one cubic metre is assumed as 1.18 m³ for the trial mix. The cement content required varies from 280 to 450 kg/m³. Generally three trial mixes are made varying the cement content in steps of 280, 365 and 450 kg/m³. The fine and coarse aggregates are used in equal proportions in the first trial mix.

First trial mix

The weights of ingredients required to produce one cubic metre of concrete are given as

Absolute volume of aggregates	=	$(1000 - 354) \times 10^3$
Volume of water	=	245×10^3
Cement	=	280 kg
Fine aggregate (0.59 × 900)	=	531 kg
Coarse aggregate (0.59 × 700)	=	413 kg
Water (Required to obtain the desired slump of 5 cm, determined by test)	=	245 kg
Volumes of ingredients		cm³
Absolute volume of cement	=	$\dfrac{280 \times 10^3}{3.15} = 89 \times 10^3$
Volume of water	=	245×10^3
Entrapped air (about 2 per cent)	=	20×10^3
		354×10^3
Absolute volume of aggregates	=	$(1000 - 354) \times 10^3$
	=	0.646 m³
Volume occupied by fine aggregate	=	0.323 m³
Volume occupied by coarse aggregate	=	0.323 m³

For the second and third trial mixes, using different cement content, the quantities of materials required can be quickly calculated using specific gravity factor (SGF).

$$\text{SGF for fine aggregate} = \left(\frac{513 \times 10^3}{0.323 \times 10^6} \right) = 1.65$$

$$\text{SGF for coarse aggregate} = \left(\frac{413 \times 10^3}{0.323 \times 10^6} \right) = 1.28$$

Second trial mix

The weights of coarse aggregate and water are maintained constant as provided in the first trial mix. The ratio of fine to coarse aggregate gets reduced due to the increased amount of cement content in the second trial mix.

Cement content	$= 365 \text{ kg/m}^3$	cm^3	

$$\text{Absolute volume of cement} = \left(\frac{365 \times 10^3}{3.15} \right) = 89 \times 10^3 \text{ cm}^3$$

Volume of water $= 245 \times 10^3 \text{ cm}^3$

Entrapped air $= 20 \times 10^3 \text{ cm}^3$

         $= 323 \times 10^3 \text{ cm}^3$

         $704 \times 10^3 \text{ cm}^3$

Absolute volume of fine aggregates $= (1000 - 704) \times 10^3$

Volume of water $= 296 \times 10^3 \text{ cm}^3$

Weight of fine aggregate $= (296 \times 1.65)$ $= 490 \text{ kg}$

Weight of coarse aggregate $= (323 \times 1.28)$ $= 413 \text{ kg}$

Third trial mix

Cement content $= 450 \text{ kg/m}^3 \text{ cm}^3$

$$\text{Absolute volume of cement} = \left(\frac{450 \times 10^3}{3.15} \right) = 143 \times 10^3$$

Volume of water $= 245 \times 10^3$

Entrapped air $= 20 \times 10^3$

         $= 323 \times 10^3$

         731×10^3

Absolute volume of fine aggregates $= (1000 - 731) \times 10^3$

Volume of water $= 269 \times 10^3 \text{ cm}^3$

Weight of fine aggregate $= (269 \times 1.65)$ $= 444 \text{ kg}$

Weight of coarse aggregate $= (323 \times 1.28)$ $= 413 \text{ kg}$

The control specimens of the three trial mixes are tested for compressive strength and the results are plotted as cement factor against strength. From the plot of results, the optimum cement factor which produces the desired strength is selected.

12.4.3 Adjustment of Trial Mix Proportions

The trial mixes designed may require adjustment from time to time to compensate for some unintentional change in the characteristics of the concrete or to make a planned change in the characteristics. The most common adjustments generally required are:

(a) To compensate for a change in the moisture content of aggregates

(b) To proportion a mix for greater to lesser cement content

(c) To effect changes in slump or air content

The method recommended in the new standard ACI: 211.2–98, is a variation on the method of absolute volumes. The specific gravity factor referred to in ACI Code corresponds to that of an initially dry lightweight aggregate. Specific gravity factors generally vary with the moisture content of aggregates[113]. The variation is usually approximately linear in the lower range of moisture contents but may digress from linearity at higher moisture contents. Fig. 12.6 shows the relationship between specific gravity factor and moisture content for lightweight aggregates as determined by the pycnometer method[126]. According to the commendations of the new standard ACI: 211.2–98, the specific gravity factor is defined as the ratio of the weight of the aggregates as introduced into the mixer to the effective volume displaced by the aggregates. It is important to note that the weight of the aggregate as introduced into the mixer included any moistures absorbed in the aggregate and any free water on the aggregates.

In calculating the effective volume displaced by lightweight aggregates in the concrete, specific gravity factors corresponding to the actual moisture content in the damp aggregates have to be used. The approximate numerical values of the compensating adjustments recommended by the ACI standard are summarized below:

(a) *Proportion of fine aggregate:* When the percentage of finer fractions of the aggregates are increased, a corresponding increase in water content will be necessary. For every one per cent increase in fine aggregate, the water content may be increased by 2 kg/m^3. To maintain constant strength, the cement content of the mix should be increased by approximately one per cent for every 2 kg/m^3 increase in water.

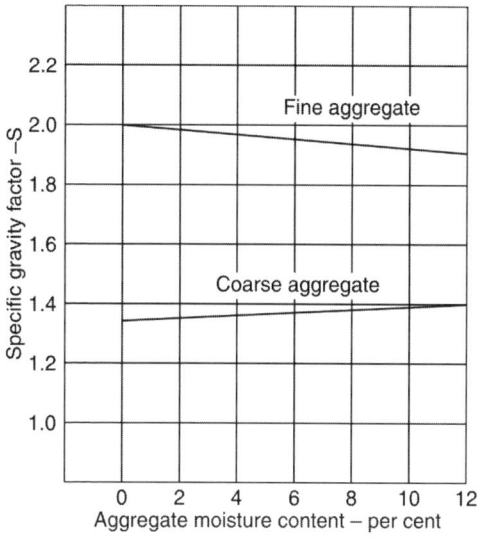

Fig. 12.6: Relationship between pycnometer SGF and moisture content

(b) *Air content*: Entrainment of air generally increases the workability of concrete, unless the water content is reduced to maintain a constant slump. Water should be reduced by approximately 3 kg/m^3 for every one per cent increase in air content. Increased air content generally results in a reduction in the compressive strength. To maintain constant strength, a corresponding increase in cement content and a reduction in fine aggregate content is made so as to maintain the required total effective displaced volume.

(c) *Slump*: The workability of concrete measured in terms of slump may be increased by increasing the water content in the mix. For each desired 2.5 mm slump, water should be increased approximately by 6 kg/m^3. To maintain constant strength, cement content is increased by approximately 3 per cent for every increase of 6 kg/m^3 in water content. The fine aggregate content is suitably decreased to maintain the required total effective displaced volume.

Table 12.4: Adjustment for changes in aggregate moisture content

	Original mix (Dry basis)		Original mix (Damp basis)	
	$m_c = 0$ per cent $S_{CO} = 1.34$	$m_f = 0$ per cent $S_{FO} = 1.99$	$m_c = 1.5$ per cent $S_{C1.5} = 1.5$	$m_t = 4$ per cent $S_{F.4} = 1.97$
	Weight (kg)	Effective displaced volume (m³)	Weight (kg)	Effective displaced volume (m³)
Cement	415	0.132	415	$\left(\dfrac{415}{10^3} \times 3.15\right) = 0.132$
Air (5.5 per cent)	–	0.055	–	0.055
Coarse aggregate	$\left(\dfrac{356}{1.015}\right) = 315$	$\left(\dfrac{351}{10^3} \times 1.34\right) = 0.262$	356	$\left(\dfrac{356}{10^3} \times 1.35\right) = 0.264$
Fine aggregate	$\left(\dfrac{571}{1.04}\right) = 549$	$\left(\dfrac{549}{10^2} \times 1.99\right) = 0.276$	571	$\left(\dfrac{571}{10^3} \times 1.97\right) = 0.290$
Added water	$(10^3 \times 0.215) = 275$	$(1.0 - 0.725) = 0.275$	259	$\left(\dfrac{259}{10^3}\right) = 0.259$
Total	1590	1.000	1601	1.000

| | Adjusted mix (Dry basis) | | Adjusted mix (Damp basis) | |
	$m_f = 0$ per cent $S_{CO} = 1.34$	$m_f = 0$ per cent $S_{FO} = 1.99$	$m_c = 5$ per cent $S_{C.4} = 1.37$	$m_t = 8$ per cent $S_{F.8} = 1.95$
	Weight (kg)	Effective displaced volume (m^3)	Weight (kg)	Effective displaced volume (m^3)
Cement	415	0.132	415	0.132
Air (5.5 per cent)	–	0.055	–	0.055
Coarse aggregate	351	0.262	$(351 \times 1.05) = 369$	$\left(\dfrac{369}{10^3} \times 1.37\right) = 0.269$
Fine aggregate	549	0.276	$(549 \times 1.08) = 593$	$\left(\dfrac{593}{10^3} \times 1.95\right) = 3.304$
Added water	275	0.275	$(10^3 \times 0.24) = 240$	$(1.0 - 0.760) = 0.240$
Total	1590	1.000	1617	1.000

m_c = Moisture content of coarse aggregate S_C = Specific gravity of coarse aggregate

m_f = Moisture content of fine aggregate S_F = Specific gravity of fine aggregate

12.4.4 Adjustment for Changes in Aggregate Moisture Condition

The moisture content of various aggregate consignments may vary for a particular work and this necessitates suitable adjustments in the field mix proportions to account for the variation in water content of the aggregates. The adjustments are made in the aggregate weights with due consideration for the variation in the specific gravity factor with the moisture content. The procedure to be followed is outlined below:

(a) The weight of cement and the effective displaced volume of cement and air are maintained constant.

(b) Using the appropriate volume of moisture content, the new weights of fine and coarse aggregates are computed such that their dry weight remains constant.

(c) The effective displaced volumes of aggregate corresponding to the appropriate moisture condition and the specific gravity factor are computed.

(d) The water content is suitably decreased to maintain the total effective displaced volume constant.

These adjustments are illustrated by an example shown in Table 12.4, in which a field mix with coarse and fine aggregates having a moisture content of 1.5 and 4 per cent, is suitably adjusted when the moisture content in the coarse and finer fractions changes to 5 and 8 per cent, respectively.

13

Design of No-Fines
Concrete Mixes

13.1 NO-FINES CONCRETE

No-fines concrete is a simple form of lightweight concrete obtained by eliminating the finer fractions of the aggregate in a normal concrete mix. The total omission of fine aggregate in the mix will result in a system of uniformly distributed voids throughout the mass of concrete, generally reducing the density of the resulting material. The main advantages of using no-fines concrete are the high degree of thermal insulation, speedy construction, low density and shrinkage. No-fines concrete is not prone to segregation and it can be dropped from a considerable height, facilitating the use of high lifts. Due to the absence of large surface area of sand particles that would have to be coated with cement paste, the cement in a no-fines concrete mix may be as little as 70 to 130 kg per cubic metre of concrete, resulting in comparatively lower costs.

The density of no-fines concrete depends on the grading of coarse aggregate use and is generally in the range of 60–75 per cent[127] of that of normal concrete. The aggregate generally used is 10–20 mm material, although the other sizes may be used. If lightweight aggregate is used, densities as low as 70 kg/m^3 can be obtained. Various types of aggregates, like crushed rock, gravel, blast furnace slag and clinker have all been used successfully to produce no-fines concrete. The use of crushed rock generally results in higher strengths than when gravel aggregates are used in the mixes.

13.2 STRENGTH CHARACTERISTICS

The compressive strength of no-fines concrete mainly depends on its density. The strength varies generally between 70 kg/cm^2 for a density of 1900 kg/m³ to 140 kg/cm² for a density of 2100 kg/m³ at 28 days. The strength continues to increase after 28 days in a manner similar to that of normal concrete. In the case of no-fines concrete mixes, water/cement ratio is not the main controlling factor and for a given aggregate/cement ratio, there is an optimum water/cement ratio, yielding the highest strength. This trend can be clearly identified in the data of McIntosh, Botton and Muir[128] shown in Fig. 13.1 in which aggregate/cement ratios ranging from 6 to 10 by volume are covered. The corresponding water/cement ratios vary from 0.37 to 0.45 by weight, while the density of the concrete varies from 1940 to 2100 kg/m³ for aggregate/cement ratios in the range from 10 to 6 by volume.

The optimum water/cement ratio touching the peaks of the aggregate/cement ratio curves is useful in proportioning no-fines concrete mixes. A water/cement ratio higher than the optimum will result in segregation of the aggregate particles while lower values adversely affect the workability of the mix resulting in improper compaction.

13.3 MIX PROPORTIONING

The design of no-fines concrete mixes is based on the required strength at a particular age. The relations between the water and aggregate/cement ratios and strength shown in Fig. 13.1 are useful

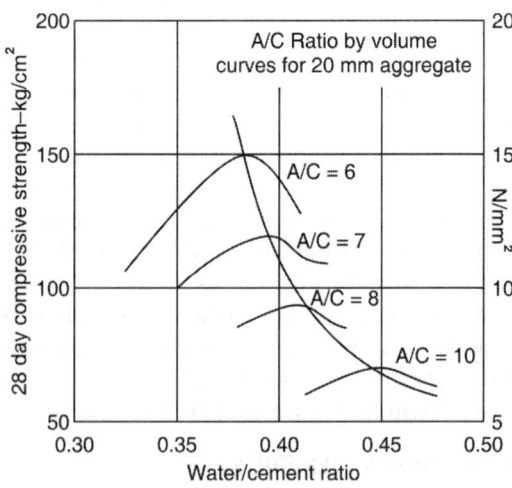

Fig. 13.1: Relation of compressive strength, water/cement ratio and aggregate/cement ratio for No-fines concrete

in this regard. The proportions are governed more by the necessity to achieve a cellular structure and coat each aggregate particle with cement grout than by a high strength requirement. Trial mixes are made using the estimated proportions of ingredients and suitable adjustments necessary to produce the required workability and strength are incorporated. If the workability is to low, it may be necessary to increase the cement content and/or the water/cement ratio. On the other hand, if the workability is too high and bleeding occurs, the water/cement ratio or the richness of mix is reduced. Normally, gravel and crushed rocks do not absorb a significant percentage of water in the mix. However, if lightweight aggregates are used to produce no-fines concrete, it is rather difficult to predict the optimum water/cement ratio since they absorb as much as 10 to 20 per cent of water (by weight) in 24 hours. In such cases, the optimum combination of ingredients in the mix should be decided by trial mixes followed by tests.

13.4 EXAMPLES OF MIX DESIGN

13.4.1 Design a No-Fines Concrete Mix with a Minimum Specified 28 Days Cube Strength of 90 kg/cm²

The degree of control is such that a control factor of 0.75 may be assumed.

Maximum size of coarse aggregate	= 20 mm
Type of coarse aggregate	= Irregular gravel
Type of cement	= Ordinary Portland
Bulk density of cement	= 1472 kg/m³
Bulk density of coarse aggregate	= 1520 kg/m³

Design

Average 28 days compressive strength $= \dfrac{90}{0.75} = 120 \text{ kg/cm}^2$

From Fig. 13.1

Optimum water/cement ratio	= 0.39 (by weight)
Aggregate/cement ratio by volume	= 7.00
Corresponding density of concrete	= 2050 kg/m³

Aggregate/cement ratio by weight $= \left(\dfrac{7 \times 1520}{1472}\right) = 7.25$

Proportion of ingredients by weight:

Cement	:	Coarse aggregate	:	Water
1	:	7.25	:	0.39

Batch quantities per m^3 of concrete

$$\text{Cement} = \frac{1}{8.64} \times 2050 = 263 \text{ kg}$$

$$\text{C.A.} = \frac{7.25}{8.64} \times 2050 = 1722 \text{ kg}$$

$$\text{Water} = \frac{0.39}{8.64} \times 1050 = 92 \text{ kg}$$

13.4.2 Design a Concrete Mix without Fine Aggregate to Satisfy the following Requirements

Average 28 days compressive strength	=	75 kg/cm^2
Type of coarse aggregate	=	Irregular gravel 20 mm maximum size
Type of cement	=	Ordinary Portland
Bulk density of cement	=	1472 kg/m^3
Bulk density of gravel	=	1600 kg/m^3

The aggregate contains 1.5 per cent moisture by weight. Estimates the field mix quantities required for 1 m^3 of concrete.

Design

Average 28 days compressive strength = 75 kg/cm^2

From Fig. 13.1

Water/cement ratio required	=	0.43
Aggregate/cement ratio by volume	=	9.5
Corresponding density of concrete	=	1955 kg/m^3
Aggregate/cement ratio by weight	=	$\dfrac{9.5 \times 1600}{1472}$
	=	10.4

Proportion of ingredients by weight:

Cement	:	Coarse aggregate	:	Water
1	:	10.4	:	0.43

Weights of ingredients:

$$\text{Cement} = \left(\frac{1}{11.83} \times 1955 \right) = 165 \text{ kg}$$

Water $\qquad = \dfrac{0.43}{11.83} \times 1955 \qquad = 70\,\text{kg}$

C.A. $\qquad = \dfrac{10.4}{11.83} \times 1955 \qquad = 1720\,\text{kg}$

Moisture content in C.A. $= \dfrac{1.5}{100} \times 1720 \qquad = 26\,\text{kg}$

Batch quantities per m^3 of concrete:
(Field mix)

Cement $=$ 165 kg
Water $= (70 - 26) \qquad = 44\,\text{kg}$
C.A. $= (1720 + 26) = 1746\,\text{kg}$

Mass Concrete Mixes

14.1 GENERAL FEATURES OF MASS CONCRETE

In the construction of dams and other massive structures using concrete, the compressive strength of concrete is usually of secondary importance, unless it is regarded as a criterion of general quality. In mass concrete construction, the mix proportions are generally selected with a view to economy of materials and reducing the tendency for the temperature of concrete to rise due to the heat developed by the hydration of cement. According to the American Concrete Institute Committee[129], "mass concrete is defined as any large volume of cast in place concrete with dimensions large enough to require that measures be taken to cope with the generation of heat and attendant volume change to minimize cracking".

Mass concrete is generally composed of cement, aggregate and water and admixtures, like pozzolans and fly ash, are sometimes used. Crushed rock aggregates are invariably used since deposits of large sized gravel aggregates are not likely to be economically available. An important feature of mass concrete made with aggregates of large maximum size is that for any given mix, there exists only a small range of water/cement ratio within which the concrete is not prone to segregation. The cohesiveness of the concrete can be improved by using rock dust in the concrete mixes in suitable proportions.

The workability of mass concrete can be improved by air-entrainment and this will also permit the use of lower water/cement ratios to achieve the same strength and improves the durability of hardened concrete. The rise in temperature in the interior of a large concrete mass due to heat developed by the hydration of cement, may lead to serious cracking as a result of thermal gradients and low tensile strength of concrete. To circumvent this problem, low heat development was first produced in the United States for specific use in large concrete gravity dams. The low rate of strength development is a disadvantage with this type of cement. The modified cement Type II successfully combines a somewhat higher rate of heat development than that of low heat with a rate of gain of strength similar to that of ordinary Portland cement. Type II cement is recommended for use in massive structure where a moderately low heat generation is desirable or where moderate sulphate attack is expected. The modified cement is extensively used in the United States and is covered by ASTM specification C: 150–11[130] and the British Standard, BS: 8500–2:2006.

14.2 CEMENT FACTOR IN MASS CONCRETE

It was the general practice during early thirties to insist on cement contents of the order of 220 to 330 kg/m^3 in mass concrete used for large dams. An objectionably high degree of cracking was noticed in Norris dam constructed by the Tennessee valley authority in 1936, where a cement factor of 223 kg/m^3 was used. Thereafter, a trend towards reducing the cement content is clearly noticeable and use of cement factors as low as 88 kg/m^3 have been reported[132]. A number of mass concrete dams constructed in the United States clearly indicate the advantage of using lower cement factors of the order of 120 to 160 kg/m^3.

The results of laboratory investigations by the ACI Committee[129], have indicated that an air-entrained mass concrete with a cement factor of 56 kg/m^3 and fly ash of equivalent solid volume of 112 kg/m^3 of cement, produced a very workable mix with a very low water content of 60 kg/m^3. The one year cylinder strength of this concrete was reported to be of the order of 210 kg/cm^2. The optimum cement factor to produce a workable mix depends upon the shape and grading of aggregates.

14.3 AGGREGATES REQUIREMENTS FOR MASS CONCRETE

In the design of mass concrete mixes with large aggregates, a suitable selection of aggregate grading is important to achieve overall economy and to ensure satisfactory properties of the fresh concrete. The grading limits for fine and coarse aggregate as recommended by the ACI Committee[129] are compiled in Tables 14.1 and 14.2. The coarse aggregates are grouped under six different sizes varying from a maximum of 150 mm to a minimum of 5 mm.

Table 14.1: Recommended grading limits for fine aggregate

Sieve size	Percentage retained (individual by weight)	Percentage retained (cumulative by weight)
10 mm	0	0
4.76 mm	0–8	0–8
2.38 mm	5–20	10–25
1.19 mm	10–25	30–50
600	10–30	50–65
300	15–30	70–83
150	12–20	90–97
Pan fraction	3–10	100

Table 14.2: Recommended grading limits for coarse aggregate

Maximum size aggregate in concrete (mm)	Percentage of coarse aggregate fractions					
	Coarse			Fine (mm)		
	Cobbles	Coarse	Medium	(5–19)	(10–19)	(5–10)
19	0	0	0	100	55–73	27–45
38	0	0	40–55	45–60	30–35	15–25
76	0	20–40	20–40	25–40	15–25	10–15
150	20–35	20–32	20–30	20–35	12–20	8–15

Figure 14.1 shows the grading limits and suggested gradings of fine aggregate, successfully used on a number of jobs involving plain and air-entrained mass concretes, according to McIntosh[133]. The shapes of the grading curves were found to be convex upwards for natural sand and convex downwards for crushed rock sand. The suggested range of combined gradings used on a number of jobs is shown in Fig. 14.2. The aggregate and water/cement ratios used in conjunction with these gradings for 80 mm aggregates varied from 6 to 9 and 0.5 and 0.7 by weight, respectively. The corresponding ratios with larger aggregate sizes of 150 mm varied from 8 to 15 and 0.48 to 0.70, respectively. The suggested proportions[134] should be considered as a rough guide to the type of gradings which are used and the final choice should invariably be based on full scale trial

mixes. The dotted line shown in Fig. 14.2 refers to the combined grading recommended by Orchard[135] when the maximum size of aggregate is 160 mm.

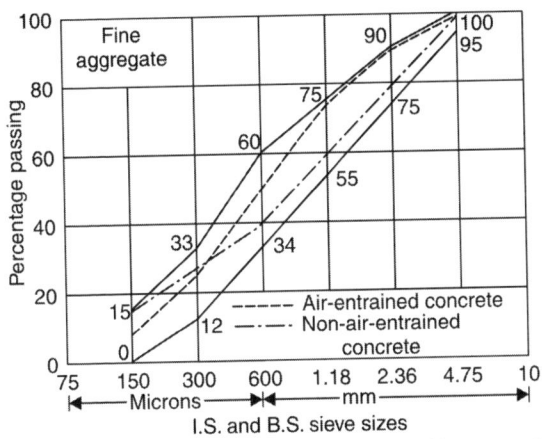

Fig. 14.1: Range of fine aggregate gradings used with suggested gradings for air-entrained and non-air-entrained concretes with aggregates of large maximum size

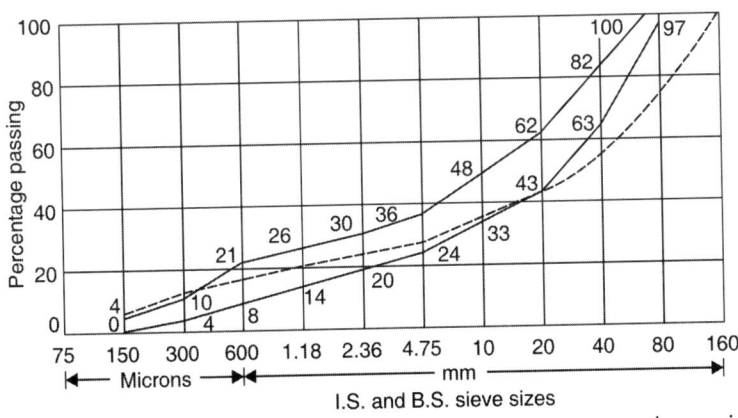

Fig. 14.2: Range of gradings used with aggregates of 80 mm maximum size

The curves in Fig. 14.3 connecting the water/cement ratio, the aggregate/cement ratio and workability, have been suggested by Orchard based on a study of the available information and extra-polation from similar curves for aggregates of smaller maximum size. It is important to note that these curves are not based on direct experimental data and hence they must be regarded as a rough guide in the selection of mix proportions for mass concrete.

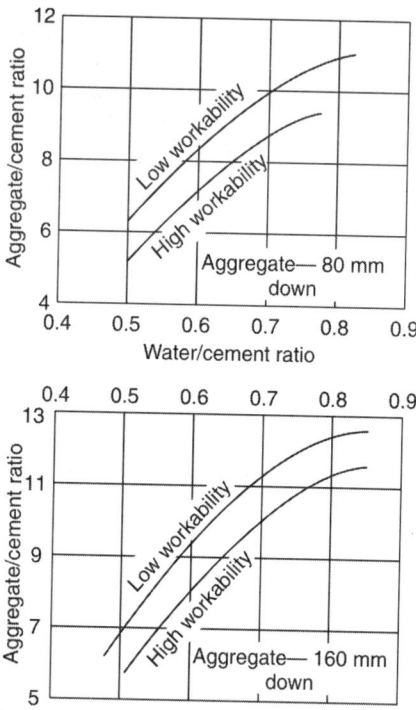

Fig. 14.3: Relation between workability aggregate and water/cement ratio for mass concrete

Investigations by Higginson et al[136] have indicated that to achieve the greatest cement efficiency, there is an optimum maximum size for each compressive strength level to be obtained, with a given aggregate and cement. The economical maximum size of aggregate is generally determined by various factors, like the design strength, plant requirements for processing, batching, mixing, transporting, placing and consolidating the concrete. Large aggregate particles of very irregular shape have a tendency to promote cracking at the interface of mortar and aggregate due to differential volume changes. Very large size aggregates have been occasionally used and in general an aggregate size of 15 cm is considered as the maximum practical size for use in mass concrete.

14.4 SELECTION OF MIX PROPORTIONS

The mix proportions of mass concrete are generally established by the trial mix method using the data based on previous experience. Since the well-established ACI method[67] provides for use of large aggregates of maximum size up to 150 mm, the method may be used to estimate the trial mix proportions. According to this method, the minimum cement content in a mix used for the exterior portions of massive structures subjected to severe exposure is of the order of 225 kg/m^3. However, for interior portions, the cement content may be considerably reduced since the water/cement ratio is fixed based on strength rather than durability requirements.

A comparative study[132–138] of the data on mix proportions used in the construction of major dams in the world indicates that the cement content varied from a minimum of 88 kg/m^3 to a maximum of 270 kg/m^3. Based on past experience, the ACI Committee[129] has suggested a cement content of 125 and 140 kg/m^3 for best shaped and angular aggregates respectively when the maximum size of aggregates is 150 mm.

The water content for air-entrained concrete using 150 mm maximum size aggregates may vary from 71 to 89 kg/m^3 for natural aggregates and 83 to 107 kg/m^3, when crushed aggregates are used. The corresponding water requirements are 20 per cent higher when 75 mm maximum size aggregates are used. The optimum proportion of fine to total aggregate depends on the grading and shape of aggregates. For 150 mm aggregate concrete containing natural sand and gravel, the ratio of fine to total aggregate by absolute volume may be as low of 21 per cent while that for crushed aggregates, the ratio is of the order of 25 to 27 per cent.

14.5 MIX DESIGN EXAMPLE

Design a concrete mix suitable for use in a gravity dam exposed to severe atmosphere. Ordinary Portland cement is used and the coarse aggregate available is graded from 5 to 150 mm.

28 days cylinder compressive strength = 150 kg/cm^2

Dry rodded weight of coarse aggregate = 1800 kg/m^3

Fineness modulus of sand = 2.8

Slump required for the job = 25 to 50 mm

Specific gravities of cement, coarse aggregate and sand are 3.15, 2.70 and 2.65, respectively.

Design

Air-entrained concrete will be used since the dam structure is exposed to severe weather conditions.

Water/cement ratio required is
(a) from strength considerations (Table 7.4) = 0.71
(b) from durability considerations (Table 7.3) = 0.48
Minimum water/cement ratio of 0.48 is adopted.
Approximate mixing water for the known slump (25–50 mm) and aggregate size (150 mm) = 120 kg/m^3 from Table 7.5.
Desirable air content   = 3 per cent

$$\text{Weight of cement required} \quad = \frac{120}{0.48} = 250 \text{ kg}$$

This higher cement content may be used for the exposed portions of the dam.
For the interior portions, where durability is not the main consideration, using the water/cement ratio based on strength (0.71)

$$\text{The weight of cement required} \quad = \frac{120}{0.71} = 169 \text{ kg}$$

Volume of coarse aggregates required (Table 7.6)   = 0.83 m^3
Weight of coarse aggregate   = 0.83 × 1800   = 1494 kg
If V = absolute volume of sand required, for the interior portions of the dam, quantity of sand required is determined by the absolute volume method.

The absolute volumes of the ingredients in one cubic metre of concrete are:

$$\left(\frac{169 \times 10^3}{3.15}\right) + (120 \times 10^3) + \left(\frac{1494 \times 10^3}{2.70}\right) + (30 \times 10^3) + V = 10^6$$

$V = (10^6 - 757 \times 10^3)$ = 243 × 103 cm^3
Weight of sand required   = (2.65 × 243) = 644 kg
Estimated batch quantities per m^3 of concrete:
Cement   = 169 kg
Water   = 120 kg
Sand   = 644 kg
Coarse aggregate   = 644 kg
The trial mix proportions can be suitably adjusted in the light of test results of fresh and hardened concrete.

High-Density Concrete Mixes

15.1 CONCRETE FOR ATOMIC RADIATION SHIELDING

Concretes of various types have been used extensively in making structures for shielding personnel and equipment from harmful nuclear particles and rays. The properties[139] desirable in concrete used for radiation shielding are that it shall have large hydrogen content to capture fast neutrons, that it shall be able to resist the thermal stresses induced by thermal neutron capture and that it shall have sufficient mass to attenuate the Gamma rays. In addition, the concrete must be able to withstand the heat radiated from the atomic pile during its operation. Since the ability of concrete to absorb Gamma rays is almost proportional to its density, the thickness of the shield can be reduced, if concrete with a higher density than the normal is used.

The density of concrete can be increased by using aggregates of high-specific gravity. Natural aggregates such as magnetite; baryte, limonite and artificial heavy aggregates like steel shot have been extensively used in the construction of concrete biological shields for atomic reactors. The percentage of hydrogen in dry concrete being very low, most of the slowing down of fast neutrons is done by elements of medium atomic weight and not by hydrogen. If the temperature gradients are not excessive, concrete has been found to have a life of at least ten years in a flux of 10^{11} neutrons/cm^2 according to Davis[140]. Concrete for a biological shield should not have any straight-through construction joints in the line of radiations.

The properties of high density concretes are similar to those of normal concretes when compared on a volume basis to take into account the higher specific gravities of heavy aggregates. It is reported[141] that ordinary concrete generally produces the most economical shield where space is not an important consideration. But in situations where space and access through the shield becomes more important as in a hot cell or nuclear reactor, the more dense mineral ore aggregates are generally preferred.

15.2 AGGREGATES FOR HIGH-DENSITY CONCRETE

Natural mineral aggregates, like barytes, hematite, magnetite and limonites, have been widely used for making high density concrete. Barium sulphate, marketed as barytes, has a specific gravity of 4.3 to 4.6 in the coarser fractions and as low as 4.0 in the finer fractions. Crushed baryte aggregates do not present any special problems as far as proportioning of mixes is concerned; but care must be exercised in handling and processing the aggregates since they are prone to degradation which causes dusting of the aggregate and breakdown of the particles during mixing.

Based on experimental investigations on baryte aggregate concrete mixes, Witte and Backstrom[142] have recommended the grading limits shown in Table 15.1 in which the percentage limits retained on different standard sieves are given for fine and coarser fractions having different nominal maximum sizes varying from 20 to 76 mm. The wet density of baryte concrete varies from 3550 to 3700 kg/m^3, depending upon the aggregate/cement and water/cement ratios.

Steel punchings and shot have been successfully used for producing concrete with a density in the range of 5000 to 6000 kg/m^3. But these artificial heavy aggregates are many times dearer than the natural aggregates.

Concrete made with steel shot having a specific gravity of 7.8 is liable to segregation during placing and compaction. To overcome this problem, prepacked method of placing concrete is generally preferred. Davis et al[143] have reported the use of prepacked method for placing high-density concrete made with steel punchings as coarse aggregate and limonite in the finer fractions.

15.3 PROPORTIONING OF HIGH-DENSITY CONCRETE USING NATURAL HEAVY AGGREGATES

The empirical method recommended by Road Note No. 4[12] for designing concrete mixes with normal density aggregates, can be used for high-density concretes made with heavy aggregates. The aggregate/cement ratios given in Tables 6.2, 6.3 and 6.4 of Road Note No. 4 are based on the gross apparent volume of the solid particles, with the specific gravity of aggregates being 2.6. While using heavy aggregates, allowances are to be made for the difference between the specific gravities of normal and dense aggregates. The aggregate/cement ratios from the tables should be multiplied by the ratio of the specific gravities of dense to normal aggregates. Furthermore, if the specific gravities of the fine and coarse aggregates are significantly different, the proportions should be such that the overall grading curve represents the absolute volume proportions rather than the weight proportions of each grade of material. The method of designing a high-density concrete mix is illustrated by the following examples.

Table 15.1: Recommended grading limits of baryte aggregate for use in high-density concrete

(a) Coarse aggregate

Maximum size of aggregate (mm)	*Percentage by weight retained*			
	4.75–10 mm	*10–20 mm*	*20–38 mm*	*38–76 mm*
20	35–50	50–65	–	–
38	20–30	30–40	40–50	–
76	15–25	20–35	20–35	20–35

(b) Fine aggregates

Maximum size of aggregate	*Percentage by weight retained*						
	4.75 mm	*2.36 mm*	*1.18 mm*	*600 micron*	*300 micron*	*150 micron*	*Smaller than 150 micron*
For all maximum sizes	0–5	10–25	20–35	15–25	10–20	7–15	10–15

Mix design example

Design a high density concrete mix to suit the following requirement; structural considerations require an average 28 days cylinder strength of 300 kg/cm^2.

Type of cement: Ordinary Portland cement

Type of coarse aggregate: Crushed baryte with a maximum size of 20 mm and specific gravity of 4.25.

Type of fine aggregate: Crushed baryte with a specific gravity of 4.0.

Desired workability: Medium

Design

Water/cement ratio (from Fig. 5.1) = 0.50

For medium workability, using angular aggregate of 20 mm maximum size, the aggregate/cement ratio corresponding to the coarsest grading (grading curve no. 1) is obtained from Table 6.3 as 4.2.

If the fine and coarse aggregates are combined to correspond to the grading curve no. 1, then the ratio of fine to coarse aggregate = $\dfrac{30}{70}$

Using normal density aggregates with a specific gravity of 2.7, the proportions of ingredients by weight are given as

Cement	:	F.A.	:	C.A.	:	Water
I	:	$\left(\dfrac{4.2 \times 30}{100}\right)$	:	$\left(\dfrac{4.2 \times 70}{100}\right)$	:	0.50
I	:	1.26	:	2.94	:	0.50

Using heavy aggregates, the corresponding proportions of ingredients by weight are obtained as

I	:	$\left(\dfrac{1.26 \times 4.0}{2.7}\right)$	:	$\left(\dfrac{2.94 \times 4.25}{2.7}\right)$	:	0.50
I	:	1.86	:	4.60	:	0.50

Quantities of materials required for one cubic metre of concrete can be computed by the absolute volume.

If 'C' is the weight of cement required for the one cubic metre of concrete, then

$$\frac{C}{3.15 \times 10^3} + \frac{1.86C}{4.0 \times 10^3} + \frac{4.60C}{4.25 \times 10^3} + \frac{0.50C}{10^3} = 1$$

Weight of cement	= 420 kg	
Fine aggregate	= (1.86 × 420)	= 780 kg
Coarse aggregate	= (4.60 × 420)	= 1940 kg
Water	= (0.50 × 420)	= 210 kg
Density of fresh concrete	= 3350 kg/m³	

15.4 RELATION BETWEEN COMPRESSIVE STRENGTH, WATER-CEMENT RATIO AND DENSITY FOR BARYTE CONCRETE

Witte and Bockstrom[142] have reported the variation of compressive strength and density of baryte aggregate concrete for a range of water/cement ratios varying from 0.53 to 0.90 by weight, for aggregates of 38 mm maximum size. The results of their investigations are shown in Figs 15.1 and 15.2. The 28 days cylinder compressive strength varied from a minimum of 250 kg/cm² to maximum of 450 kg/cm² for water/cement ratios varying from 0.90 to 0.53. These results indicate that the range of compressive strengths attainable are significantly higher with baryte aggregates than with normal density aggregates for the same range of water/cement ratios. However, Jordan[144] has reported that concrete made with baryte has a lower ultimate compressive strength than that made with ordinary aggregates.

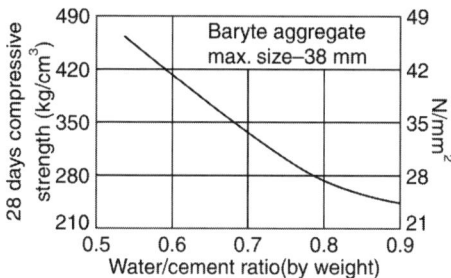

Fig. 15.1: Relation between compressive strength and water/cement ratio for baryte aggregate concrete

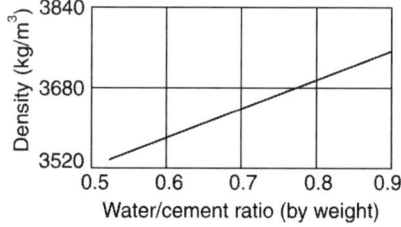

Fig. 15.2: Relation between water/cement ratio and density of baryte aggregate concrete

The density of baryte concrete varies more or less linearly with the water/cement ratio from a minimum of 3550 kg/m³ to a maximum of 3700 kg/m³. The density increased with the water/cement ratio as a result of improved workability and compaction of the mixes with increased water content. The recommended mix proportions using three different maximum sizes of aggregate for conventional baryte concrete[142] is given in Table 15.2, which is based on full scale experi-mental investigations. It is always preferable to make trial mixes using the available materials, followed by suitable adjustments based on the results of test specimens.

Table 15.2: Recommended mix proportions for 20 mm, 38 mm and 76 mm maximum size, conventional baryte aggregate concrete

	Unit	Maximum size of aggregate		
		20 mm	38 mm	76 mm
Water content	kg/m^3	185	167	145
Cement content	kg/m^3	318	288	250
Aggregate content (Specific gravity 4.1)	kg/m^3	3050	3180	3320
Water/cement ratio (by weight)	–	0.58	0.58	0.58
Fine aggregate (Per cent of total aggregate)		50	42	35
Slump	cm	8.7	8.7	7.6
Plasticizer (Per cent wt. of cement)	–	1	1	1
Unit weight (Fresh concrete)	kg/m^3	3550	3600	3700

15.5 HIGH-DENSITY CONCRETE USING STEEL AGGREGATES

Artificial heavy aggregates, like steel punchings and steel shot, have been used to produce concrete with very high density in the range of 4000 to 6000 kg/m^3. Davis et al[143] have achieved densities of 4540 kg/m^3 using steel punchings and limonite as coarse and fine aggregate, respectively. The compressive strength of standard cylinders at 28 days averaged about 220 kg/cm^2 for a mix containing Type II cement and a water/cement ratio of 0.58.

Fiesenheiser and Wasil[145] have developed empirical graphs for the design of high density concrete containing steel punchings and steel shot as aggregates. The specific gravities of the various ingredients used is given in Table 15.3. The steel shot was graded to a fineness modulus of 4. A shot factor of 50 per cent produced the highest dry packed density as shown in Fig. 15.3. The relation between compressive strength and density as a function of age is given in Fig. 15.4. The water/cement ratio and the cement factor necessary to produce a concrete of required density are obtained from the combined relationships shown in Fig. 15.5.

Table 15.3: Specific gravities and density of ingredients for high density concrete using steel aggregates

Material	Specific gravity	Specific weight kg/m³
Water	1.00	1000
Cement	3.10	3100
Steel punchings⁺	7.56	7560
Steel shot*	7.46	7460

⁺ Maximum thickness = 10 mm, Maximum diameter = 25 mm

* Graded to a fineness modulus of 4.00

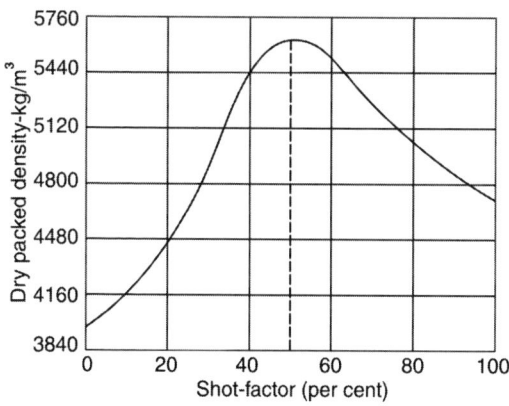

Fig. 15.3: Relationship of shot factor to dry packed density

It should be noted from the strength-density relationships that for any one strength, there is one density and vice versa. In an actual design problem; the density is selected as the minimum that is required or a higher value than the minimum, if the strength controls the design. There is a choice of either fixing the cement factor and determining the water/cement ratio or fixing the water/cement ratio and determining cement factor. It is, however, recommended that in no case should the cement factor be less than 400 kg/m³, nor should the cement factor is reduced resulting in savings in the cost of cement.

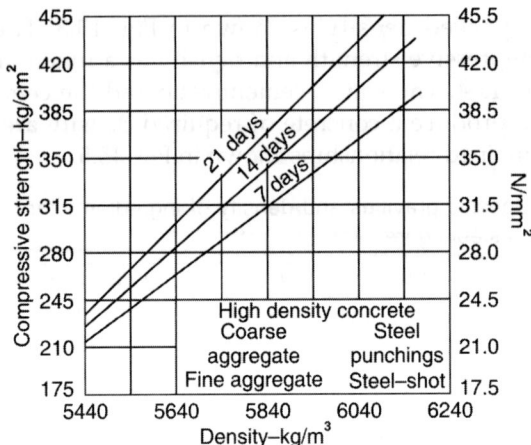

Fig. 15.4: Combined relationships of 7, 14 and 21 days compressive strengths to theoretical density

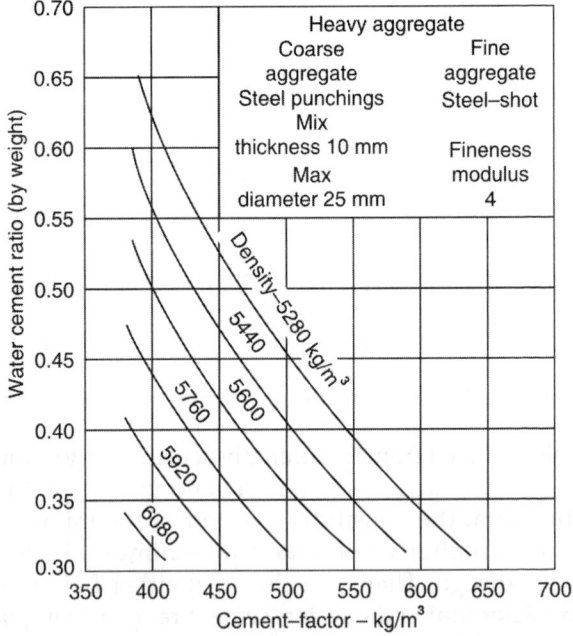

Fig. 15.5: Relationship of water/cement ratio, cement factor and density

However, it may be necessary to use more steel shot and punchings in such mixes. An economical mix is obtained by examining a few combinations of water/cement ratio and cement factors to produce a unit volume of concrete.

It is well known that the strength of ordinary concrete increases with an increase in the cement factor. However, with high-density concretes, the opposite is true because the specific gravity of cement is lower than that of steel shot and punchings but greater than that of aggregates in ordinary concrete.

The design of a high-density concrete mix containing steel aggregates is illustrated by the following example.

Mix design example

Design the trail mix proportions of high density concrete using steel aggregates to suit the following requirements:

Minimum 7 days cylinder compressive strength = 250 kg/cm^2

Minimum density required = 5760 kg/m^3

Specific gravity and specific weight of water, cement, steel shot punchings are as given in Table 15.3.

Design

Density corresponding to a strength of 250 kg/cm^2 at 7 days

(from Fig. 15.4) = 5570 kg/cm^3

But minimum density required = 5760 kg/m^3

Hence the minimum density criterion controls the design.

Compressive strength corresponding to the minimum required density of 5760 kg/m^3 = 290 kg/cm^2

From Fig. 15.5, there are many combinations of water/cement ratio and cement factors producing the required density. Three different combinations are examined for a comparative analysis with different water/cement ratios.

	Mix 1	Mix 2	Mix 3
Water/cement ratio	0.44	0.40	0.32
Cement factor (kg/m^3)	400	425	500
Weight of water per m^3 of concrete (kg)	176	170	160

The quantities of materials required for the three different mixes are compiled in Table 15.4. If the cost of steel shot, punchings and cement is know, the total costs of the three different mixes are evaluated and the most economical mix is selected.

Table 15.4: Trial mix proportions for high density concrete

Material	Mix 1		Mix 2		Mix 3	
	Weight kg/m³ of yield	Volume m³	Weight kg/m³ of yield	Volume m³	Weight kg/m³ of yield	Volume m³
Steel shot	2592	0.3470	2583	0.3450	2550	0.3400
Steel punchings	2592	0.3430	2583	0.3430	2550	0.3365
Cement	400	0.1340	425	0.1420	500	0.1635
Water	176	0.1760	170	0.1700	160	0.1600
Concrete	5760	1.0000	5761	1.0000	5760	1.0000

15.6 EXPERIMENTAL INVESTIGATIONS ON HIGH-DENSITY CONCRETE USING STEEL PUNCHINGS

1. Introduction

The properties of high-density concrete made with sheared mild steel bars as coarse aggregate and river sand as fine aggregate were studied through experimental investigations[184]. Sheared mild steel bars of 6 to 10 mm diameter and of equal length were used as coarse aggregate. The specific gravity of the coarse aggregate was 7.85 while the dry unit weight and fineness modulus were 4800 kg/m³ and 7.834, respectively. Gurpur river sand with a particle size distribution corresponding to zone-II of IS: 383–1970 was used as fine aggregate. The fineness modulus of sand was 2.89. The grading curves of the coarse and fine aggregates used in the present work are shown in Fig. 15.6.

2. Mix Proportions

Ordinary Portland cement of C-33 grade conforming to IS: 269–1976 was used in the investigation. For the production of high-density concrete, optimum density is the main criterion in determining the relative proportions of the ingredients. Hence the proportion of the fine to the total aggregate was determined by mixing the fine and coarse aggregates in the dry state and selecting the proportions which gives the highest density. The fine and coarse aggregates were filled in successive layers and tamped well in a container to determine the dry packed density. The relation between dry packed density and the ratio of fine to total aggregate content determined experimentally is shown in Fig. 15.7. This helps in establishing the range of theoretical density of heavy concrete for various mix proportions of fine to total aggregate ranging from 0 to 100 per cent. The details of various mix proportions used in the investigations reported by Krishna Raju[184] are compiled in Table 15.5.

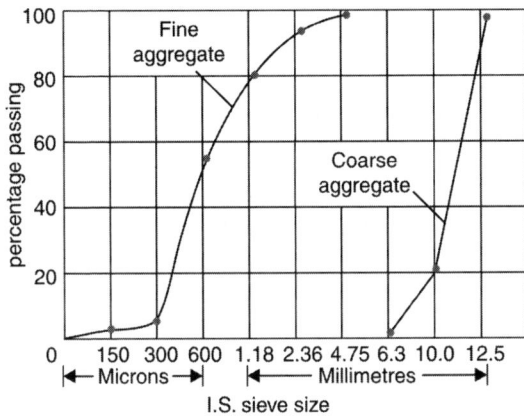

Fig. 15.6: Grading curves for fine and coarse aggregates used in investigation

Table 15.5: Mix proportions of high-density concrete

Cement content (kg/m³)	F.A./C.A. ratio	A/C ratio	W/C ratio	Cement	:	F.A.	:	C.A.
300	10 : 90	16.67	0.608	1	:	1.67	:	15.00
	20 : 80	14.00	0.637	1	:	2.80	:	11.20
	30 : 70	11.33	0.630	1	:	3.40	:	7.93
	40 : 60	10.00	0.670	1	:	4.00	:	6.00
	50 : 50	9.33	0.720	1	:	4.67	:	4.67
400	10 : 90	12.50	0.450	1	:	1.25	:	11.25
	20 : 80	10.50	0.470	1	:	2.10	:	8.40
	30 : 70	8.50	0.415	1	:	2.55	:	5.95
	40 : 70	7.50	0.480	1	:	3.00	:	4.501
	50 : 50	7.00	0.590	1	:	3.50	:	3.50
500	10 : 90	1.00	0.420	1	:	1.00	:	9.00
	20 : 80	8.40	0.446	1	:	1.68	:	6.72
	30 : 70	6.80	0.448	1	:	2.04	:	4.76
	40 : 70	6.00	0.530	1	:	2.40	:	3.60
	50 : 50	5.60	0.680	1	:	2.80	:	2.80

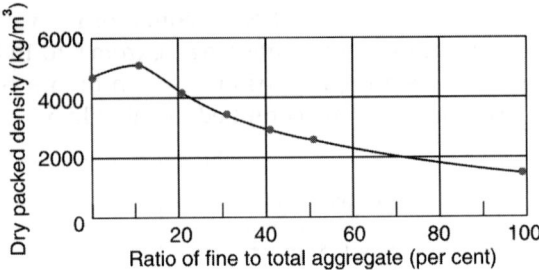

Fig. 15.7: Variation of dry packed density with percentage of fine to total aggregate.

The cement content in the mix was varied from 300 to 500 kg/m³ while the ratio of fine to total aggregates was varied from 10 to 50 per cent for each quantum of cement content. The amount of water added in each mix was such as to produce a workable mix which was neither too stiff nor too wet, corresponding to a narrow range of workability from low to medium.

3. Workability Characteristics

The workability characteristics of the various mixes were determined by the slump, compaction factor and Vebe tests conducted according to IS: 1199–1959. The high-density concrete did not exhibit any appreciable slump being less than 5 mm for all cement contents and fine to total aggregate ratios. Hence the slump test cannot be considered as a measure of the workability of high-density concrete.

The compacting factor was found to be more or less constant, having values in the narrow range of low to medium workability. The Vebe time varied from 75 to 25 seconds as the cement content increased from 300 to 500 kg/m³. The test results indicate that the Vebe test is very sensitive to changes in the various parameters in the mix and can be considered as a suitable workability test for high-density concrete with steel aggregates. Fig. 15.8 shows the typical Vebe test results for an aggregate/cement ratio of 10 as a function of the cement content in the mix.

4. Casting and Curing of Test Specimens

Six standard test specimens comprising 150 mm cubes were cast for each mix using steel moulds and the concrete was compacted using a platform vibrator. The specimens were cured in a water tank for 28 days before the start of tests.

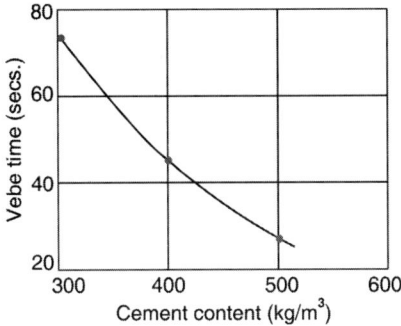

Fig. 15.8: Variation of Vebe time with cement content

5. Properties of Hardened Concrete

The compressive strength tests were conducted in a 2000 kN compression testing machine.

(a) *Compressive strength:* The results of 28 days compressive strength and the corresponding water/cement ratio for a constant aggregate/cement ratio of 10 is shown in Fig. 15.9. The results conform to the pattern of decreasing strength with increasing water/cement ratio. The relationship between 28 days compressive strength and cement content for the various fine to total aggregate ratios used in the tests is shown in Fig. 15.10. For all the ratios, the compressive strength increases more or less linearly with the cement content. The maximum compressive strength of 55 N/mm² was achieved with a cement content of 500 kg/m³ and a fine to total aggregate/cement ratio of 10 per cent with a water/cement ratio of 0.42.

(b) *Density:* The variation of density with cube compressive strength was found to be linear. Based on the test results, an empirical relation expressing compressive strength as a function of density is derived as

$$f_{ck} = 0.9 \, D_c$$

where,

f_{ck} = 28 days cube compressive strength (N/mm²)

D_c = density of heavy concrete (kN/m³)

The present test results confirm the earlier findings of Fiesenheiser and Wasil[145] who reported a linear relation between compressive strength and density for high-density concrete made with steel punchings as coarse aggregate and steel shot as fine aggregates.

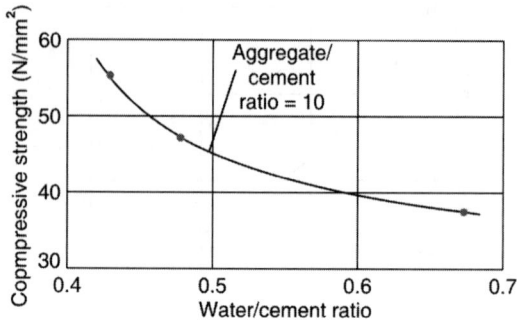

Fig. 15.9: Relation between compressive strength and water/cement ratio

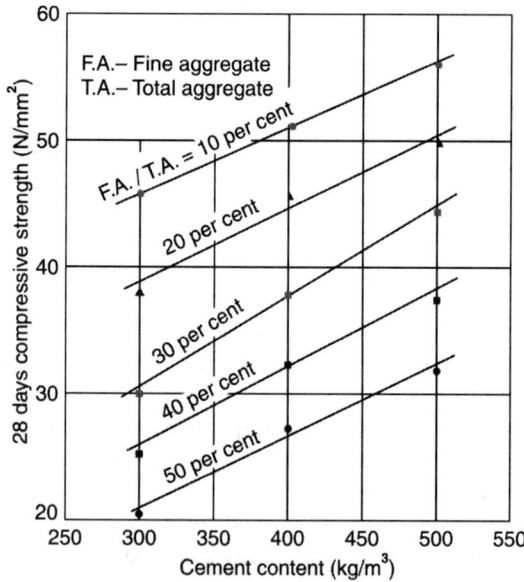

Fig. 15.10: Relationship between compressive strength and cement content

6. Conclusions

The main conclusions resulting from the tests on high-density concrete using sheared steel bars as coarse aggregate can be summarised as follows:

(a) High-density concrete can be made with sheared steel bar punchings as coarse aggregate to give densities ranging from 34 to 56 kN/m³ and compressive strength ranging from 20 to 55 N/mm².

(b) The Vebe and compacting factor tests are suitable for the measurements of workability of fresh concrete but the slump test is not sensitive for high-density concrete.

(c) A linear relation is proposed for prediction of the compressive strength expressed as a fraction of the density of concrete.

15.7 EXPERIMENTAL INVESTIGATIONS ON HIGH-DENSITY CONCRETE USING HAEMATITE AGGREGATES

1. Introduction

The properties of high-density concrete made with haematite aggregates and C-53 grade cement were studied through experimental investigations[185]. Several concrete mixes were designed using the Indian Standard Code IS: 10262–1982[179] and Road Note No. 4[12] methods by varying the aggregate/cement ratios from 6 to 10 and water/cement ratios in the range of 0.38 to 0.52 Conplast-430 super plasticizing admixture was used to improve the workability of the concrete mixes. The resulting high-density concrete indicated a characteristic compressive strength of 27 to 50 N/mm^2 having densities in the range of 35 to 37 kN/m^3.

2. Materials and Test Specimens

Birla super C-53 grade Portland cement conforming to IS: 12269–82 code was used for all the mixes. Haematite aggregates mined at Hospet, Karnataka state and supplied by Mineral Sales Private Limited was used in the fine and coarser fractions with the maximum size of the aggregate being 20 mm and having the grading analysis shown in Table 15.6.

Table 15.6: Grading of fine and coarse aggregates (haematite)

I.S. sieve size	Percentage passing	
	F.A.	C.A.
40 mm	–	100.00
20 mm	–	99.00
10 mm	–	22.0
4.75 mm	99.00	1.50
2.36 mm	83.3	–
1.18 mm	61.6	–
600 micron	50.2	–
300 micron	30.0	–
150 micron	2.3	–

The grading of the aggregates conformed to the requirements specified in IS: 383–1970. The fineness moduli of fine and coarse aggregates were 2.74 and 6.8, respectively. The fine aggregate grading corresponded to the grading zone-II of IS grading requirements. The specific gravity and bulk density of the aggregate being 4.35 and 25.4 kN/m^3, respectively. The test specimens comprised of 150 mm standard cubes for the determination of compressive strength.

3. Concrete Mix Design

Concrete mixes corresponding to the grade M-25 to M-40 were designed using the Indian Standard Code IS: 10262 and the Road Note No. 4 methods. The degree of workability assumed was very low since compaction by vibration was available and super plasticizing admixture was used to improve the workability.

Three different fine to coarse aggregate ratios of 0.429, 0.471 and 0.538 were used for A/C ratios varying from 6 to 10. These ratios corresponded to 30, 32 and 35 per cent fine aggregate in the combined aggregate grading of the standard grading curves of Road Note No. 4. The details of various mixes used in this investigation are compiled in Table 15.7. For each of the mixes, control specimens comprising six standard cubes were cast for standard tests at 28 days.

The control specimens were water cured for a period of 28 days. The slump, compaction factor and Vebe tests were conducted on fresh concrete according to the method specified in IS: 1199–1959. The strength tests were conducted on standard cubes using a 2000 kN compression testing machine.

4. Workability Characteristics

The variation of slump, compaction factor and Vebe time with W/C and A/C ratios are shown in Figs 15.11 to 15.13, respectively. For a given A/C ratio, the workability of the high-density concrete increased with increasing ratio of fine to coarse aggregate. Also the workability generally decreased with increasing A/C ratio from 6 to 10. Based on pilot tests with varying dosages of superplasticizer content from 0 to 1.25 per cent, it was decided to use 1 per cent admixture in all the mixes to improve the workability of concrete. Fig. 15.14 shows the variation of slump with cement content.

5. Compressive Strength

The results of 28 days compressive strength tests on hardened concrete are compiled in Table 15.8. The variation of cube compressive

Table 15.7: Details of high-density concrete mixes

Mix no.	Proportions (by weight)	W/C ratio	A/C ratio	F.A./C.A. ratio	Cement content (kg/m³)
A-1	1 : 1.8 : 4.2			0.429	
A-2	1 : 1.9 : 4.1	0.38	6.0	0.471	482
A-3	1 : 2.1 : 3.9			0.538	
B-1	1 : 2.1 : 4.9			0.429	
B-2	1 : 2.2 : 4.8	0.41	7.0	0.471	427
B-3	1 : 2.5 : 4.5			0.538	
C-1	1 : 2.4 : 5.6			0.429	
C-2	1 : 2.6 : 5.4	0.45	8.0	0.471	382
C-3	1 : 2.8 : 52			0.538	
D-1	1 : 2.7 : 6.3			0.429	
D-2	1 : 2.9 : 6.1	0.49	9.0	0.471	345
D-3	1 : 2.5 : 6.5			0.538	
E-1	1 : 3.0 : 7.0			0.429	
E-2	1 : 3.2 : 6.8	0.52	10.0	0.471	320
E-3	1 : 3.5 : 6.5			0.538	

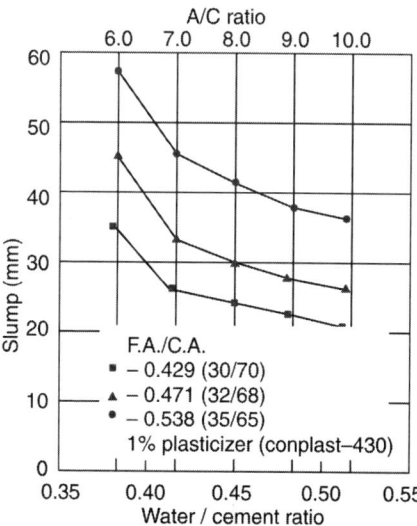

Fig. 15.11: Relation between W/C ratio and slump for various A/C ratios

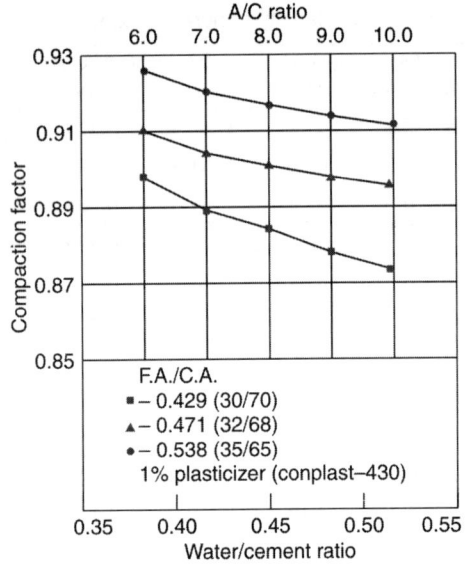

Fig. 15.12: Relation between W/C ratio and compaction factor for various A/C ratios

Table 15.8: Compressive strength of high-density concrete

Mix no.	F.A./C.A. ratio	W/C ratio	A/C ratio	28 days cube strength f_{ck} (N/mm²)
A-1	30/70			49.0
A-2	32/68	0.38	6.0	49.6
A-3	35/65			50.6
B-1	30/70			42.0
B-2	32/68	0.41	7.0	42.9
B-3	35/65			43.8
C-1	30/70			42.0
C-2	32/68	0.45	8.0	38.2
C-3	35/65			38.7
D-1	30/70			34.1
D-2	32/68	0.49	9.0	34.8
D-3	5/65			35.3
E-1	30/70			31.3
E-2	32/68	0.52	10.0	31.7
E-3	35/65			32.4

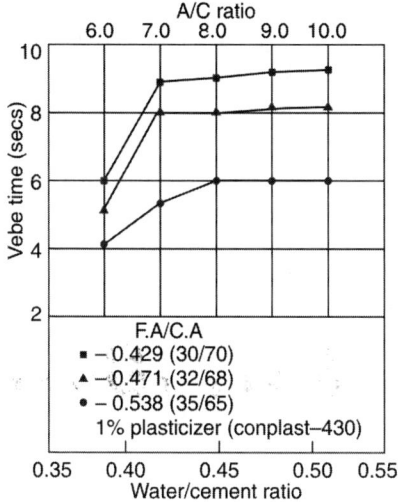

Fig. 15.13: Relation between W/C ratio and Vebe time for various A/C ratios

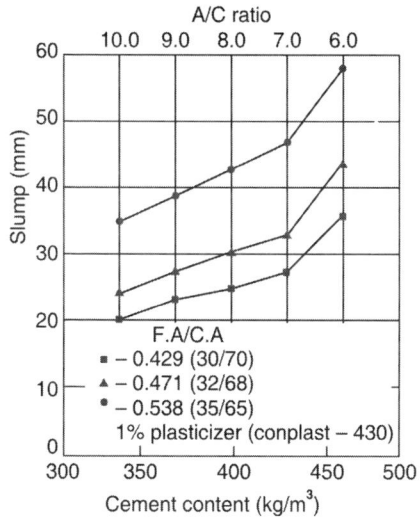

Fig. 15.14: Relation between cement content and slump for various A/C ratios

strength with W/C ratio is shown in Fig. 15.15. The general trend of results is in agreement with Duff Abram's water/cement ratio law of decreasing compressive strength with increasing W/C ratio.

Maximum compressive strength of 50 N/mm^2 was recorded for W/C ratio of 0.38 and A/C ratio of 6. The lowest compressive

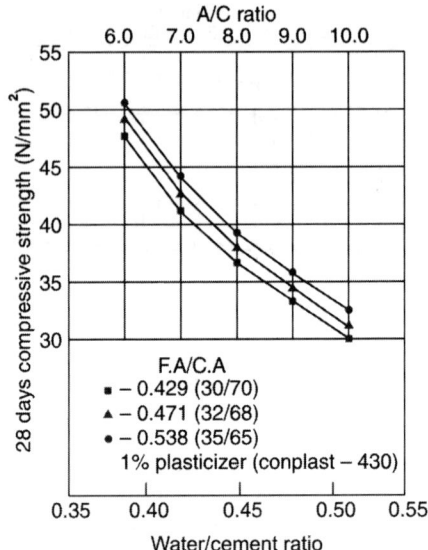

Fig. 15.15: Relation between W/C ratio and compressive strength for various A/C ratios

strength of 32 N/mm² was observed for W/C ratio of 0.52 and an A/C ratio of 10. For each of the W/C and A/C ratios used, the highest compressive strength was recorded for fine to coarse aggregate ratio of 0.538. The variation of compressive strength with cement content is shown in Fig. 15.16. The compressive strength increased from 30 to 50 N/mm² when cement content was increased from 320 to 480 kg/m².

6. Density

The variation of density of concrete with cube compressive strength is shown in Fig. 15.17. Since density is highly influenced by the aggregate content in the mix, higher densities were observed for mixes with higher A/C ratio of 10.

The density varied within the narrow range of 35 to 37 kN/m³, corresponding to A/C ratios of 6 to 10. Higher fine to coarse aggregate ratios generally resulted in higher density for each of the A/C ratios used in this investigation. Higher W/C ratios corresponding to higher A/C ratios generally resulted in higher density. Maximum density of nearly 37 kN/m³ was recorded for mixes with W/C ratio of 0.52 and an A/C ratio of 10.

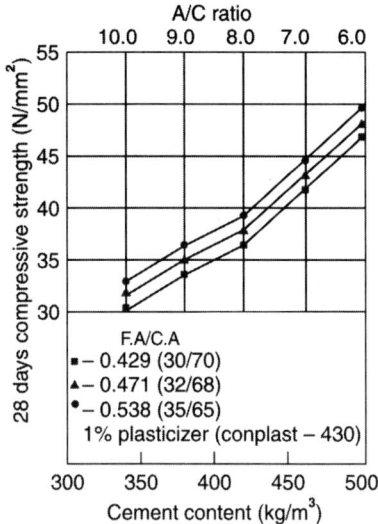

Fig. 15.16: Relation between cement content and 28 days cube compressive strength

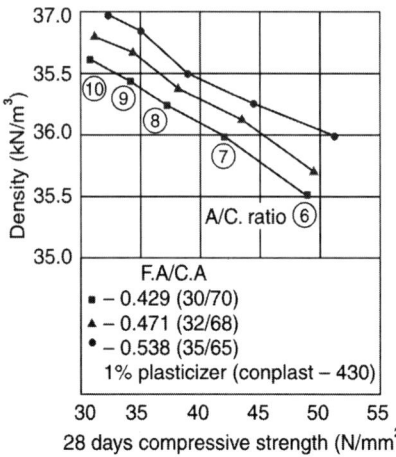

Fig. 15.17: Relation between cube compressive strength and density

7. Conclusions

The following conclusions are valid to the parameters covered in the tests on high-density concrete using haematite aggregates of specific gravity 4.35 in the fine and coarser fractions.

(a) The workability of high-density concrete expressed as slump and compaction factor increased with the ratio of fine to coarse aggregate and with cement content in the mix.

(b) The compressive strength of concrete increased from 30 N/mm^2 for a W/C ratio of 0.38.

(c) High-density concrete with a compressive strength varying from 30 to 50 N/mm^2 can be produced by varying the cement content from 320 to 482 kg/m^3 with an A/C ratio of 10. The density varied more or less linearly with increasing A/C ratio and at a decreasing rate with W/C ratio.

16

Fly Ash Cement
Concrete Mixes

16.1 INTRODUCTION

Pulverized fuel ash or commonly fly ash is the residue of the combustion of finely ground coal used in thermal power stations. The pozzolanic properties of fly ash were first reported in the 1914 edition of Engineering News, USA[146]. In 1930, the term fly ash was coined and its properties were investigated by McMillan and Power (USA)[147]. In 1939, Davis[148] and his associates published the results of the most comprehensive investigations into the use of fly ash in concrete and the era of fly ash concrete originated in USA. These earlier studies established that fly ash could be transformed from a waste product to a useful by-product for use in concrete as a companion to Portland cement.

The results of Davis formed the basis for many of the standard specifications on fly ash in use worldwide. First major application of fly ash in concrete was in the construction of Hungry Horse Dam (USA) in 1948[149]. Worldwide demand for power has resulted in the increase of thermal power stations and consequent increase in the production of fly ash resulting in environmental problems. Research work by Malhotra and Berry[150] has clearly established that the use of fly ash in concrete results in significant benefits.

16.2 ESTIMATES OF FLY ASH PRODUCTION

The installation of super power thermal stations like Raichur, the estimated annual worldwide production of fly ash is of the order of 1000 million tonnes at present and it is likely to exceed 2000 million tonnes by early 21st century. The dumping of fly ash in open fields results in ecological and environmental problems. The example of the havoc caused by the breach of the huge fly ash pond of Vijayawada thermal power station in the year 1990, indicates the magnitude of environmental hazards of storing huge quantities of fly ash.

16.3 TYPES OF FLY ASH

For the general purpose of use in concrete, fly ash is grouped under the following two major categories:

(a) *Low lime fly ash:* This type of fly ash contains less than 10 per cent calcium oxide and produced from anthracite and bituminous coals. The fly ash possesses truly pozzolanic properties, i.e. needs an activator to start reactions and produce cementatious properties. This is classified as Type-F fly ash according to ASTM-618 and the chemical constituents $(SiO_2 + Al_2O_3 + Fe_2O_3) < 70\%$.

(b) *High-lime fly ash:* This type contains greater than 10 per cent calcium oxide and it is produced from subbituminous and lignite coals. This type of fly ash possesses some cementatious properties itself in addition to pozzolanic properties. This type is classified as Type-C fly ash according to ASTM-618 and the chemical constituents comprising $(SiO_2 + Al_2O_3 + Fe_2O_3) < 50\%$. Fly ash produced in most countries generally satisfies the requirements of ASTM-618.

16.4 BENEFITS OF USING FLY ASH

Research investigations over the last three decades have indicated three major benefits in utilising fly ash.

(a) *Economical benefits:* Cement production requires large quantities of energy. Replacement of cement results in energy savings since fly ash does not need additional energy input before use. Larger the quantity of fly ash replacing cement, the energy saved is also proportionately more. Thermodynamic computations indicate that 1 tonne of cement replacement saves at least 6000 mJ of energy which is equivalent to a barrel of oil or 0.25 tonnes of coal. In Normandy Dam (USA), 72000 m³ of fly ash concrete containing 10000 tonnes of fly ash was used resulting in savings of 225000 dollars. In the construction of Watts Bar Nuclear Plant, 330000 m³ of fly ash concrete was used resulting in savings of 825000 dollars.

Environmental Protection Agency (EPA) in USA has formulated regulations requiring all agencies in USA utilising Federal funds in construction should compulsorily utilise fly ash in construction unless such use can be shown to be technically improper.

(b) *Technical benefits*: The use of fly ash in concrete results in the following technical benefits:
 1. Reduced bleeding and segregation
 2. Improved finishability and flow properties
 3. Increased cohesiveness leading to excellent pumpability
 4. Reduced heat of hydration
 5. Increased resistance to cracking
 6. Increased durability
 7. Increased resistance to chemical attack

(c) *Environmental benefits*: Worldwide production of fly ash by 2000 AD is estimated to be around 2500 to 3000 million tonnes, which poses enormous problems of disposal. In this background, utilisation of fly ash in civil engineering constructions can ease the environmental problems to a great extent. At present, very small percentage of fly ash is used due to the following reasons:
 (i) Failure to provide quality assured products
 (ii) Poor marketing methods
 (iii) Absence of effective technology transfer
 (iv) Transport and storage problems
 (v) Lack of effective legislation and standardisation in the country.

16.5 PHYSICAL AND CHEMICAL PROPERTIES OF FLY ASH

Fly ash suitable for use in concrete consists mostly of glassy, hollow spherical particles generally referred to as cenospheres. In contrast, Portland cement particles are generally angular. The salient properties are briefly outlined in the following sections.

(a) *Fineness:* The particle size distribution of fly ash depends upon the production process. Fly ash removed from flue gases using electrostatic precipitators has a finer particle size distribution with a specific surface in the range of 2700 to 3500 cm^2/gm. Fly ash resulting from mechanical collectors is comparatively coarser. For fly ash, the criterion of specific surface is not suitable due to the wide range of density and porosity of individual particles. Hence the percentage retention on a single sieve size of 45 micron is considered as a measure of fineness. Based on this criterion, progressive classification of fly ash for efficient use in concrete is shown in Table 16.1.

Table 16.1: Classification of fly ash

Grade	Fineness (per cent retained on 45 micron sieve)	Observations
Premium	< 5	Excellent properties
1	5–20	Good properties
2	20–35	Good to adequate properties
3	35	Doubtful economic viability

(b) *Density*: The average relative density of fly ash ranges between 1.9 and 2.4 which is around two-thirds that of ordinary Portland cement.
Loose bulk density of fly ash   = 800 kg/m^3
Loose bulk density of Portland cement = 1470 kg/m^3

(c) *Colour:* The colour of fly ash ranges from almost cream to dark grey essentially depending upon the proportion of unburnt carbon present and the iron content.

(d) *Pozzolanic activity index:* This index is a measure of the reactivity of fly ash with Portland cement in the presence of water to form complex compounds. The single most factor determining the quality of fly ash for use in concrete is fineness (expressed as percentage retained on 45 micron sieve) which influences the pozzolanic activity index.

The index decreases from 100 to 50 per cent as the fineness increases from 5 to 35. This indicates that finer the particle size, the reactivity is more resulting in improved workability and strength properties.

(e) *Chemical composition:* The typical chemical compositions of two major types of fly ash are compared with the ordinary Portland cement in Table 16.2.

16.6 STANDARD SPECIFICATIONS FOR FLY ASH

Many countries have evolved their own specifications for fly ash based on extensive research investigations spread over a period of

Table 16.2: Typical bulk oxide composition

Oxides	Mass percentage		
	O.P.C.	Low-lime fly ash (F)	High-lime fly ash (C)
SiO_2	20	50	40
Al_2O_3	5	28	18
Fe_2O_3	3	3	8
CaO	64	3	20
MgO	2	1	4
SO_3	2	1	2
Others	4	8	8

several decades. The standard specifications of Australia, Austria, Canada, India, Japan, Korea, Turkey, UK, USA and USSR are compiled in Table 16.3.

16.7 APPLICATIONS OF FLY ASH CONCRETE

Fly ash concrete has found extensive applications in mass concrete, precast concrete, concrete used for pavements and roller compacted concrete with the added advantages of increased workability, pumpability, resistance to chemical attack and with increased durability in comparison with ordinary Portland cement.

(a) *Mass concrete:* First application of fly ash was in mass concrete construction of dams and it is widely accepted that fly ash in conjunction with Portland cement is ideally suited for mass concrete constructions. Reduction of cement content in mass concrete due to the addition of fly ash, reduces the temperature rise due to heat of hydration. Hence use of fly ash by direct replacement of cement controls the rise of temperature and overcomes the risk of excessive bleeding tendency to segregate and increased permeability.

(b) *Precast concrete:* Fly ash is well established for use in precast concrete masonry units (solid and hollow cement blocks). The main benefits of using fly ash in the production of concrete masonry units are improved cohesion in the relatively harsh mixes and better finished products as well as reduction in drying shrinkage. However, the level of use of fly ash in structural precast members comprising reinforced and prestressed concrete is relatively low mainly due to the slow gain of strength.

Table 16.3: Standard specifications of p.f.a. for use in concrete

Requirement	Australia	Austria	Canada F	Canada C	India	Japan	Korea	Turkey	UK	USA F	USA C	USSR
1. Loss on ignition (Max %)	8.0	5.0	12.0	6.0	12.0	5.0	12.0	10.0	7.0	12.0	6.0	10.0
2. SO_3 (Max %)	2.5	–	5.0	5.0	3.0	–	5.0	5.0	2.5	5.0	5.0	3.0
3. MgO (Max %)	–	5.0	–	–	5.0	5.0	5.0	5.0	4.0	5.0	5.0	–
4. Na_2O (Max %)	–	–	–	–	1.5	–	–	–	–	–	1.5	1.5
5. SiO_2 (Min %)	–	42/60	–	–	3.5	45	–	–	–	–	4.0	–
6. $SiO_2 + Al_2O_3$ (Min %)	–	–	–	–	70	70	70	70	–	70	50	–
7. Moisture (Max %)	1.5	1.0	3.0	3.0	–	1.0	3.0	3.0	0.5	3.0	3.0	–
8. Water requirement (Max %)	–	–	–	–	–	100	102	105	95	105	105	–
9. Pozzolanic activity index With cement (Max %)	–	80	75	75	–	60/70	85	70	85	75	75	85
With lime (Min %)	–	–	–	–	3.9	–	5.5	–	–	5.5	5.5	5.5
10. Fineness Retained on sieve 45 mm (Max %)	50	–	34	34	–	–	–	–	12.5	34	34	–
Specific surface (cm²/gm)	–	–	–	–	3200	2700	–	–	–	–	–	–

(c) *Pavement concrete:* First trials of fly ash concrete for use in pavements in USA showed good results in 1950. Since then, the use of fly ash concrete in road pavements has gradually been accepted. The problem of air entrapment in fly ash concrete is more important to develop resistance to frost attack. In fly ash concrete, the dosage of air entraining agent required for a given amount of air content is usually much higher in comparison with the ordinary Portland cement concrete.

(d) *Roller compacted concrete:* Roller compacted concrete also known as Rollcrete, is a very lean and dry concrete, compacted in place using vibrating rollers. From 1970 onwards, roller compacted concrete is widely used for pavements.

Fly ash is well suited since the most critical requirement of reinforced cement concrete is that for effective compaction, it must be dry enough to support the weight of the vibrating equipment and yet fluid enough to permit distribution of the paste throughout the mass during mixing and compaction. The use of fly ash reduces the mix water demand, improves workability, placeability and pumpability of mix and resulting in a better finish. Another advantage is a higher proportion of fly ash can be incorporated and the resulting concrete is called high fly ash content concrete (HFCC). Great deal of research and development has established fly ash concrete as a useful material in dams and pavements.

16.8 METHODS OF DESIGNING FLY ASH CONCRETE MIXES

Experimental investigations by Smith[151], Cannon[152] and Dhir et al[155] clearly indicate that it is possible to design fly ash concrete mixes of comparable strength to Portland cement concrete covering a strength range up to 50 N/mm² at 28 days. The salient steps involved in these methods are herewith outlined with numerical design examples.

16.8.1 Smith's Method of Designing Fly Ash Concrete

A rational method of mix design by which trial mixes of fly ash concretes could be produced with an accuracy equivalent to that obtained when applying Road Note No. 4[12] to the design of orthodox concretes, has been formulated by Smith. The method is based on extensive experimental investigations on concretes with fly ash from over 25 generating stations. The mixes designed covered the normal strength range of structural concrete. The usefulness of the method is shown by comparison with a large number of tests on fly ash cement concretes and with the results of earlier investigations [153,154].

The cementing efficiency method proposed by Smith gives the required values of strength and placeability to the fly ash concrete mixes. In particular, the strength of fly ash concrete is shown to depend only on the relative proportions of ash, cement and water. The method can be applied equally well to those concretes in which fly ash was considered as replacing sand [156], as in those where it replaced cement.

The design procedure to be followed comprises of the following steps:

1. The water/cement ratio (W/C) of an orthodox concrete mix to produce the desired strength is selected from Fig. 6.1 of Road Note No. 4[12].

2. From a knowledge of the aggregate characteristics and behaviour, the aggregates/cement ratio (N), which in a normal concrete of water/cement ratio (W/C), would give the required degree of workability.

$$(F/C) = \frac{(W/C)_s - (W/C)_w}{(3.15/G)(W/C)_w - K(W/C)_s}$$

For most of the ashes, the specific gravities lie between 1.9 and 2.3. An average value of 2.1 is assumed for the specific gravity to ash (C). 'K' is the cementing efficiency of an ash relative to cement as measured by the effect of the ash on the ratio $(W/C)_s$. In its effect on this ratio, a weight F of ash will be equivalent to a weight 'KF' of cement. The value of 'K' based on experiments has been found to be 0.25 and the water/cement ratio $(W/C)_w$ is taken as 0.4. Using these values, the optimum ratio becomes

$$(F/C) = \left[\frac{(W/C)_s - 0.4}{0.6 - 0.25 (W/C)_s} \right]$$

The value chosen for the water/cement ratio $(W/C)_w$ has an effect on the optimum quantity of ash in the fly ash concrete. The ash quantity decreases with an increasing value of $(W/C)_w$. For this reason, the value of $(W/C)_w$ should be as low as possible. The suggested practical minimum value is 0.4, although the designer is free to choose any value.

4. The water/cement ratio (W/C) for the fly concrete mix is calculated from the relation,

$$(W/C) = (W/C)_s (1 + 0.25 \, F/C)$$

5. The aggregate/cement ratio (A/C) for the fly ash concrete mix is evaluated from the relation,

$$(A/C) = N \left[\frac{(W/C)}{(W/C)_w} \right]$$

If $(W/C)_w$ is taken as 0.4, then

$$(A/C) = N \left[\frac{(W/C)}{0.4} \right]$$

where, 'N' is the aggregate/cement ratio of a normal concrete of water/cement ratio of 0.4.

6. The fly ash concrete mix proportions are given by:

Water	:	Fly ash	:	Cement	:	Aggregate
(W/C)	:	(F/C)	:	1	:	(A/C)

Design example

Design a fly ash concrete mix to suit the following data.

28 days mean compressive strength $= 210 \text{ kg/cm}^2$

Workability: Medium

Type of cement: Ordinary Portland

Fly ash specific gravity $= 2.1$

Type of aggregate: Irregular gravel of 20 mm maximum size and natural sand graded to curve no. 3 or Road Note No. 4.

Design

1. The required water/cement ratio for an average strength of 210 kg/cm² at 28 days, estimated from Fig. 6.1 of Road Note No. 4 = 0.71 = $(W/C)_s$.

2. Using irregular aggregate of 20 mm maximum size and natural sand graded to curve no. 3 (Fig. 2.2), the Table 6.3 shows that in a normal concrete of water/cement ratio, 0.4 the aggregate/cement ratio = 3.5 = N.

$$(F/C) = \left[\frac{(0.71 - 0.4)}{0.6 - (0.25)(0.71)} \right] = 0.735$$

3. The water/cement ratio for the fly ash concrete mix is calculated as,
$$(W/C) = (0.71)(1 + 0.25 \times 0.735) = 0.84$$

4. The aggregate/cement ratio is computed by the relation,
$$(A/C) = 3.5 \ (0.84/0.4) \qquad = 7.35$$

5. Hence the mix proportions by weight are

Cement	:	Fly ash	: Aggregate	:	Water
1	:	0.735	: 7.35	:	0.84

The quantities of materials required to produce a unit volume of concrete can be calculated by the absolute volume method.

16.8.2 Cannon's Method of Proportioning Fly Ash Concrete Mixes

The procedure for proportioning fly ash concrete mixes has been evolved from extensive investigations by the Tennessee Valley Authority; as a result of using fly ash in all classes of concrete for over a decade. However, the method is intended only for proportioning cement and fly ash and does not deal with the proportioning of aggregates or the determination of basic water requirements. The method assumes that the quantity and gradation of the coarse aggregate is the same in comparable mixes and that the difference in yield due to the large volume of cementations material in the fly ash mix is balanced by a reduction of the sand content.

The economy resulting from the use of fly ash depends entirely on the relative cost of fly ash to cement, the quality of ash and the strength requirements of the concrete. Figure 16.1 shows the relation between water/cement ratio and 28 and 90 days compressive strength of concrete of control mixes in which Type III cement and lime stone aggregates are used.

The economic proportions of fly ash to be used for different strength ranges are given in Figs 16.2 and 16.3 and these depend upon the relative cost of fly ash to cement. The percentage increase or decrease in water content of the fly ash mix in comparison with control mix, is shown in Figs 16.4 and 16.5, for different water/cement ratios of the control mix and for different percentages of fly ash content in the mix. Figures 16.6 and 16.7 give the comparative cement requirements for various fly ash proportions to produce concrete of equal strength at 28 and 90 days of age to mixes without fly ash. These curves have been developed from the strength versus water/cement plus fly ash curves by making allowances for the differences in water requirements at the various levels of strength using Figs 16.4 and 16.5.

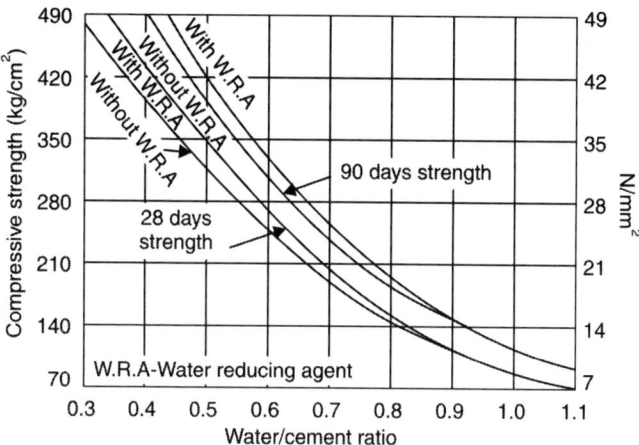

Fig. 16.1: Relation between water/cement ratio and strength of control mixes for an average type III cement limestone and sand

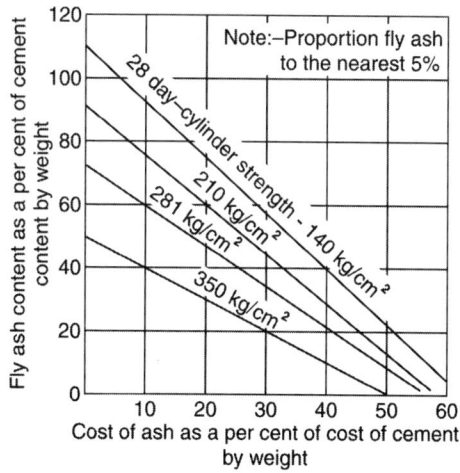

Fig. 16.2: Economic proportions of fly ash for 28 days strength concrete

The investigations showed that fly ash should be proportioned in concrete on the basis of economy and equal strength requirements and not as a straight substitute for cement either on a weight or volume basis. As the strength requirements decrease, the use of fly ash in concrete becomes more economical and marginal savings in

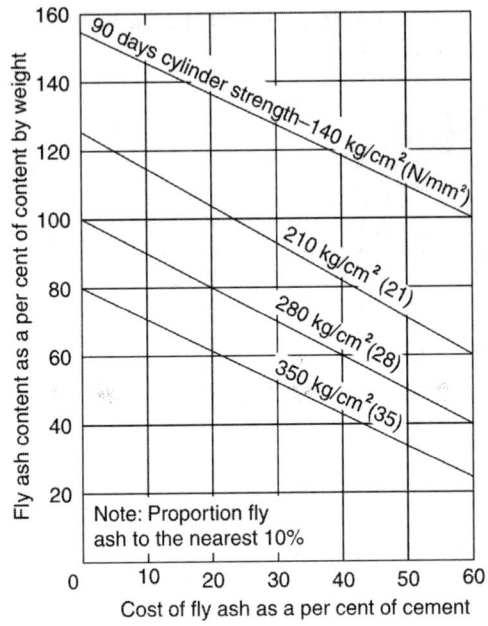

Fig. 16.3: Economic proportions of fly ash concrete for 90 days strength concrete

material coasts may be achieved even at high strength ranges of up to 350 kg cm² at 28 days. The design procedure to be followed is outlined in the following steps.

1. The water/cement ratio required to produce the desired 28 or 90 days strength is selected from Fig. 16.1 for the control mix.
2. The approximate water requirements for the maximum size of aggregate to be used and the required slump is estimated from Table 7.2 (ACI-method).
3. The volume of coarse aggregate per unit volume of concrete is selected from Table 7.6 of ACI: 211.1–99. In making this selection, the fineness modulus of the sand should be reduced by 0.20 to allow for the effect of the larger volume of cementatious material in the fly ash mix.
4. The economical proportion of fly ash to be used is determined from Figs 16.2 and 16.3, using the appropriate relative cost of fly ash and the required strength.
5. Using the water/cement ratio and the fly ash proportion, the increase or decrease in water content for the fly ash mix is determined from Figs 16.5 and 16.6.

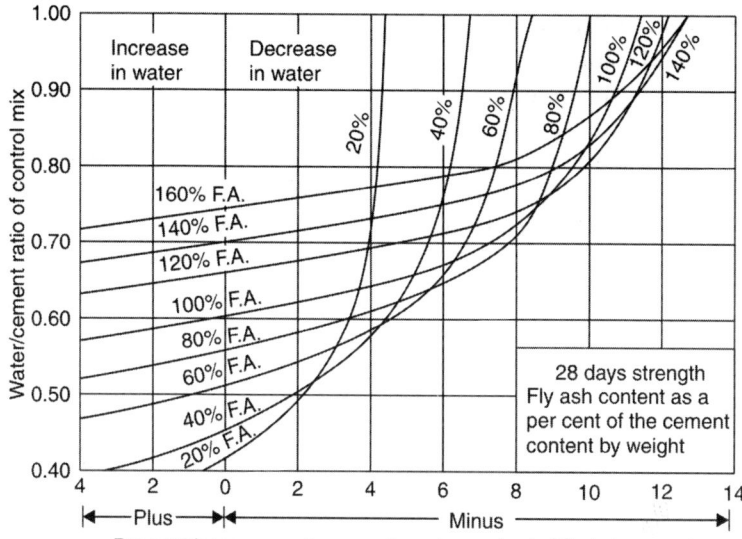

Fig. 16.4: Comparison of water requirements of concrete with and without fly ash equally proportioned for 28 days strength identical slump and air contents

6. The water requirements of the fly ash mix are determined by using the estimated water requirements of the control mix and the increase or decrease in water content.

7. The cement requirements of the control mix are determined by dividing the control mix water requirements by the water/cement ratio.

8. The proportionate cement requirement of the fly ash mix is selected from Figs 16.6 and 16.7 using the water/cement ratio of the control mix and the fly ash proportion.

9. The weight of sand required is determined by the absolute volume method.

10. The trial mix should be checked for slump and air content and suitable adjustments are made in the water, cement and fly ash requirements to achieve the desired results.

Design example

A fly ash concrete mix is required to suit the following requirements:

28 days specified minimum compressive strength = 210 kg/cm^2

Control ratio   = 0.82

Maximum size of aggregate	= 40 mm
Desirable slump	= 25–50 mm
Desirable air content	= 4.5 per cent
Fly ash coast	= 25 per cent of cement cost
Dry rodded weight of coarse aggregate	= 1620 kg/m³
Specific gravity of cement	= 3.15
Specific gravity of fly ash	= 2.3
Specific gravity of sand	= 2.65
Specific gravity of coarse aggregate	= 2.67
Fineness modulus of sand	= 2.80

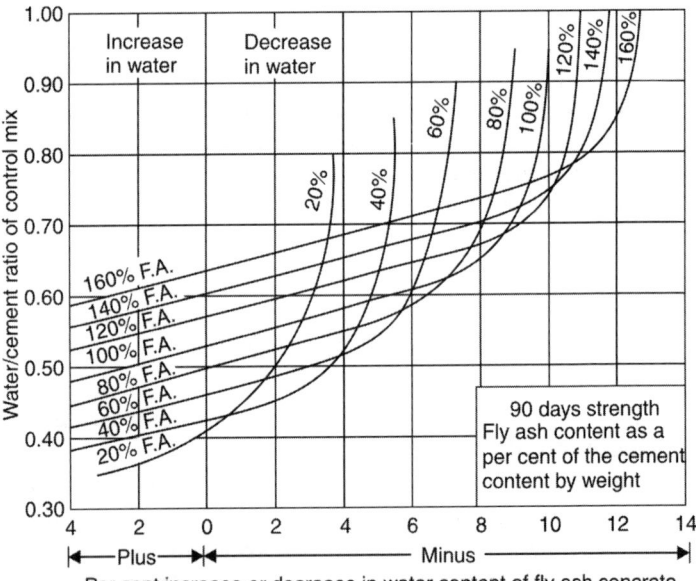

Fig. 16.5: Comparison of water requirements of concrete with and without fly ash equally proportioned for 90 days strength identical slump and air contents

Design

1. Mean compressive strength = (210/0.82) = 260 kg/cm² water/cement ratio (from Fig. 16.1) = 0.59
2. Approximate water requirement using 40 mm aggregate for a slump of 25–50 mm, using air-entrained concrete (Table 7.5) = 145 kg/m³

3. The volume of coarse aggregate per cubic metre of concrete using 40 mm aggregates and corresponding to a fineness modulus of sand of (2.80–0.20) is obtained from Table 7.6 as 0.74 m^3 weight of coarse aggregate = (0.74 × 1620) = 1200 kg.

4. The economical proportion of fly ash, for fly ash coast at 25 per cent of cement coast is (Fig. 16.2) = 50 per cent

5. For a water/cement ratio = 0.59 and fly ash proportion = 50 per cent water reduction in the fly ash mix (Fig. 16.4) = 4 per cent

6. Water content of fly ash mix = (145 × 0.96) = 139 kg
   Control mix cement content = (145/0.59) = 245 kg

7. Cement content of fly ash (Fig. 16.6) = 82 per cent of that of control mix.

8. Fly ash mix cement = (0.82 × 245) = 201 kg
   Fly ash content = (0.5 × 201) = 101 kg

Determination of sand content by absolute volume method

Ingredient	Weight (kg)	Absolute volume (m^3)	
Coarse aggregate	120	$\left(\dfrac{1200}{2.67 \times 10^3}\right)$	= 0.400
Cement	200	$\left(\dfrac{200}{3.15 \times 10^3}\right)$	= 0.063
Fly ash	101	$\left(\dfrac{101}{2.30 \times 10^3}\right)$	= 0.043
Water	128	–	= 0.138
Air	–	–	= 0.045
		Total	= 0.689 m^3

Volume of sand = (1.000 – 0.689)   = 0.311 m^3
Weight of sand = (2.65 × 0.311 × 10^3)   = 830 kg

Weights of ingredients for one cubic metre of fly ash concrete are given as

Cement	= 20 kg
Fly ash	= 101 kg
Coarse aggregate	= 1200 kg
Sand	= 830 kg
Water	= 139 kg
Density of fresh concrete	= 2470 kg/m^3

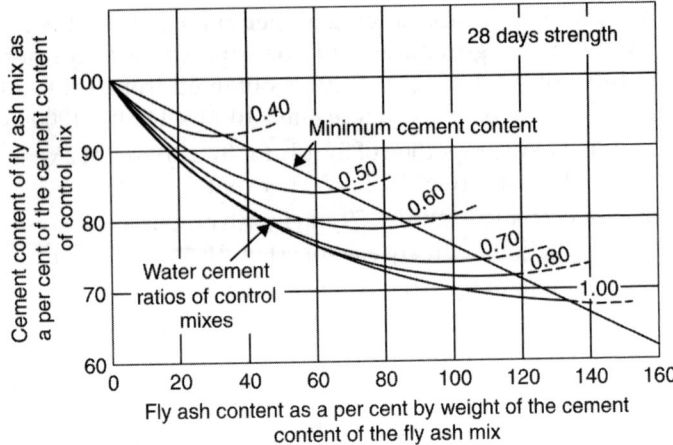

Fig. 16.6: Relative cement requirements of various fly ash proportions of concrete equally proportioned for 28 days strength identical slump and air contents

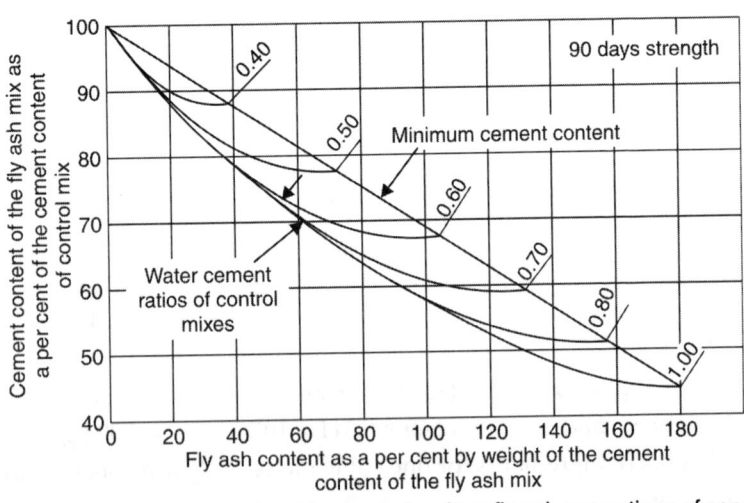

Fig. 16.7: Relative cement requirement of various fly ash proportions of concrete equally proportioned for 90 days strength identical slump and air contents

16.8.3 Dhir's Method of Designing Fly Ash Concrete Mixes

Dhir's method of mix design is useful for fly ashes which are of finer variety with percentage retained on 45 micron sieve not greater than 35 per cent.

The design procedure to be followed comprises of the following steps:

1. *Determination of free W/C ratio*: For a given target mean strength, the appropriate W/C ratio is obtained using Fig. 16.8.

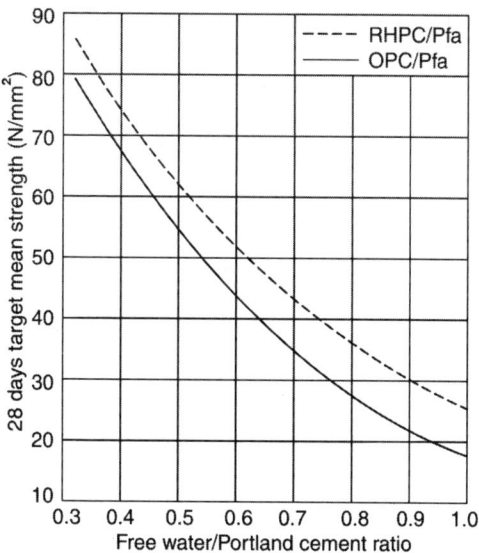

Fig. 16.8: Relationship between 28 days strength and free water/Portland cement ratio

2. *Determination of fly ash/composite cement ratio [F/(F + C)]*: From Fig. 16.9, the optimum ratio of fly ash/composite cement ratio is obtained corresponding to the free water/cement ratio for different grades of fly ash, designated as *P*, 1 and 2 as classified in Table 16.1.

   For coarser grades of fly ash (generally resulting from thermal power stations in India, e.g. Raichur) in which the percentage material retained on 45 micron sieve is greater than 35 per cent (Grade-3 fly ash) extrapolated curves suggested by Krishna Raju et al[156] are used.

3. *Determination of free water content (W)*: For the specified workability and the type of coarse aggregate, the appropriate value of the water content (W) is read out from Table 16.3 for the various grades of fly ash. The approximate free water content for fly ash concrete containing approximately 30 per cent fly ash is determined from Table 16.4 for grades of fly ash designated as

P, 1 and 2. For Grade-3 fly ash of coarser variety, the free water content is obtained from Table 16.5.

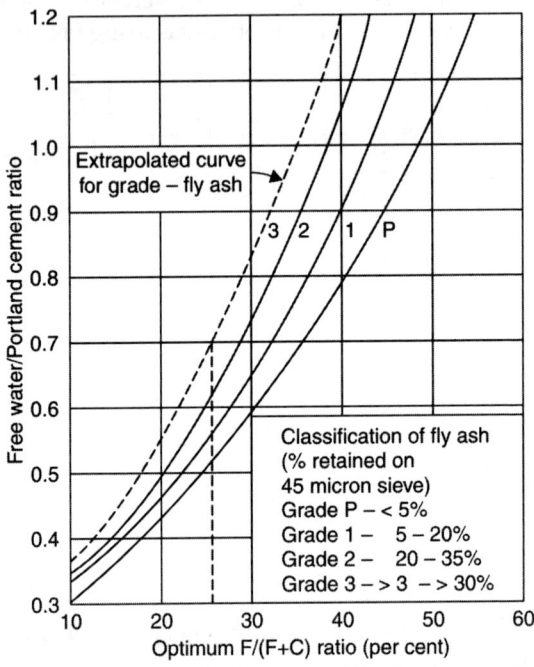

Fig. 16.9: Relationship between free water/Portland cement ratio and fly ash/total composite cement content

4. *Determination of Portland cement (C) and fly ash (F) contents:*

The cement content $= C = \left[\dfrac{W}{W/C}\right]$

The fly ash content $= F = \left(\dfrac{CX}{100}\right) \div \left(1 - \dfrac{x}{100}\right)$

where, $X =$ percentage ratio $\left(\dfrac{F}{F+C}\right)$

5. *Determination of total aggregate content (A):* The plastic density of fly ash concrete (p), corresponding to the free water content and relative density of combined aggregates are obtained from Fig. 16.10. Then the total aggregate content (A) is expressed as

$$A = p - (C + F + W)$$

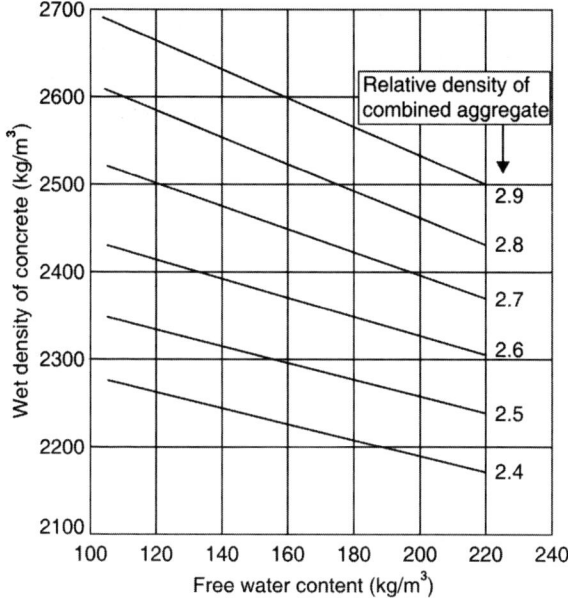

Fig. 16.10: Estimated plastic density of compacted Pfa concrete

Table 16.4: Approximate free water contents for fly ash concrete containing approximately 30% fly ash*

Maximum size of C.A. (mm)	Type of aggregate	Grade of fly ash	Free water content (L/m³) for slump (mm)		
			10–30	*30–60*	*60–12*
10	Gravel	Premium	160	175	190
		1	170	185	200
		2	180	195	210
	Crushed	Premium	180	195	210
		1	190	205	220
		2	200	215	230
20	Gravel	Premium	140	155	170
		1	150	165	180
		2	160	175	190
	Crushed	Premium	160	175	190
		1	170	185	200
		2	180	195	210

* For other fly ash contents, add (for lower fly ash contents) or subtract (for higher fly ash contents) about one litre/m³ of water per 2 per cent of fly ash

The proportion of fine aggregate (A_f) corresponding to the desired workability, maximum size of aggregate and the total composite cement content is obtained from Fig. 16.11. Then the coarse aggregate content can be computed as

$$A_c = (A - A_f)$$

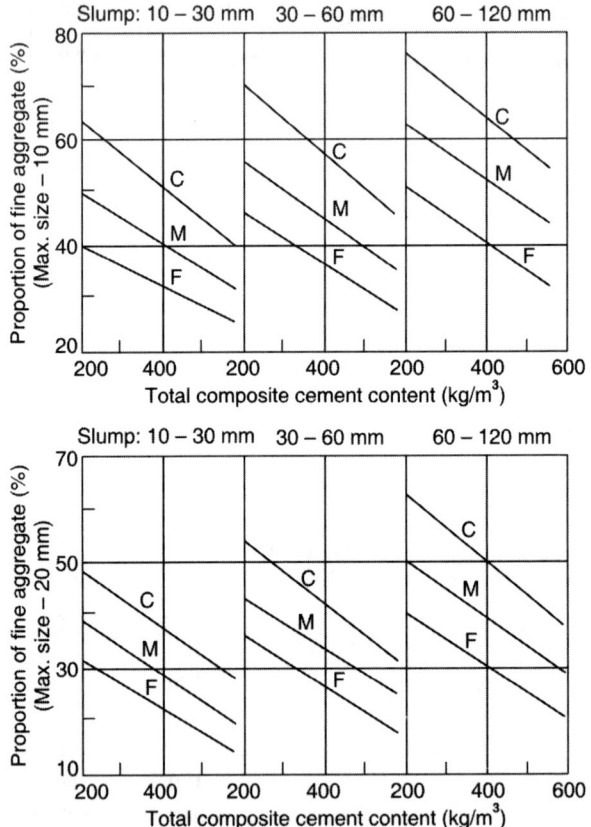

Fig. 16.11: Determination of fine aggregate proportion (top) 10 mm maximum size and (bottom) 20 mm maximum size aggregate

6. *Trial mix*: A trial mix is prepared using the mix proportions determined from the design procedure. If workability differs from the designed value, the water content is suitably adjusted and if the strength differs, the water/cement ratio is slightly modified to achieve the desired strength and workability of fly ash concrete.

Table 16.5: Approximate water content for fly ash containing approximate 30% fly ash* of Grade-3[156]

Maximum size of aggregate	Type of aggregate	Grade of fly ash	Free water content (L/m³) for slump (mm)		
			10–30	30–60	60–12
20	Gravel	3	145	160	175
	Crushed		160	180	195

The following design example illustrates the use of modified Dhir's method for designing fly ash concrete mix using coarser (Grade-3) fly ash for a given strength and workability.

7. *Design example*: Design a concrete mix with ordinary Portland cement and fly ash to suit the following data:

Cement: C-43 grade conforming to IS: 8112–1976

Fly ash: Grade-3 with percentage material retained on

45 micron sieve	= 43.5
Specific gravity	= 2.10
Specific surface	= 3200 cm²/gm

Coarse aggregate: Crushed granite, 20 mm maximum size

Specific gravity	= 2.68
Bulk density	= 1600 kg/m³

Fine aggregate: River sand with medium grading

Bulk density	= 1590 kg/m³
Workability: Design slump	= 60–120 mm
Grading of concrete	= M-15

Degree of control: Very good (s = 2.5 for M-15 grade concrete according to IS: 10262)

Durability: Mild exposure plain concrete,

Maximum water/cement ratio	= 0.7
Minimum cement content	= 220 kg/m³ (IS: 456–1978)

Design the fly ash/cement concrete mix and set out the quantities of materials required for one cubic metre of concrete using the modified Dhir's method.

Mix design

1. Target mean strength = F_{ck} =]15 + (1.65 × 2.5)] 20 N/mm²

2. From Fig. 16.8, read out the free water/cement ratio corresponding to the target mean strength (for RHPC) which is

greater than 1.0. But from durability consideration (mild exposure), according to IS: 456, maximum permissible W/C ratio = 0.7 for plain concrete and minimum cement content = 220 kg/m^3.

3. From Fig. 16.9, for W/C ratio = 0.7 and Grade-3 fly ash, the ratio $[F/(F + C)]$ = 25 per cent.

4. For workability (slump) = 60–120 mm and using crushed granite aggregate, the approximate free water content required is obtained from Table 16.4 as W = 195 kg/m^3.

5. Cement content $= C = \left[\dfrac{W}{W/C}\right] = \left(\dfrac{195}{0.7}\right) = 278$ kg/m^3.

6. Fly ash content is computed from the ratio $[F/(F + C)] = 0.25$

$$\left[\dfrac{F}{F + 2.78}\right] = 0.25$$

Solving, $\qquad F = \left[\dfrac{278 \times 0.25}{1 - 0.25}\right] = 92$ kg/m^3.

7. From Fig. 16.10, the plastic or wet density (p) of concrete for relative density of combined aggregate expressed as

$$p = \left[\dfrac{268 + 2.60}{2}\right] = 2.64 \text{ is obtained as } p$$

$$= 2360 \text{ kg/m}^3$$

8. The fine aggregate content is obtained from Fig. 16.11. For medium sand (M) and slump corresponding to 60–120 mm and the total composite cement content = (278 + 92) = 370 kg, the corresponding fine aggregate content for maximum coarse aggregate size of 20 mm is read out as 40 per cent of the total aggregate.

$\therefore$ Fine aggregate $\quad = (0.4 \times 1975) = 718$ kg

$\qquad$ Coarse aggregate $\quad = (1795 - 718) = 1077$ kg

9. The final design quantities of the various ingredients in the concrete mix are evaluated as

Cement $\qquad\qquad\qquad = 278$ kg

Fly ash $\qquad\qquad\qquad = 92$ kg

Coarse aggregate $\qquad = 1077$ kg

Fine aggregate	= 718 kg
Water	= 195 kg
Density	= 2360 kg/m^3

10. The mix proportions by weight are expressed as

Cement	:	Fly ash	:	Coarse aggregate	:	Fine aggregate	:	Water
1	:	0.3	:	3.87	:	2.58	:	0.70

17

Ultra-High-Strength Concrete

17.1 RANGE OF COMPRESSIVE STRENGTH

The normal strength concrete with its range of compressive strength from 100 to 300 kg/cm^2 is widely used in the building industry for different types of constructions. The development of prestressing techniques in 1930, emphasized the importance of high-strength concrete to withstand the high compressive stresses developed in prestressed concrete structural elements. By proper selection and proportioning of ingredients and improved methods of manufacture, concrete with a compressive strength ranging from 300 to 700 kg/cm^2 can be produced, using the natural aggregates and the common types of cements commercially available in the market. With the development of vibration technique as a means of compacting concrete, even extremely dry and stiff mixes, having lower water/cement ratios, have been successfully used to produce good concrete of high strength. Recent developments in the technology of concrete have conclusively shown the possibility of producing ultra-high-strength concrete, which has any desired 28 days cube compressive strength ranging from 700 to 1000 kg/cm^2, by using the modern cements and the natural aggregates. However, compressive strength exceeding 1000 kg/cm^2 can be achieved only by resorting to special materials and manufacturing methods or by using polymerisation techniques in conjunction with ordinary concrete.

17.2 METHOD OF PRODUCTION OF ULTRA-HIGH-STRENGTH CONCRETE

The FIP commission[157] on methods of producing ultra-high-strength concrete has reported several approaches, based on improved compaction, improved adhesion of cement matrix to aggregate and application of triaxial stress to achieve higher compressive strengths. However, the methods which are readily practicable are the use of high-quality coarse aggregate, synthetic, aluminous fine aggregate and Ciment Fondu. By using these materials, it is possible to produce very high-strength concretes, resorting to the normal concrete making techniques.

In the range of high compressive strength, with the use of classic materials and methods, adhesion of matrix to aggregate becomes critical. Alexander et al[158] have conclusively shown the compressive strength of concrete is a function of aggregate matrix bond, after exhaustive experimental investigations. Kleiger has reported from USA, the aggregates formed of Portland cement clinker in a mix with a water/cement ratio of 0.3, have resulted in strengths of 1200 kg/cm^2 at 90 days. Harris[159] has reported that in the United Kingdom, compressive strengths of 1300 kg/m^2 have been achieved by Robson, using Ciment Fondu with a synthetic aggregate derived from Ciment Fondu available commercially. Also, compressive strengths of up to 1240 kg/cm^2 were achieved with 1: 2 : 3 mix by using good granite coarse aggregate known as ALAG in the fine fraction only. The synthetic resin concretes, made by using expensive epoxy resin as a matrix, result in an exceedingly viscous mix and large quantities of resin 1000 kg/cm^2, has a very low modulus of elasticity.

Concrete being a granular material, it is possible to increase the strength of the mixture by greater cohesion, which can be realised by decreasing the voids between particles by compaction. In general, compaction has been found to improve almost all desirable qualities of concrete. The technique of compaction by pressure and vibration used by Freyssinet[160] for the production of high-strength concrete poles. The resulting concrete strengths were observed to be in the range of 1030 to 1240 kg/cm^2. Experiments by Lawerence[161] on cement compacts, which were precompressed under high pressure, revealed high compressive strength of the order of 372 N/mm^2. There appears to be an optimum frequency for each type of cement to achieve maximum compaction. Bennett and Gokhale[162] have reported that the optimum frequency for rapid hardening Portland cement as 800 Hz, resulting in strengths of 145 N/mm^2.

Recently, Parrott[163,164] has investigated the limits of strength attainable with classic materials and using vibration. The experimental investigations revealed that it is possible to produce concrete having a mean compressive strength in the range of 88 to 100 N/mm^2, using crushed rock aggregates such as lime stone, basalt and dolerite along with natural sand as fine aggregate. The salient details of mix proportions used and the compressive strengths measured as shown in Table 17.1. The optimum water/cement and aggregate/cement ratios were observed to be 0.28 and 2.0 respectively to achieve good compaction. Due to the high cement content in the mixes, the fine aggregate content is kept low and a satisfactory grading recommended, comprises of 10 per cent of fine aggregate passing 4.75 mm sieve, combined with 90 per cent of 10 mm to 4.75 mm, single-sized coarse aggregate.

Table 17.1: Details of mixes of ultra-high-strength concrete

Aggregate type	Aggregate cement ratio by weight	Ratio of fine to total aggregate	Water cement ratio (by weight)	Workability vebe time (secs)	28 days cube strength (kg/cm)^2
Limestone	2.0	0.1	0.28	16	830
Basalt	2.0	0.1	0.28	12	990
Dolerite	2.0	0.1	0.28	12	1000

Very high strengths may be achieved by the increased compaction and activation[165] of cement by the electrohydraulic impulse method. Activation of cement method, which is still in the development stage, consists of interrupted high-intensity electric discharges which create a plasma sphere within the grout. Collapse of this sphere causes a breakdown of the cement particles and results in the formation of water films on the grain surface and ionization of the grout. These lead to be increased magnitude of gel formation. The resulting grout is both denser and stronger due to the increased surface area of the gel. Certain chemical wetting agents are also effective in increasing the activation of the cement and strength of the paste.

17.3 POLYMER CONCRETE

Compressive strengths of concrete exceeding 1000 kg/cm^2 have been achieved by resorting to special techniques, like polymerisation. Experimental investigations conducted at the United States Bureau of Reclamation and the Brookhaven National Laboratory[166,167] have shown that substantial increase in strengths and other properties can be achieved by impregnation of dried concrete with a monomer followed by treatment by irradiation or thermal-catalytic process. The monomers which have proved most effective are methyl methacrylate and styrene. Vinyl acetate and ethylene gas dissolved in sulphur dioxide have also been used. The concrete is usually thermally dried and subjected to a vacuum of about 8 cm of mercury. Then the precast concrete element is soaked in monomer with a nitrogen blanket pressure, generally in the range of 0.4 kg/cm^2. This is followed by polymerisation by irradiation treatment of Cobalt-60 or thermal treatment at 75°C for about 4 hours.

The increase in compressive strength depends on the magnitude of polymer, which is impregnated and polymerised in concrete. A typical relation between polymer loading and the corresponding compressive strength is shown in Fig. 17.1[168]. In specimens containing 6.4 per cent by weight of methyl methacrylate, polymerised by radiation, the compressive strength and tensile strength increased by nearly four times that of the control specimens. The bond strength was observed to increase by three times, while there was a fourfold increase in the freeze-thaw resistance, with the permeability almost decreasing to zero. Significant increase in hardness was also noted. In general, an increase in the compressive strength of the material is found to have a beneficial effect on many of the other properties.

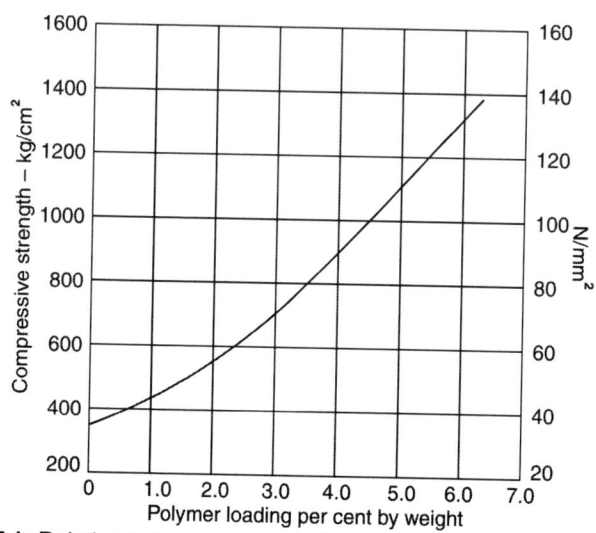

Fig. 17.1: Relation between compressive strength and polymer loading

The range of strength of ultra-high-strength concrete justifies a working stress of over 300 kg/cm^2, readily applicable for compression members. The new material finds extensive applications in prestressed concrete structures and it is possible to extend the use of existing types of structures in which present limitations of concrete strength necessitates a change to another type of material.

17.4 SILICA FUME CONCRETE

Condensed silica fume obtained as a by product from electric arc furnaces producing silicon metal or alloys, when used as an admixture with ordinary Portland cement, has been found to significantly increase the compressive strength of concrete. Silica fume consisting of very fine particles with average size of 0.1 to 0.2 micron has a very low bulk density of 200 to 250 kg/m^3. Consequently, it causes difficulty in handling and disposal. Pelletizing and densification operations increase the bulk density to 500 kg/m^3 so that it can be transported in bulk silos mounted on trucks. Mixed with water, condensed silica fume slurry with a specific gravity of 1.3 to 1.4 can be pumped into tankers and transported.

Silica fume particles have a very large specific surface of the order of 33000 to 77000 cm^2/gm which is nearly an order magnitude of large than that of the ordinary Portland cement. Norwegian standard-3474 and Canadian standard Can3A-23.5 specify that the condensed silica fume used as an admixture in concrete should contain at least 85 per cent SiO$_2$ and it should be in amorphous form. The maximum permissible quantity of silica fume is restricted to 10 per cent by weight of ordinary Portland cement.

Research investigations during the last four decades indicate that the use of silica fume as a pozzolanic admixture in concrete leads to technical, economical and environmental benefits. Silica fume concrete exhibits reduced thermal cracking and heat of hydration and shows improved durability and resistance to sulphate and acid waters. Tests have also shown that silica fume when used in conjunction with ordinary Portland cement, significantly improves the compressive strength of concrete.

Investigations by Malhotra[191] has indicated that concrete containing 0 to 15 per cent by weight of replacement of cement and silica fume ratio from 0.4 to 0.6 in both air-entrained and non-air-entrained concretes showed significant increase in the 28 days compressive strength of concrete.

Ultra-high-strength concrete has been produced by Wolsiefer[194] having the following specification:

C = Ordinary Portland cement = 593 kg/m³

S = Condensed silica fume = 119 kg/m³

Limestone aggregates of 10 mm maximum size

Combined water/cementatious ratio = $\left(\dfrac{W}{C+S}\right)$ = 0.22

Superplasticizer was used to improve workability.

The compressive strengths (f_{ck}) recorded at the end of 14 days and 4 months were 100 and 125 N/mm², respectively.

Tests on ultra-high-strength concrete by Bache[200] have indicated very high-strength values with the following specifications:

Ordinary Portland cement = 50 per cent ⎤

Condensed silica fume = 50 per cent ⎥ By volume

Water/cementatious ratio = 0.13 to 0.15 ⎦

Aggregate: Calcined bauxite

Large dosage of plasticizer was used.

28 days compressive strength (f_{ck}) = 250 N/mm²

The resulting concrete was found to be very brittle ultra-high-strength concrete with silica fume was produced and used for a high rise building in Montreal, Canada with the following specifications:

Average compressive strength of concrete cylinder = 60 N/mm² at 3 days, 75 N/mm² at 7 days 90 N/mm² at 28 days.

Ultra-high-strength concrete was developed and used for the repair work of the stilling basin of Kinzua dam in Pennsylvania by US Army Corps of Engineers in 1983[204].

The various chemical constituents, classification and properties of silica fume concrete are presented in detail by the author in Chapter 22 of this book.

18

Introduction to Concrete Mix Proportioning Using Digital Computer

18.1 APPLICATION OF MATHEMATICAL PROGRAMMING TECHNIQUES TO CONCRETE MIX DESIGN

The primary aim of mix design is to determine the relative quantities of ingredients, so as to achieve the most economical mix resulting in concrete processing certain minimum properties, like workability, strength and durability. The cost of concreting includes the cost of materials, plant and labour and generally in the design of concrete mixes, the leanest possible mix is preferred since it not only reduces the cost but also provides considerable technical advantages. Basically, the process of mix design may be looked upon as a mathematical programming problem in which that cost of concrete, is minimized subject to certain given constraints, like the compressive strength, workability, durability and aggregate characteristics.

The development of high-speed digital computers has revived interest in the optimum seeking methods. Significant advances have been made in the field of optimization techniques[169,170]. If the objective function is a linear function of the various constraints, which by themselves are linear, then the programming problem is said to be a linear programming problem. The solution to a linear programming problem is obtained by the simplex method developed by Dantzig[171], as early as in 1947. In fact the real impetus to the growth of interest in programming techniques came only after this work.

The simplex algorithm for solving the general linear programming problem is an iterative procedure which yields an exact optimal solution in a finite number of steps. One of the most powerful techniques for solving non-linear programming problems is to transform the problem by some means into a form which permit applications of the simplex algorithm. Thus the simplex algorithm turns out to be one of the most powerful computational device for solving non-linear as well as linear programming problems. There are many algorithms available to seek the solution of a non-linear programming problem which are classed as direct and indirect methods[172]. Direct methods are those where the constraint equations are explicitly handled in the algorithm, while indirect methods cast a constrained minimization problem as a sequence of unconstrained minimization problem.

A significant class among the direct methods of approach on general non-linear inequality constrained optimization problems, is referred to as the methods of feasible direction. Two well-known procedures which embody the philosophy of the method of feasible directions are Rosen's gradient projection algorithm[173] and Zouten-dijk's[174] procedure.

18.2 DESIGN VARIABLES AND CONSTRAINTS

The optimized concrete mix proportions having the least cost should satisfy the criteria of strength and durability of the hardened concrete and workability of the fresh concrete. The fine and coarse aggregates should be proportioned in such a way that the combined aggregate is within the prescribed grading limits since grading affects the workability of the mix. Aggregates of different maximum sizes, having different properties and unit costs and their combinations can be handled by introducing recycling capability in the programme.

The design variables to be considered in the computerised mix design process, necessarily include the various concrete ingredients, such as cement, aggregates, water and admixtures, if any used to achieve some desired properties or to economies on the use of cement. The design constraints which are generally considered comprise of:

(a) Compressive strength
(b) Workability
(c) Durability
(d) Aggregate characteristics

The objective function to be minimized is generally the total cost of concrete per unit volume, expressed as the sum of the costs of individual materials.

Design Constraints

The constraints controlling the design variables in concrete mix design generally fall into the category of behaviour constraints rather than side constraints.

(a) *Compressive strength*: It is well established that the strength of concrete at any given age is a function of the water/cement ratio, according to the law established by Abram[2]. The non-linear relationship between strength and water/cement ratio can be linearised by suitable transformation and incorporated in the programme. Experimental investigations by Cordon and Gillespia[175], Walker and Bloem[176] and Bloem and Gaynor[177] have indicated that the strength of concrete is also affected by the maximum size of aggregate in addition to the water/cement ratio. Suitable linear relationships can be established between strength and water/cement ratio as a function of the aggregate size for use as design constraint.

(b) *Workability*: The degree of workability suitable for the job in hand is generally categorized as very low, low, medium and high with the specified values of slump, compacting factor and Vebe time to cover these ranges. However, for a design constraints to be used in the optimization programme, it should be expressible in terms of the design variables. A survey of the literature indicates that the expressions developed by Murdock[13] relating compacting factor with the water/cement ratio, aggregate/cement ratio, surface and angularity index of aggregates and the standard consistency of cement, can be conveniently used as workability constraints in the optimization process.

(c) *Durability*: In the design of concrete mixes, the durability criterion is generally satisfied by proper selection of cement and by using air entrainment in concrete exposed to severe weather conditions. Practical recommendations[50,67] generally include a ceiling on the water/cement ratio and a minimum cement content to be used in concrete to improve its resistance to cycles of freezing and thawing. Depending upon the type of structure and exposure conditions, a maximum permissible water/cement ratio constraint is introduced. Sometimes a minimum water/cement ratio constraint is also imposed from practical considerations to avoid very stiff mixes.

(d) *Aggregate characteristics*: The shape, size and grading of aggregates influence the water content required to produce a workable mix. Various methods have been proposed to select the proportions of aggregates in a concrete mix. The ACI method[67] is based on the fineness modulus approach while the Road Note No. 4[12] method recommends the use of grading curves.

The method suggested by Murdock[83] involves the use of surface and angularity index of aggregates, which depend upon the size, shape and grading characteristics of the aggregate. An examination of these various methods indicates that the standard grading curves of Road Note No. 4 are well suited for constraint formulation since the maximum and minimum permissible grading limits of the combined aggregates are specified for different maximum sizes of aggregates ranging from 10 to 40 mm. If ten standard sieves from 40 mm down to 75 microns are used, there will be a total of twenty constraints corresponding to the maximum and minimum grading limits.

18.3 DEVELOPMENT OF SOFTWARE FOR CONCRETE MIX DESIGN

(a) *Introduction*: Concrete mix design being a laborious and tedious process involving several variable parameters and different methods of design, it will be highly convenient if software are prepared for each of the well-established and time-tested methods of mix design.

A software for concrete mix design developed in Fortran-77 by Dhananjaya[178] is very useful for concrete mix designers. The programme incorporates the Indian Standard Code method[179]. The software gives quantities of ingredients and cost per cubic metre of concrete for the desired grade of concrete. The input data comprises the important properties of cement and aggregate, cost of materials, workability requirements and the degree of quality control at site.

(b) *Flowchart for concrete mix design*: A typical flowchart developed for the design of concrete mix using the IS: 10262–1982 Code method is presented in the conventional form at the end of the chapter.

(c) *Software for concrete mix design*: The programme for the design of concrete mix was developed in Fortran-77 programming language by incorporating the various rules and procedures prescribed in the Indian Standard Code, IS: 10262–1982.

The salient features of the software developed are:

1. The programme accepts the characteristics strength (28 days strength) of cement tested according to IS: 4031–1968 which varies from 31.9 to 61.3 N/mm².
2. The input data includes the workability of concrete either in the form of slump or compacting factor.
3. Results of grading of aggregates (sieve analysis) either in the form of weights retained or percentage passing through Indian Standard sieves.
4. The following design tables and figures/charts are incorporated in the software:
   (i) Tables 1 to 6 and Figures 1 to 3 of IS: 10262–1982.
   (ii) Tables 2 and 4 of IS: 383–1970.
   (iii) Table 30 of SP: 23–1982.
   (iv) Table 19 of IS: 456–2000.
5. Quantity of cement in the mix is checked for the minimum and maximum contents depending on exposure and durability requirements as prescribed in IS: 456–2000 for RCC and IS: 343–2010 for P.S.C.

   The software for concrete mix design is given at the end of the chapter.

   (d) *Design example (Comparison of results using manual computations and computer method)*

Design Data

1. Characteristic compressive strength (28 days) $= f_{ck}$
   $= 40 \text{ N/mm}^2$
2. Desired degree of workability (compacting factor) $= 0.8$
3. Degree of quality control $=$ Good
4. Type of exposure $=$ Mild
5. Compressive strength of cement $= 53 \text{ N/mm}^2$
6. Specific gravity of cement $= S_c$
   $= 3.15$
7. Specific gravity of aggregates:
   Fine aggregate (S_{fa}) $= 26$
   Coarse aggregate (S_{ca}) $= 2.68$

8. Water absorption of fine aggregate = 1 per cent
9. Water absorption of coarse aggregate = 0.2 per cent
10. Free (surface) moisture of fine aggregate = Nil
11. Free (surface) moisture of coarse aggregate = 0.3
12. Result of sieve analysis (grading of aggregates)

IS sieve size	Percentage passing	
	F.A.	C.A.
40 mm	–	100
20 mm	–	100
10 mm	100	26.5
4.75 mm	97.5	1.55
2.36 mm	92.0	0.30
1.18 mm	48.0	0.10
600 micron	16.0	–
300 micron	5.0	–
150 micron	0.1	–

13. Cost of cement per bag = Rs 700
14. Cost of fine aggregate (sand) per m^3 = Rs 50
15. Cost of coarse aggregate (granite jelly) per m^3 = Rs 450

Mix design by manual computation (IS: 10262-1982)

Target mean strength F_{ck} = $[f_{ck} + t \times s]$
= $[40 + 1.65 \times 6.6]$
= 50.89 N/mm^2

(for values of t and s refer Tables 1 and 2 of IS: 10262–1982).

From Fig. 2 of IS: 10262–1982, for F_{ck} = 50.89 N/mm^2, read out water/cement ratio, W/C = 0.35.

The results of sieve analysis show that sand conforms to the standard grading requirements of Zone I and that for coarse aggregate shows that the nominal maximum size of coarse aggregate is 20 mm.

From Table 5 of IS: 10262–1982, for 20 mm nominal maximum size of the coarse aggregate and sand conforming to Zone II, the following values are taken:

Water content per cubic metre of concrete = 180 litres

Sand content as percentage of total aggregate by volume = 25

Since the sand sample grading conforms to Zone I, the following corrections are made in accordance with Table 6 of IS: 1026–1982.

Required sand content as percentage of total aggregate = sand content as percentage of total aggregate in the zone II + Corrections needed for zone difference + corrections needed at the rate of + 1 per cent for each 0.05 increase or decrease in free water/cement ratio.

$$= \left[25 + 1.5 + \frac{0.35 - 0.35}{0.05} \times 1 \right] = 26.5$$

Hence, $p = \dfrac{26.5}{100} = 0.265$

Corrections for water content are not required since compacting factor assumed is 0.8.

Estimation of air content: For 20 mm nominal maximum size aggregate, 2 per cent air content is assumed as per Table 3 of IS: 10262–1982.

Net volume of concrete = (Gross volume – Air content)
$$V_{net} = 1 - 0.05 = 0.98 \text{ m}^3$$

Determination of quantities of ingredients

$$\text{Cement content} = C = \left[\frac{\text{Water content } (w)}{\text{Water/cement ration } (w/c)} \right]$$

$$= \left[\frac{180}{0.35} \right] = 514.3 \text{ kg/m}^3$$

The cement content is adequate for mild exposure conditions according to Table 9 of IS: 1343–1980.

$$V_{net} = \left[w + \frac{C}{S_c} + \frac{f_a}{p \times S_{fa}} \right] \times \frac{1}{1000}$$

$$0.98 = \left[180 + \frac{513.4}{3.15} + \frac{f_a}{0.265 \times 2.6} \right] \times \frac{1}{1000}$$

Solving, $f_a = 438.73 \text{ kg}$

Computation of coarse aggregate content (C_a)

$$V_{net} = \left[w + \frac{C}{S_c} + \frac{C_a}{(1 - p) S_{ca}} \right] \times \frac{1}{1000}$$

$$0.98 = \left[180 + \frac{513.4}{3.15} + \frac{C_a}{(1 - 0.265)2.68} \right] \times \frac{1}{1000}$$

Solving, $C_a = 1254.25 \text{ kg}$

Therefore, mix proportions required before corrections for water absorption and moisture content in fine and coarse aggregates is

$$[C \ : \ f_a \ : \ C_a \ : \ w] \quad = \quad [1 \ : \ 0.85 \ : \ 2.44 \ : \ 0.35]$$

Actual quantities of ingredients

The final quantities of ingredients after applying the corrections for water absorption and moisture content in fine and coarse aggregates are evaluated as shown below:

(i) Required quantity of cement = 513.4 kg

(ii) Corrections for water content in aggregates for water/ cement ratio of 0.35, quantity of water required = 180 litres. Extra quantity of water to be deducted for free moisture present in fine aggregate at 0.2 per cent by mass is given by $(0.2/100) (1254.25) = 2.51$ litres

Quantity of water to be deducted for free moisture present in fine aggregate at 0.3 per cent by mass is given by $(0.3/100)$ $(438.73) = 1.316$

Quantity of water to be added for water absorption in case of fine aggregate at 1 per cent by mass is computed as $(1/100)$ $(438.73) = 4.387$ litres

Therefore, actual quantity of water required for field mix is

$$w \ = \ (180 + 2.51 - 1.316 + 4.387) = 185.58 \text{ litres}$$

(iii) Actual quantity of sand required after allowing for mass of free moisture is computed as

$$f_a \ = \ (9438.73 + 1.316) \qquad = 440.05 \text{ kg}$$

(iv) Actual quantity of coarse aggregate required is

$$C_a = \ (1.254.25 - 2.51) \qquad = 1251.74 \text{ kg}$$

Actual quantities of ingredients required for one cubic metre of M-40 grade concrete using 53 grade cement are as follows:

(i) Cement content $\quad\quad\quad = C = 514.3$ kg

(ii) Fine aggregate content $\quad = f_a = 440.05$ kg

(iii) Coarse aggregate content $= C_a = 1251.74$ kg

(iv) Water content $\quad\quad\quad\quad = w = 185.58$ litres

The final mix proportions of ingredients required for M-40 grade with 53 grade cement are computed as,

$$[C \ : \ f_a \ : \ C_a \ : \ w] \quad = \quad [1 \ : \ 0.84 \ : \ 2.39 \ : \ 0.36]$$

Total cost of concrete per cubic metre = Rs 1991.30

A comparative analysis of mix proportion results obtained from manual computations and by using the software developed indicates that the percentage differences in the quantities are negligible small. Thus the software developed can be conveniently used to avoid lengthy and tedious manual computations required for concrete mix design. The software developed can be suitably modified to include the American Concrete Institute method and the British method.

18.4 CONCLUDING REMARKS

The computer aided mix design is very useful in achieving considerable savings in the cost of production, where large quantities of concrete are involved. In the case big projects, like concrete dam works and precast concrete plants, a comparative analysis of the costs of concretes using materials from different sources can be made to identify the optimum solution. The cost of cement in general being the maximum in comparison with other ingredients in concrete, leanest mixes generally produce the most economical solutions. But while selecting the leanest possible mix, it is important to satisfy constraints like minimum cement content or the workability requirements suitable for the job. As the utility of digital mix proportioning depends on the integrity of the design constraints and variables, a clearer understanding and identification of several basic variables will help in selecting the optimum mix proportions.

FLOWCHART FOR CONCRETE MIX DESIGN

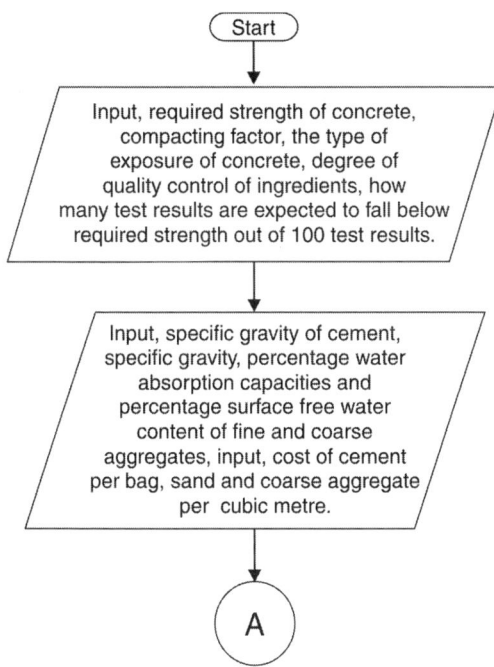

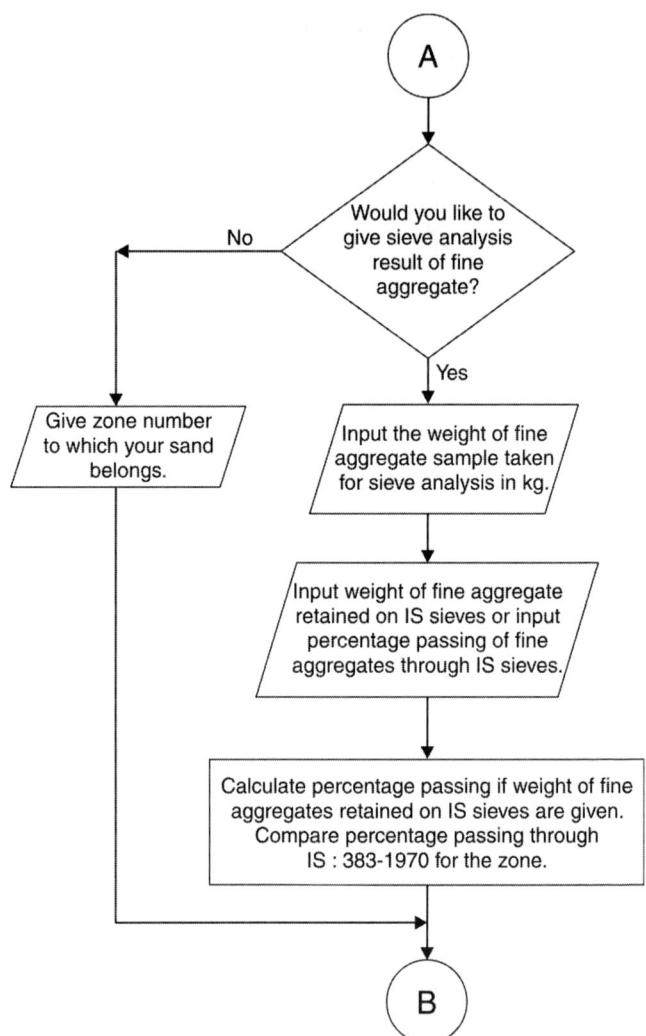

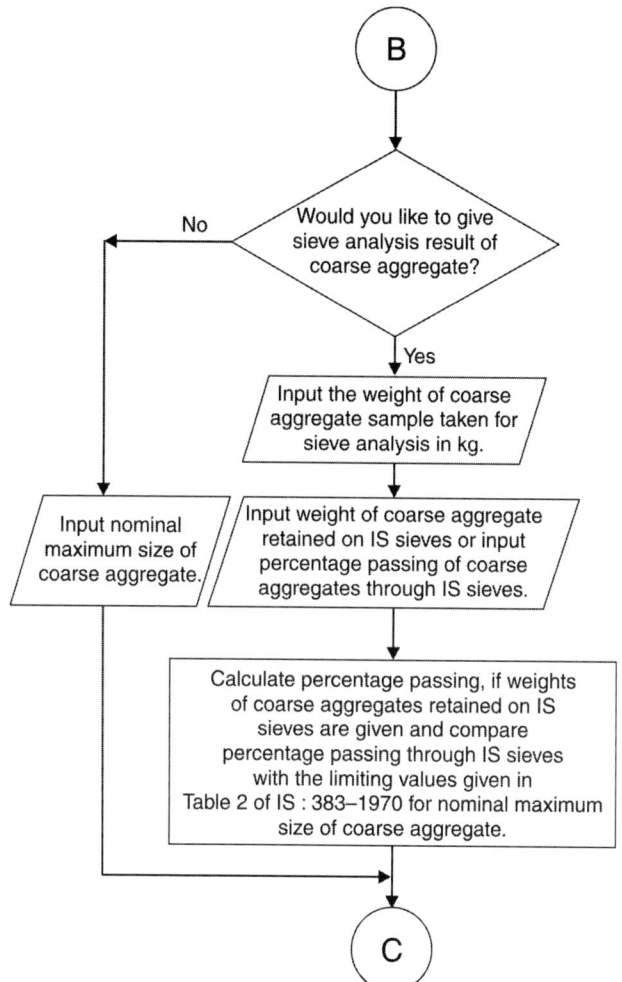

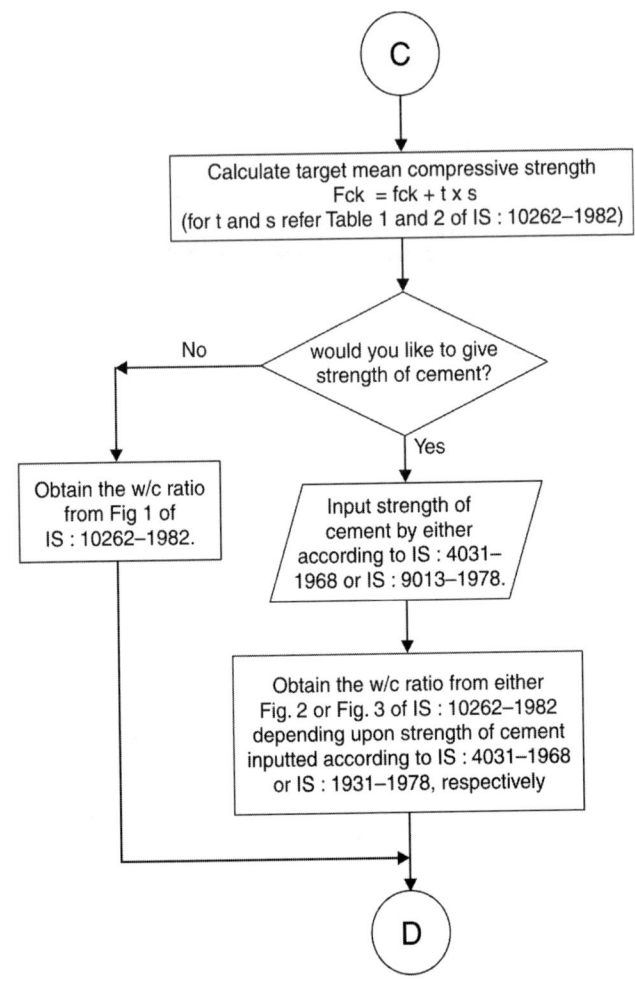

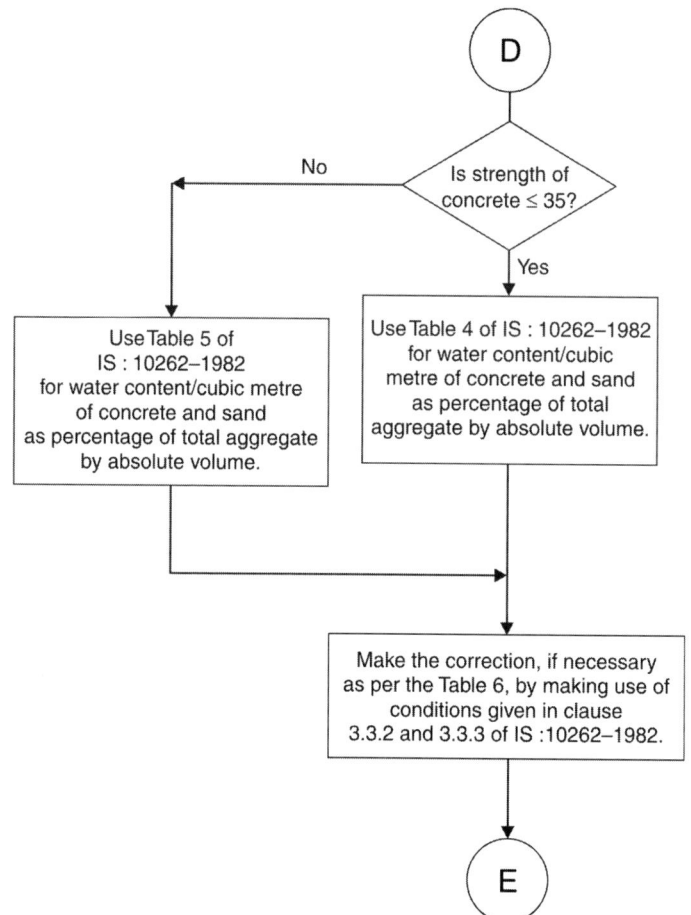

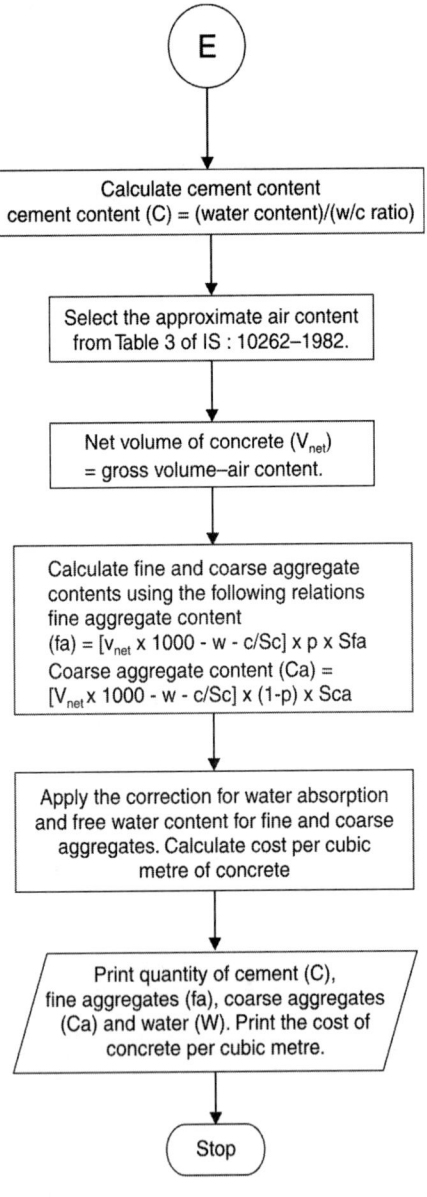

A SOFTWARE FOR CONCRETE MIX DESIGN

```
      dimension sd(20,5), rfi (0:10), zone (40,5), zzco (40,5), esfi (10)
      dimension esco (10), rco (0:10)
      real k
      character y
      open (unit = 3, file = 'cmd18.dat')
      write (*,13)
13    format (////////// )
      write (*, 23)
      write (*,*)                'You are most welcome to'
      write (*,12)
12    format (/, 5x,             'concrete mix design package'
      write (*,23)
      write (*,13)
      do 1000 i = 1,10000
1000  continue
      write (*,*) 'would you like to use this package ? Y/N'
      read (*,101)y
```

```
101   format (a)
      j = 0
      if (y.ne. 'y') go to 110
158   write (*,156)
156   format (////////, 75(1h-),/)
157   format (/,75(1h-),////////)
      write (*,*) 'Give the volume in cubic metre of concrete for which'
      write (*,*) 'proportions of ingredients required'
      write (*,157)
      read (*,/)v1
      write (*,156)
      write (*,*) 'Give the strength of concrete required'
      write (*,157)
      read (*,*)fck
      write (*,156)
      write (*,*) 'Give how many test results you are expected to fall'
      write (*,*) 'below strength of concrete out of 100 cube tests'
      write (*,157)
      read (*,*)c
      write (*,156
      write (*,*) 'Give assumed compaction factor of the concrete'
      write (*,157)
      read (*,*)cf
      write (*,156)
      write (*,*) 'Give the type of aggregate applicable to'
      write (*,*) 'Your coarse aggregate'
      write (*,*) '1 for crushed or angular aggregate'
```

```
write (*,*) '2 for rounded aggregates'
write (*,157)
read (*,*)tco
write (*,156)
write (*,*) 'Give the type of quality control of ingradients'
write (*,*) '1 → for Very good quality control of ingradients'
write (*,*) '2 → for Good quality control of ingradients'
write (*,*) '3 → for Fair quality control of ingradients'
write (*,157)
read (*,*)dqc
write (*,156)
write (*,*) 'Give the type of exposure of concrete at site'
write (*,*) '1 → for Mild exposer of concrete structure'
write (*,*) '2 → for Moderate exposer of concrete structure'
write (*,*) '3 → for Severe exposer of conc ete structure'
write (*,157)
read (*,*)texp
write (*,156)
write (*,*) 'Give specific gravity of cement'
write (*,157)
read (*,*)gce
write (*,156)
write (*,*) 'Give specific gravity of fine aggregate'
write (*,157)
read (*,*)gfi
write (*,156)
```

```
      write (*,*) 'Give specific gravity of coarse aggregate'
      write (*,157)
      read (*,*)gco
      write (*,156)
      write (*,121)
121   format ('Give percentage water absorption capacity of coarse aggregates')
      write (*,157)
      read (*,*)wco
      write (*,156)
      write (*,122)
122   format ('Give percentage water absorption capacity of fine aggregates')
      write (*,157)
      read (*,*)wfi
      write (*,156)
      write (*,123)
123   format ('Give percentage surface moister content of coarse aggregates')
      write (*,157)
      write (*,*)fco
      write (*,156)
      write (*,124)
124   format ('Give percentage surface moister content of fine aggregate')
      write (*,157)
```

```
      read (*,*)ffi
      write (*,156)
      write (*,*) 'Give the cost of cement per bag'
      write (*,157)
      read (*,*)coce
      write (*,156)
      write (*,*) 'Give the cost of fine aggregates per cubicmeter'
      write (*,157)
      read (*,*)cofi
      write (*,156)
      write (*,*) 'Give the cost of coarse aggregates per cubicmeter'
      write (*,157)
      read (*,*)coco
      write (*,156)
      read (3,*) ((sd(i,j), j=1,3), i=1,11)
      read (3,*) ((zone(i,j), j=1,2), i=1,28)
      read (3,*) ((zzco(i,j), j=1,2), i=1,28)
      read (3,*) (esfi(i), i=1,8)
      read (3,*) (esco (i), i=1,7)
      write (*,*) 'Would you like to give analysis of fine aggregates'
      write (*,*) 'on IS sieves ? Y/N'
      write (*,157)
      read (*,101)y
      write (*,156)
      if (y.eq.'y')then
      write (*,21)
```

```
21      format (5x, 'Sieve analysis for fine grained aggregates.')
        write (*,23)
        write (*,*) 'would you like to give % passing of fine aggregates'
        write (*,*) 'through IS sieves ? Y/N
        write (*,157)
        read (*,101)y
        write (*,156)
        if (y.eq.'y') then
        do 39 i=1, 8
        if (esfi(i).eq.0.0)then
        write (*,*) 'Give percentage of fine aggr. passing through pan'
        elseif(esfi(i).ge.150.0)then
        write (*,104)esfi(i)
104     format ('Give percentage of fine aggregate passing through
        IS sieve 'f8.2, 'microns')
        else
        write (*,105)esfi(i)
105     format ('Give percentage of fine aggregates passing through
        IS sieve 'f8.2, 'mm')
        endif
        write (*,157)
        read (*,*)rfi(i)
        write (*,156)
39      continue
        call fine(rfi,zone,zon)
        else
```

```
      write (*,*) 'Would you like to give weight of fine aggregates'
      write (*,*) 'retained on IS sieve ? Y/N
      write (*,157)
      read (*,101)y
      write (*,156)
      if (y/eq/'y')then
      write (*,*) 'Give then wt. of fine aggregate sample taken for sieve'
      write (*,*) 'analysis in Kgs'
      write (*,157)
      read (*,*)w1
      write (*,156)
      do 42 i=1,8
      if (esfi(i).eq.0.0)then
      write (*,*) 'Give fine aggregates in gms. retained on pan'
      elseif(esfi(i).ge.150.0)then
      write (*,102)esfi(i)
102   format ('Give fine aggregates in gms. retained on IS sieve f8.2, 'microns')
      else
      write (*,103)esfi(i)
103   format ('Give fine aggregates in gms. retained on IS sieve f8.2, 'mm')
      endif
      write (*,157)
      read (*,*)rfi(i)
      write (*,156)
```

```fortran
42    continue
      rfi(0) = 100.0
      do 18i=1,8
      rfi(i) = rfi(i)/(w1*1000.0)*100.0
      rfi(i) = rfi(i-1)-rfi(i)
c     write (*,*)rfi(i)
18    continue
      call fine (rfi,zone,zon)
      gogo 40
      else
      write (*,*) 'Lack of information, Sorry mix proportion can not be'
      write (*,*) 'given. Suggestion: Give either % passing through or'
      write (*,*) 'weight of fine aggregates retained on IS sieve'
      write (*,157)
      goto 110
      endif
      endif
      else
      write (*,*) 'Give zone no. to which your fine aggregates belongs'
      write (*,*) 'Give 1 for zone 1'
      write (*,*) 'Give 2 for zone 2'
      write (*,*) 'Give 3 for zone 3'
      write (*,*) 'Give 4 for zone 4'
      write (*,157)
      read (*,*)zon
      write (*,156)
      endif
```

```fortran
40      write (*,*)  'Would you like to give analysis of coarse aggregates'
        write (*,*)  'on IS sieves ? Y/N
        write (*,157)
        read (*,101)y
        write (*,156)
        if (y.eq.'y')then
        write (*,22)
22      format (5x, 'sieve analysis for coarse grained aggregates')
        write (*,23)
23      Format (5x, 45(1h⁻))
c       write (*,156)
        write (*,*)  'Would you like to give percentage passing of coarse'
        write (*,*)  'aggregates through IS sieves : Y/N
        write (*,157)
        read (*,101)y
        write (*,156)
        if (y.eq.'y')then
        do 44 i=1,7
107     write (*,107)esco(i)
        format ('Give percentage of coarse aggregates passing through'
        IS sieve', f8.2, 'mm')
        write (*,157)
        read (*,*)rco(i)
        write (*,156)
44      continue
        call coarse(rco,zzco,dmax)
        goto 49
```

```
      else
      write (*,*) 'would you like to give weight of coarse aggregates'
      write (*,*) 'retained on IS sieves ? Y/N
      write (*,157)
      read (*,101)y
      write (*,156)
      write (*,*) 'Give the weight of coarse aggregate sample taken for'
      write (*,*) 'Sieve analysis in kgs'
      write (*,157)
      read (*,*)w2
      write (*,156)
      if (y.eq.'y')then
      do 46 i=1,7
      write (*,108)esco(i)
108   format ('Give coarse aggregates in gms. retained on IS sieve',
     f8.2, 'mm')
      write (*,157)
      read (*,*)rco(i)
      write (*,156)
46    continue
      rco(0)=100.0
      do 28 i=1,7
      rco(i)=rco(i)/(w2*1000.0)*100.0
      rco(i)=rco(i-1)-rco(i)
28    continue
      call coarse (rco,zzco,dmax)
      goto 49
```

```
      else
      write (*,*) 'Lack of information, sorry mix proportion can not be'
      write (*,*) 'given. Suggestion: Give either % passing through or'
      write (*,*) 'weight of coarse aggregates retained on IS sieve'
      write (*,157)
      goto 110
      endif
      endif
      else
      write (*,*) 'Give nominal maximum size of coarse aggregates taken'
      write (*,157)
      read (*,*)dmax
      write (*,156)
      endif
      if(c.le.50) k=0.00
      if (c.le.16) k=1.00
      if(c.le.10) k=1.28
      if(c.le.5) k=1.65
      if(c.le.2.5) k=1.96
      if(c.le.1.0) k=2.33
      if(c.le.0.5) k=2.38
      if(c.le.0.0)goto 100
      m=(fck-10)/5+1
      do 10 i=1,11
      if(i.eq.m)then
      do 20 j=1,3
```

49

```
         if(j.eq.dqc)sigma=sd(i,j)
20       continue
         endif
10       continue
         tfck=fck+sigma*k
         write (*,*)  'tfck='tfck
         write (*,*)  'Would you like to give strength of cement after'
         write (*,*)  '28 days normal way of curing[IS: 4031-1962]? Y/N
         write (*,157)
         read (*,101)y
         write (*,156)
         if(y.eq.'y')then
         write (*,125)
125      formate ('Give the strength of cement after 28 days curing')
         write (*,157)
         read (*,*)fck1
         write (*,156)
         call normal (fck1,tfck,wc)
         else
         write (*,*)  'Would you like to give strength of cement'
         write (*,*)  'based on boiling water method(IS: 9013-1978) ? Y/N
         write (*,157)
         read (*,101)y
         write (*,156)
         if(y.e.'y')then
         write (*,*)  'Give strength of cement based on boiled water method'
         write (*,157)
```

```
      read (*,*)fck1
      write (*,,156)
      call boiled(fck1.tfck.wc)
      else
      write (*,*) 'Would you like not to give strength of cement : Y/N
      write (*,157)
      read (*,101)y
      write (*,156)
      if(y.eq.'y')then
      wc=0.000181*tfck*tfck-0.02196*tfck+0.956752
      else
      write (*,*) 'Give strength of cement and run the program from'
      write (*,*) 'from the begining'
      write (*,157)
      goto 110
      endif
      endif
      endif
      write (*,*) 'w/c from grapbh=', wc
      wc1=0.65
      do 30 i=1,3
      if(i.eq.texp)wc=min(wc,wc1)
      wc1=wc1-0.1
30    continue
c     Selection of water content and fine to total aggregate ratio:
c     - - - - - - - - - - - - - - - - - - - - - - - - - - - - - - - -
      if(fck.le.35)then
      if(dmax.eq.10)then
```

```
        w=208.0
        fitotl=40
      elseif(dmax.eq.12.5)then
        w=202.5
        fitotl=38.75
      elseif(dmax.eq.16)then
        w=195
        fitotl=37
      elseif(dmax.eq.20)then
        w=186
        fitotl=35
      else
        w=165
        fitotl=30
      endif
      else
      if(dmax.eq.10)then
        w=200.0
        fitotl=28
      elseif(dmax.eq.12.5)then
        w=195
        fitotl=27.25
      elseif(dmax.eq.16)then
        w=188
        fitotl=26.25
      elseif(dmax.eq.20)then
        w=180
```

```
      fitotl=25
      endif
      endif
      if(dmax.eq.10)ac=3
      if(dmax.eq.12.5)ac=2.75
      if(dmax.eq.16)ac=2.4
      if(dmax.eq.20)ac=2
      if(dmax.eq.40)ac=1
      if(zon.eq.1)fitotl=fitotl+1.5
      if(zon.eq.3)fitotl=fitotl-1.5
      if(zon.eq.4)fitotl=fitotl-3.0
      if(fck.le.35)then
      fitotl=fitotl+(wc-0.06)*1/0.05
      else
      fitotl=fitotl+(wc-0.35)*1/0.05
      endif
      w=w+(cf-0.8)*3/0.1*w/100.0
      if(tco.eq.2)then
      fitotl=fitotl-7
      w=w-15
      endif
      Determination of ingradients
      ce=w/wc
      if(texp.eq.1)ce=max(ce,250.0)
      if(texp.eq.2)ce=max(ce,290.0)
      if(texp.eq.3)ce=max(ce,360.0)
      fi=((1.0-ac/100)*1000-w-ce/gce)*fitotl*gfi/100
```

c

```
14        co=((1.0-ac/100)*1000-w-ce/gce)*(1-fitotl/100)*gco
          format (5x, '[Cement : sand : Coarse agg; : water= 'f4.2,' : 'f4.2,' :
         f4.2,' : 'f4.2,' ']' /)
c      Actual quantities
          wa=w+wco/100*co-ffi/100*fi+wfi/100*fi
          fia=fi+ffi/100*fi
          coa=co-wco/100*co
          cel=ce/ce
          fil=fia/ce
          col=coa/ce
          wl=wa/ce
          write (*,6)
6         format (5x, 'Actual quantities for One cubicmeter of concrete:'
         /5x,51(1h⁻))
          write (*,8)
8         format (5x,55(1h⁻))
          write (*,1)ce
1         format (5x, 'Quantity of cement required=' f10.2, 'kg')
          write (*,2)fia
2         format (5x, 'Quantity of fine aggregate required =' f10.2, 'kg')
          write (*,3)coa
3         format (5x, 'Quantity of coarse aggregate required=' f10.2, 'kg')
          write (*,4)wa
4         format (5x, 'Quantity of water required=', f10.2, 'Litre.')
          tcost=ce/50.0*coce+(fia/gfi*cofi+coa/gco*coco)*0.001
          write (*,15)tcost
15        format (5x, 'Total cost of concrete/cubic meter in Rs =', f10.2)
```

```
      write (*-,14)cel,fil,col,w1
      write (*,8)
      if(v1.ne.0.0)then
9     write (*,9)v1
      format(//5x, 'Actual quantities for', f6.3, 'cubicmeter of'
     'concrete : '/5x.51(1h-))
      write (*,8)
      ce=v1-fia
      fia=v1-fia
      coa=v1*coa
      wa=v1*wa
      write (*,1)ce
      write (*,2)fia
      write (*,3)coa
      write (*,4)wa
      write (*,8)
      endif
      goto 110
100   write (*,5)
5     format(5x, 'Too much expectation!')
110   write (*,159)
159   format(//5x, 'Would you like to use this package again? Y/N
      read (*,101)y
      j=j+1
      if(j.ge.12)goto 998
      if(y.eq.'y')goto 159
998   write (*,13)
```

```fortran
      write (*,23)
      write (*,*)'
      write (*,12)
      write (*,23)
      write (*,13)
      close(unite=3)
      stop
      end
c     ------------------------------------------------------------
      subroutine fine(rfi,zone,zon)
      dimension rfi(0:10),zone(40,5)
      m=7
      j=1
      kk=1
      l=0
40    do 50 i=kk,m
      l=l+1
      if(rfi(l).it.zone(i,1).or.rfi(l).gt.zone(i,2))goto 51
50    continue
      zon=j
      write (*,12) j
12    format(5x,'The given fine aggregates falls into the zone', i2,'catagory')
      write (*,78)
      goto 180
51    kk=m+1
      m=m+7
      l=0
```

```
         j=j+1
         if (m.gt.28) then
         write (*,*)  'The given fine aggregate is not falling into any zone'
         write (*,*)  'Suggestion: change fine aggregates'
         write (*,78)
78       format(//)
         stop
         endif
         goto 40
180      return
         end
c        - - - - - - - - - - - - - - - - - - - - - - - - - - - - - - - - - - - -
         subroutine coarse(rco,zzco,dmax)
         dimension rco(0:10),zzco(40,5)
         m1=7
         k1=1
         j=0
41       do 58 i=k1,m1
         j=j+1
         if(rco(j).lt.zzco(i,1).or.rco(j).gt.zzco(i,2))goto 81
58       continue
         if(k1.eq.1)dmax=40
         if(k1.eq.8)dmax=20
         if(k1.eq.15)dmax=16
         if(k1.eq.22)dmax=12.5
         write (*,17)dmax
17       format(5x, 'Maximum nominal size of aggregate IS = f6.2, 'mm')
         goto 190
```

```
81      k1=m1+1
        m1=m1+7
        j=0
        if(m1.ft.28)then
        write(*,*) 'Given coarse aggregate is not falling into any zone'
        write(*,*) 'Suggestion: change coarse aggregates'
        stop
        endif
190     goto 41
        return
        end
c       - - - - - - - - - - - - - - - - - - - - - - - - - - - - -
        subroutine normal (fck1,tfck,wc)
        if(fck1.le.61.3) wc=0.000093*tfck*tfck-0.015961*tfck+0.939598
        if(fck1.le.56.4) wc=0.000103*tfck*tfck-0.015961*tfck+0.923468
        if(fck1.le.51.5) wc=0.000131*tfck*tfck-0.018834*tfck+0.933590
        if(fck1.le.46.6) wc=0.000158*tfck*tfck-0.020524*tfck+0.922955
        if(fck1.le.41.7) wc=0.000217*tfck*tfck-0.024192*tfck+0.939611
        if(fck1.le.36.8) wc=0.000265*tfck*tfck-0.024192*tfck+0.931615
        if(fck1.lt.31.9)then
        write(*,112)
112     format ('Strength of cement is very low. So it can't be used')
        stop
        endif
        return
        end
c       - - - - - - - - - - - - - - - - - - - - - - - - - - - - -
```

```fortran
      subroutine boiled (fck1,tfck,wc)
      if(fck1.le.29.9)  wc=0.000090*tfck*tfck-0.015956*tfck+0.948512
      if(fck1.le.27.0)  wc=0.000090*tfck*tfck-0.015922*tfck+0.917364
      if(fck1.le.24.0)  wc=0.000134*tfck*tfck-0.018730*tfck+0.935242
      if(fck1.le.21.1)  wc=0.000177*tfck*tfck-0.022031*tfck+0.950271
      if(fck1.le.18.1)  wc=0.000255*tfck*tfck-0.026388*tfck+0.968945
      if(fck1.le.15.2)  wc=0.000405*tfck*tfck-0.034428*tfck+1.034427
      if(fck1.lt.12.3)then
      write(*,113)
113   format ('Strength of cement is very low. So it can't be used')
      stop
      endif
      return
      end
c
```

19

Design of Concrete Mixes According to Indian Standards

19.1 GENERAL FEATURES

The Indian Standard Code[179] IS: 10262–1982 presents guidelines for mix design which include design of normal concrete mixes (non-air-entrained) both for medium and high strength concretes. The basic assumption made in mix design is that the compressive strength of workable concrete is by and large governed by the water/cement ratio. In this method of mix design, the water content and proportion of fine aggregate corresponding to a maximum size of aggregate are first determined for reference values of workability, water/cement ratio and grading of fine aggregate in any particular case from the reference values.

The batch weight of materials per unit volume of concrete is finally calculated by the absolute volume method. The specific design data developed in this method is based on the exhaustive experimental work conducted at the Cement Research Institute of India and also on the basis of data on concrete being designed and produced in the country[180,181]. However, there are various other factors which affect the property of the concrete such as the quality and quantity of cement, water, aggregate batching, and size and shape of aggregate. Therefore, the specific guidelines recommended in proportioning concretes mixes should be considered only as a basis of trial, subject to modifications.

19.2 MIX DESIGN PROCEDURE

The mix design procedure recommended in the IS Standard is detailed above.

1. The target mean strength is first determined as follow:

$$\overline{f_{ck}} = f_{ck} + t.s$$

where, $\overline{f_{ck}}$ = Target mean compressive strength at 28 days

f_{ck} = Characteristic compressive strength at 28 days

s = Standard deviation

t = A statical value depending upon the accepted proportion of low results and the number of tests as shown in Table 19.1.

Table 19.1: Values of *t*

Percentage of results below the characteristic strength	Values of t
50	0
16	1.00
10	1.28
5	1.65
2.5	1.96
1.0	2.33
0.5	2.58
0.0	Infinity

The suggested values of standard deviation(s) as recommended in the IS: 10262–1982, for different degrees of control are complicated in Table 19.2.

2. The water/cement ratio for the target mean strength is selected from Fig. 19.1. The water/cement ratio selected is checked against the limiting water/cement ratio for the requirements of durability using the Tables 19.3 and 19.4 and the lower of the two values adopted. A more precise estimate of the preliminary water/cement ratio corresponding to the target average strength is made from the relationships shown in Fig. 19.2.

3. The air content (amount of entrapped air) is estimated from Table 19.5, for the maximum size of aggregate used.

4. The water content and percentage of sand in total aggregate by absolute volume are next selected from Tables 19.6 and 19.7, for medium and high strength concretes, respectively for the following standard reference conditions;

(i) Crushed (angular) coarse aggregate

(ii) Fine aggregate consisting of natural sand conforming to grading zone II of Table 4 of IS: 383–1970, in saturated surface dry conditions.

(iii) Water/cement ratio of 0.60 and 0.35 for medium and high strength concretes, respectively.

(iv) Workability corresponding to a compacting factor value of 0.80.

Table 19.2: Suggested value of standard deviation

Grade of concrete	Standard deviation for different degree of control in N/mm²		
	Very good	Good	Fair
M-10	2.0	2.3	3.3
M-15	2.5	3.5	4.5
M-20	3.6	4.6	5.6
M-25	4.3	5.3	6.3
M-30	5.0	6.0	7.0
M-35	5.3	6.3	7.3
M-40	5.6	6.6	7.6
M-45	6.0	7.0	8.0
M-50	6.4	7.4	8.4
M-55	6.7	7.7	8.7
M-60	6.8	7.8	8.8

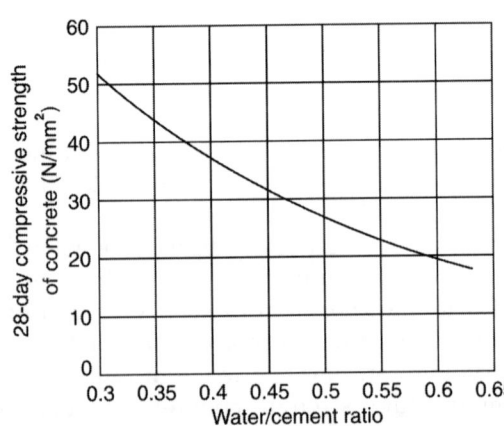

Fig. 19.1: Generalised relationship between free water/cement ratio and compressive strength of concrete

5. For other conditions of workability, water/cement ratio, grading of fine aggregate and for rounded aggregates, adjustments in water content and percentage of sand in total aggregate are made as shown in Table 19.8.
6. The cement content is calculated from the water/cement ratio and the final water content is computed after adjustment. The cement content so calculated is checked against the minimum cement content from the requirements of durability (Tables 19.3 and 19.4) and the greater of the two values adopted.
7. With the quantities of water and cement per unit volume of concrete and the percentage of sand in the total aggregate already determined, the coarse and fine aggregate content per unit volume of concrete are calculated from the following equations:

$$V = \left[W + \frac{C}{S_c} + \frac{1}{p}\left(\frac{f_a}{S_{fa}} \right) \right] \frac{1}{1000}$$

$$V = \left[W + \frac{C}{S_c} + \left(\frac{1}{1-p} \right) \frac{C_a}{S_{ca}} \right] \frac{1}{1000}$$

where,

V = Absolute volume of fresh concrete

  = Gross volume (1 m³) minus the volume of entrapped air

S_c = Specific gravity of cement

W = Mass of water (kg) per m³ of concrete

C = Mass of cement (kg) per m³ of concrete

p = Ratio of fine aggregates to total aggregate by absolute volume

f_a and C_a = Total masses of fine aggregate and coarse aggregate (kg) per m³ of concrete, respectively

S_{fa} and S_{ca} = Specific gravities of saturated surface dry fine aggregate and coarse aggregate, respectively

The application of the Indian Standard Code method for the design of concrete mixes is illustrated by the following examples.

19.3 DESIGN EXAMPLE

Design concrete mixes of low, medium and high strengths to suit the following data:

Characteristic cube strength   =   M-15
$$M\text{-}30$$
$$M\text{-}15$$

Type of cement: Ordinary Portland, fine aggregate natural river sand conforming to grading zone II of Table 4, of IS: 383–1970.

Table 19.3: Minimum cement content required in cement concrete to ensure durability under specified conditions of exposure

Exposure	Plain concrete		Reinforced concrete	
	Minimum cement content	Maximum water cement ratio	Minimum cement content	Maximum water cement ratio
(1)	(2) (kg/m³)	(3)	(4) (kg/m³)	(5)
Mild: For example, completely protected against weather, or aggressive conditions, except for a brief period of exposure to normal weather conditions during construction	220	0.7	250	0.65
Moderate: For example, sheltered from heavy and wind driven rain and against freezing, whilst saturated with water; buried concrete continuously under water	250	0.6	290	0.55
Severe: For example, exposed to sea water, alternate wetting and drying and to freezing whilst wet, subject to heavy condensation or corrosive fumes	310	0.5	360	0.45

Note 1. When the maximum water/cement ratio can be strictly controlled the cement content in the above Table may be reduced by 10 per cent

Note 2. The minimum cement content is based on 20 mm aggregate. For 40 mm aggregate, it should be reduced by about 10 per cent; for 12.5 mm aggregate, it should be increased by about 10 per cent

Table 19.4: Requirements for concrete exposed to sulphate attack

Class	Concentration of sulphates expressed as SO_3		Type of cement	Requirements for dense fully compacted concrete made with aggregate complying with IS: 383–1970*	
	In soil	In ground water			
	Total SO_3 (per cent)	SO_3 in 2 : 1 water extract g/l	(part per 100000)		Minimum cement content	Maximum free water/ cement ratio
1	2	3	4	5	6	7
1.	Less than 0.2	–	Less than 30	Ordinary Portland cement or Portland slag cement or Portland pozzolana cement	280	0.55
2.	0.2 to 0.5	–	30 to 120	Ordinary Portland cement or Portland slag cement or Portland pozzolana cement	330	0.50
				Supersulphate cement	310	0.50
3.	0.5 to 1.0	1.9 to 3.1	120 to 250	Supersulphate cement	330	0.50

Note 1. This table applies to concrete made with 20 mm aggregates complying with the requirements of IS: 383–1970* placed in near-natural ground waters of pH 6 to 9, containing naturally occurring sulphates but not contaminants such as ammonium salts. For 40 mm aggregate, the value my be reduced by about 15 per cent and for 12.5 mm aggregate the value may be increased by about 15 per cent . Concrete prepared from ordinary Portland cement would not be recommended in acidic conditions (pH 6 or less). Supersulphated cement gives and acceptable life provided that the concrete is dense and prepared with a water/cement ratio of 0.4 or less, in mineral acids, down to pH 3.5.

Note 2. The cement contents given in Class 2 are the minimum recommended. For SO_3 contents near the upper limit of Class 2, cement contents above these minimum are advised.

Note 3. Where the total SO_3 in col 2 exceeds 0.5 per cent, then a 2 : 1 water extract may result in a lower site classification, if much of the sulphate is present as low solubility calcium sulphate.

Note 4. For severe conditions, such as thin sections under hydrostatic pressure on one side only and sections partly immersed, considerations should be given to a further

(Contd...)

(Contd...)
reduction of water/cement ratio, and if necessary, an increase in the cement to ensure the degree of workability needed for full compaction and thus minimum permeability.
Note 5. Portland slag cement conforming to IS: 455–1976[†] with slag content more than 50 per cent exhibits better sulphate resisting properties.

Note 6. Ordinary Portland cement with the additional requirements that C_3A content be not more than 5 per cent and $2\,C_3A + C_4AF$ (or its solid solution $4CaO$, Al_2O_3, Fe_2O_3 + $2CaO$, Fe_2O_3) not be more than 20 per cent may be used in place of supersulphated cement.

* Specification for coarse and fine aggregates from natural sources for concrete (*second revision*).

† Specification for Portland slag cement (third revision).

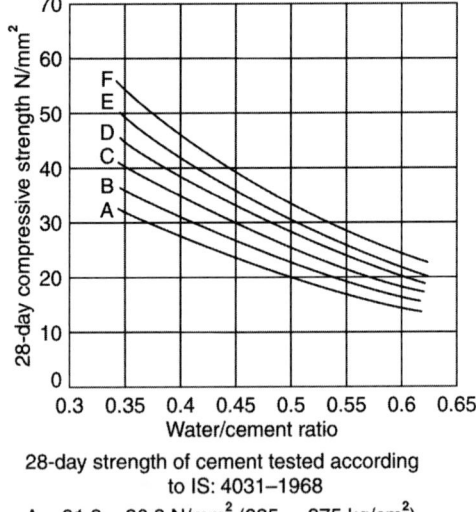

28-day strength of cement tested according
to IS: 4031–1968
A = 31.9 – 36.8 N/mm² (325 – 375 kg/cm²)
B = 36.8 – 41.7 N/mm² (375 – 425 kg/cm²)
C = 41.7 – 46.6 N/mm² (425 – 475 kg/cm²)
D = 46.6 – 51.5 N/mm² (475 – 525 kg/cm²)
E = 51.5 – 56.4 N/mm² (525 – 575 kg/cm²)
F = 56.4 – 61.3 N/mm² (575 – 625 kg/cm²)

Fig. 19.2: Relation between free water/cement ratio and concrete strength for different cement strengths

Coarse aggregate: Crushed (angular), coarse aggregate of 20 mm maximum size conforming to IS: 383 code requirements. Specific gravities of cement, sand and coarse aggregate are 3.14, 2.63 and 2.61, respectively.

Type of exposure: Mild
Degree of quality control: Very good
Degree of workability   = 0.85 for M-15
                             0.80 for M-30
                             0.75 for M-15

Design Procedure

1. Target mean strength

The target mean strength f_{ck} is calculated assuming the degree of control as 'very good' and the value of standard deviation for various grades of concrete are as follow:

Grade of concrete	Standard deviation(s) Degree of control, very good
M-15	2.5
M-30	5.0
M-45	6.0

Assuming 5 per cent of results to be below the characteristic strength, the value of t from Table 19.1 is 1.65. Hence the target mean strength for different mixes are as follows:

SI no.	Mix grade	Target mean strength (N/mm²) $\overline{f_{ck}} = f_{ck} + t.s$
1	M-15	$\overline{f_{ck}} = (15 + 1.65 \times 2.5) = 19.13$
2	M-30	$\overline{f_{ck}} = (30 + 1.65 \times 5.0) = 38.25$
3	M-45	$\overline{f_{ck}} = (49 + 1.65 \times 6.0) = 54.90$

2. Selection of water/cement ratio

The preliminary free water/cement ratio by weight corresponding to the target mean strength at 28 days for different mixes is selected from Fig. 19.1, are as follows:

Mix	Target mean strength (N/mm²)	W/C ratio (by weight)
M-15	19.13	0.60
M-30	38.25	0.39
M-45	54.90	0.30

From Table 19.3, for durability requirements, the maximum water/cement ratio for moderate exposure for plan concrete = 0.70. Hence lower values of water/cement ratio as indicated above are selected.

3. Air content

From Table 19.5, for a nominal maximum sizes of 20 mm aggregate, the entrapped air is 2 per cent of the volume of concrete.

Table 19.5: Approximate entrapped air cotent

Nominal maximum size of aggregate (mm)	Entrapped air as per cent of volume of concrete
10	3.0
20	2.0
40	1.0

4. Water content and fine to total aggregate ratio

For a nominal maximum size of 20 mm aggregate, from Table 19.6, for concrete of grade up to M-35, the water content for mixes M-15 and M-30 is 186 kg/m^3 of concrete and the sand as percentage of total aggregate by absolute volume is 35.

Table 19.6: Approximate sand and water contents per cubic metre of concrete

W/C = 0.60
Workability = 0.80 CF
(Applicable for concrete up to grade M-35)

Maximum size of aggregate (mm)	Water content per cubic* metre of concrete (kg)	Sand as per cent of total aggregate by absolute volume
10	208	40
20	186	35
40	165	30

* Water content corresponding to saturated surface dry aggregate.

Table 19.7: Approximate sand and water contents per cubic metre of concrete

W/C = 0.35
Workability = 0.80 CF
(Applicable for concrete up to grade M-35)

Maximum size of aggregate (mm)	Water content per cubic* metre of concrete (kg)	Sand as per cent of total aggregate by absolute volume
10	200	28
20	180	25

* Water content corresponding to saturated surface dry aggregate.

Similarly from Table 19.7, applicable for grades above, M-35, the water content for M-45 mix is 180 kg/m³ of concrete and the sand as percentage of total aggregate by absolute volume is 25.

Table 19.8: Adjustment of values in water content and sand percentage for other conditions

Change in conditions stipulated for Tables 19.6 and 19.7	Adjustments required in	
	Water content	Per cent sand in total aggregate
For sand conforming to grading zone I, zone III or zone IV of Table 4 of IS: 383–1970	0	+ 1.5% for zone I – 1.5% for zone III 3.0% for zone IV
Increase or decrease in the value of compacting factor by 0.1	3 per cent	0
Each 0.05 increase or decrease in free water cement ratio	0	1 per cent
For rounded aggregate	15 kg/m³	– 7 per cent

5. Adjustment of values in water content and sand percentage

Referring to Table 19.8, the following adjustments are required.

Change in conditions (Refer Table 19.8)	Adjustments required in	
	Water content per cent	Per cent sand in total aggregate
1. M-15	+ 1.5%	0
For increase in compacting factor of (0.85–0.8) = 0.05		
2. M-30		
For decrease in water/cement ratio by (0.60–4.39) = 0.21 No correction since compacting factor is 0.80	0	– 4.2%
3. M-45		
For decrease in compacting factor of (0.80–0.75) = 0.05	– 1.5%	0
For decreases in free water ratio by (0.35–0.30) = 0.05	0	– 1%

	Total adjustments	
Mix	Adjustments required in	
	Water/cement %	Sand in total aggregate %
M-15	+ 1.5%	0%
M-30	0	– 4.2%
M-45	– 1.5%	– 1.0%

Mix	Sand content as percentage of total aggregates by absolute volume (%)	
M-15	(35 – 0)	= 35%
M-34	(35 – 4.2)	= 30.8%
M-45	(25 – 1.0)	= 24.0%

6. Final water content after adjustments

(a) For M-15 grade mix, water content required is

$$\left(186 + \frac{186 \times 1.5}{100} \right) = 188.79 \text{ kg/m}^3$$

(b) For M-30 grade mix, water content required is

$$(186 + 0) \qquad = 186 \text{ kg/m}^3$$

(c) For M-45 grade mix, water content required is

$$\left(180 + \frac{1.5 \times 180}{100} \right) = 177.3 \text{ kg/m}^3$$

7. Determination of cement content

Mix	W/C (kg/m³)	Water (kg/cm³)	Cement
M-15	0.60	188.79	314.65
M-30	0.39	186.00	476.92
M-45	0.30	177.3	591.00

8. Check for minimum cement content

From Appendix-A is IS: 456–2000. The minimum cement content required for durability for plain concrete under mild exposure conditions is 220 kg/m³. The values of cement content for all the three grades are greater than the minimum value.

9. Determination of coarse and fine aggregates content

The amount of entrapped air for 20 mm maximum size aggregate from Table 19.5, is 2 per cent. For M-15 mix, fine aggregate required is calculated as

$$0.98 \text{ m}^3 = \left[188.79 + \frac{314.65}{3.14} + \frac{1 \times f_a}{0.35 \times 2.63} \right] \frac{1}{100}$$

$$\therefore \quad f_a \quad = \quad 636.06 \text{ kg/m}^3$$

Coarse aggregate required is calculated as

$$0.98 \text{ m}^3 = \left[188.79 + \frac{314.65}{3.14} + \frac{1 \times C_a}{(1 - 0.35) \, 2.61} \right] \frac{1}{1000}$$

$$C_a \quad = \quad 1169.76 \text{ kg/m}^3$$

The fine and coarse aggregates for M-30 and M-45 grades are calculated in a similar manner. Hence the aggregate requirements are as follows:

Mix	Fine aggregate (kg/m³)	Coarse aggregate (kg/m³)
M-15	636.06	1169.76
M-30	520.14	1156.38
M-45	387.86	1218.88

10. Total quantities of ingredients and mix proportions

Mix	Cement	Fine aggregate (kg)	Coarse aggregate (kg)	Water (kg)
M-15	314.65	636.06	1160.76	188.79
M-30	476.92	520.14	1156.38	186.00
M-45	591.00	387.86	121.88	177.30

Mix proportions

Mix	Cement	F.A.	C.A.	Water
M-15	1	2.01	3.71	0.60
M-30	1	1.09	2.42	0.369
M-45	1	0.65	2.06	0.30

Design of Concrete Mixes According to British Standards

20.1 GENERAL FEATURES

The latest British method[182] replaces the traditional Road Note No. 4 method. In the new method, the use of specific grading curves is discarded along with the mix design tables correlating water/ cement ratio, aggregate/cement ratio, maximum size of aggregate, type of aggregate, degree of workability and over all grading curves of combined aggregates.

In the new method, only two types of aggregates, namely crushed and uncrushed, are recognised. The water content required to give various levels of workability expressed as slump and vebe time can be determined for the two types of aggregates, namely crushed and uncrushed with different maximum sizes varying from 10 to 40 mm.

The degrees of workability 'Very low', 'Low', 'Medium' and 'High' have now been referred in terms of specific values of slump and vebe time. The new British method of mix design results in expressing the mix proportions in terms of quantities of materials per unit volume of concrete in line with European and American practice.

The method is applicable to ordinary and rapid hardening Portland cements and to sulphate resisting Portland cement used with normal weight aggregates or with air-cooled slag aggregates but not with lightweight aggregates. Three maximum sizes of aggregates are recognised, viz. 40, 20 and 10 mm.

20.2 MIX DESIGN PROCEDURE

The following procedure is adopted for the design of a concrete mix.

1. The target mean strength is calculated based on the characteristic strength, standard deviation and the statistic t which depends upon the accepted proportion of low results and the number of tests.

$$\overline{f_{ck}} = f_{ck} + t \times s$$

where, $\overline{f_{ck}}$ = Target mean strength

f_{ck} = Characteristic mean strength

2. From Table 20.1, a value is obtained for the compressive strength of a mix made with a free water/cement ratio of 0.5, according to the specified age, the type of cement and the type of aggregate used.

Table 20.1: Approximate compressive strength of concrete mixes made with water/cement ratio of 0.5 (reference 182)

Type of cement	Type of coarse aggregate	Compressive strength (N/mm²) age (days)			
		3	7	28	91
Ordinary Portland cement or	Uncrushed	18	27	40	48
Sulphate resisting Portland cement	Crushed		33	47	55
Rapid hardening Portland cement	Uncrushed	25	34	46	53
	Crushed	30	40	53	60

3. This strength value is then plotted on Fig. 20.1, and a curve is drawn from the point and parallel to the printed curves until it intercepts a horizontal line passing through the ordinate representing the target mean strength. The corresponding value of the free water/cement ratio can then be read from the abscissa.

4. The free water content, required depending upon the type and maximum size of aggregate to give a concrete of the specified slump or vebe time is obtained from Table 20.2.

5. Knowing the water/cement ratio and water content, the cement content is obtained as

$$\text{Cement content} = \left(\frac{\text{Water content}}{\text{Free water/cement ratio}} \right)$$

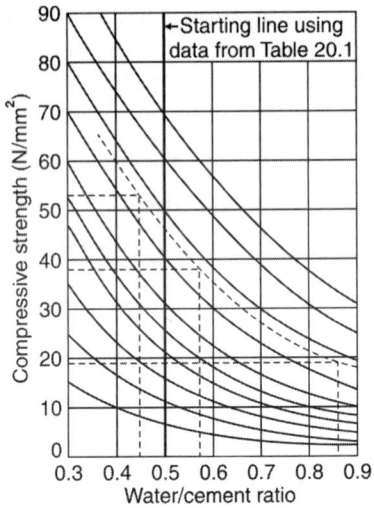

Fig. 20.1: Relationship between compressive strength and water/cement ratio

Table 20.2: Approximate water contents (kg/m³) required to give various levels of workability

| Slump (mm) | | 0–10 | 10–30 | 30–60 | 60–180 |
Vee bee (secs)		12	6–12	3–6	0–3
Maximum size of aggregate (mm)	Type of aggregate				
10	Uncrushed	150	180	205	225
	Crushed	180	205	230	250
20	Uncrushed	135	160	180	195
	Crushed	170	190	210	225
40	Uncrushed	115	140	160	175
	Crushed	155	175	190	205

Note: When coarse and fine aggregates of different types are used, the water content is estimated by the expression given by

$$\left(\frac{2}{3} \, W_f + \frac{1}{3} \, W_c\right)$$

where, W_f = Water content appropriate to the type of fine aggregate

W_c = Water content appropriate to the type of coarse aggregate

6. An estimate of the wet density of the fully compacted concrete is obtained from Fig. 20.2, depending upon the free water

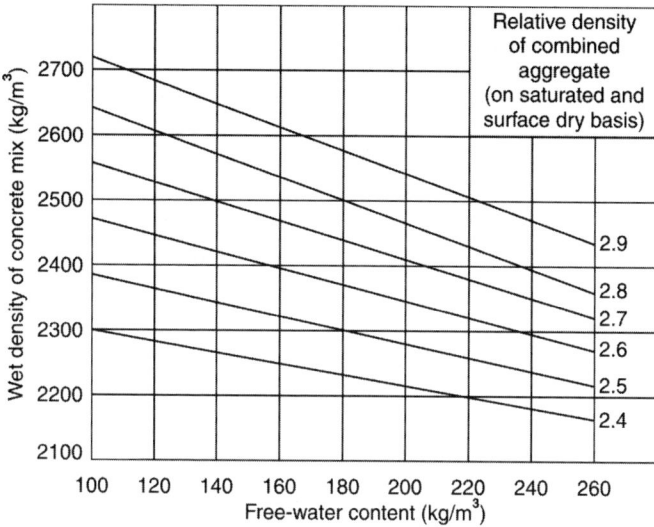

Fig. 20.2: Estimated wet density of fully compacted concrete

content and the specific gravity of the combined aggregate. From this estimated density of the concrete, the total aggregate content is determined from the following relations:

Total aggregate content = $D - W_c - W_{FW}$

where, D = The wet density of the concrete (kg/m³)

W_c = The cement content (kg/m³)

W_{FW} = The free water content (kg/m³)

7. The recommended proportion of fine aggregate depending upon the maximum size of aggregate, the workability level and the free water/cement ratio are shown in Figs 20.3 to 20.5, for sand of different zones 1 to 4, as specified in British code BS: 882 and for maximum size of aggregate of 10, 20 and 40 mm. The fine aggregate content is determined as a percentage of the total aggregate from these figures.

Fine aggregate content = (Total aggregate content)
× (Proportion of fines)

Coarse aggregate content = (Total aggregate content)
− (Fine aggregate content)

20.3 DESIGN EXAMPLE

Design a concrete mix of low, medium and high strengths to suit the following data:

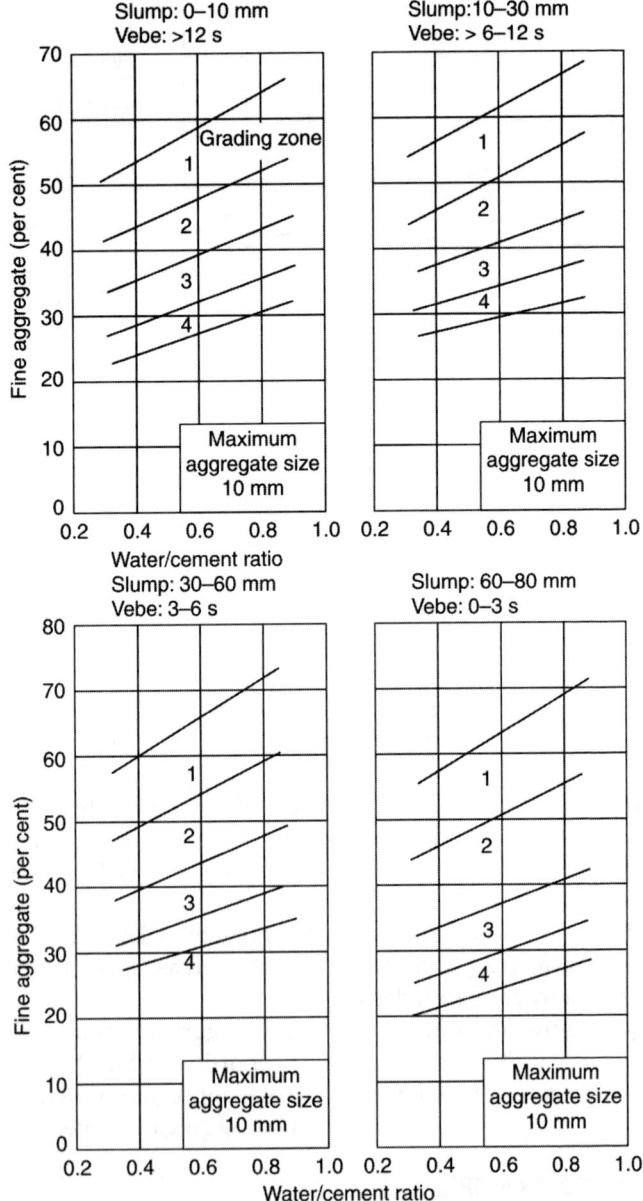

Fig. 20.3: Recommended weight of fine aggregate as a function of free water/cement ratio for various workabilities and maximum sizes

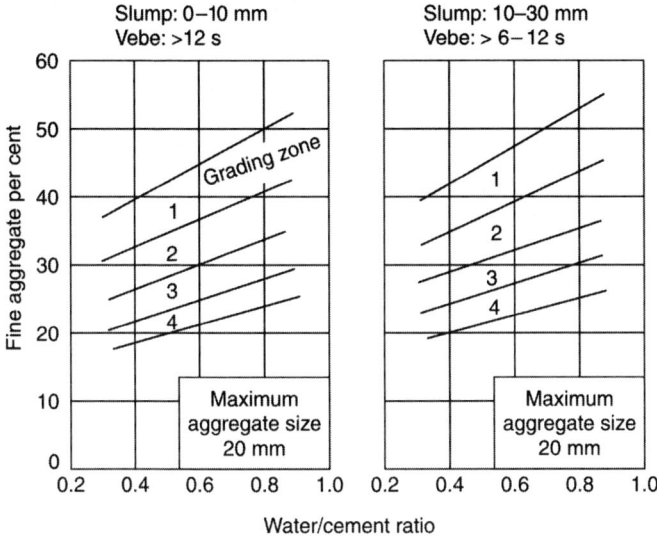

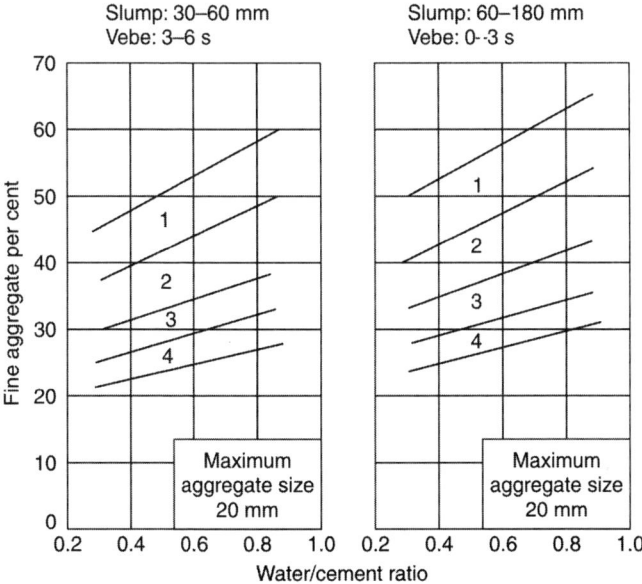

Fig. 20.4: Recommended weight of fine aggregate as a function of free water/cement ratio for various workabilities and maximum sizes

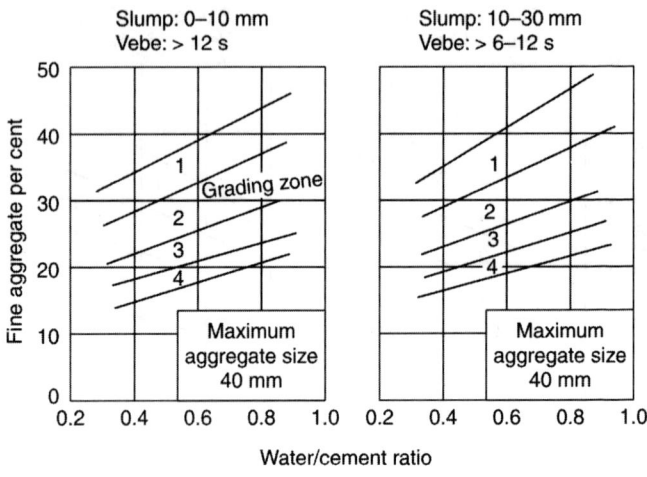

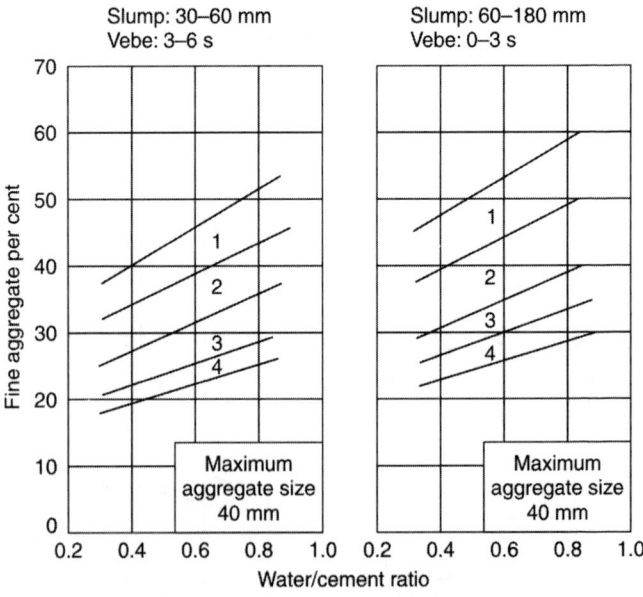

Fig. 20.5: Recommended weight of fine aggregate as a function of free water/cement ratio for various workabilities and maximum sizes

Characteristic cube strength = M-15

M-30

M-15

Type of cement: Ordinary Portland

Fine aggregate: Natural river sand
Conforming to grading zone III of Table 4, of IS: 383–1970.
Type of cement: Ordinary Portland
Coarse aggregate: Crushed (angular)
Conforming to IS: 383 code requirements. Specific gravities of cement, sand and coarse aggregate (crushed granite) are 3.14, 2.63 and 2.61, respectively.
Type of exposure: Mild
Degree of quality control: Very good
Slump: 3–6 mm for M-15
      6–12 mm for M-30
      12 mm for M-45
Design the mix proportions for the various grades of concrete.

Design Procedure

1. Target mean strength

SI no.	Mix grade	Target mean strength (N/mm²)
		$\overline{f_{ck}} = f_{ck} + t.s$
1	M-15	$\overline{f_{ck}} = (15 + 1.65 \times 2.5) = 19.13$
2	M-30	$\overline{f_{ck}} = (30 + 1.65 \times 5.0) = 38.25$
3	M-45	$\overline{f_{ck}} = (49 + 1.65 \times 6.0) = 54.90$

2. Selection of free water/cement ratio

From Table 20.1, the approximate 28 days compressive strength of concrete mixes made with a free water/cement ratio 0.5 using ordinary Portland cement with crushed type aggregate is 47 N/mm².

This strength value is plotted in Fig. 20.1 and a curve is drawn from this point and parallel to the printed curves until it intercepts a horizontal line passing through the ordinate representing the target mean strength. The corresponding value on the abscissa represents the free water/cement ratio. Thus from Fig. 20.1, the free water/ cement ratios obtained for the corresponding target mean strength of the different mixes are as follows.

Mix	Target mean strength (N/mm²)	Free water/cement ratio
M-15	19.13	0.86
M-30	38.25	0.575
M-45	54.9	0.440

3. Selection of free water content

From Table 20.2, the approximate free water contents (kg/m^3) required for various levels or workability are as follow:

Mixes	M-15	M-30	M-45
VB (secs)	3–6	6–12	12

Free water content for:
  (a) Uncrushed type of fine (W_f) aggregate of maximum size = 10 mm
       For   M-15 grade = 210
              M-30 grade = 190
              M-45 grade = 170
  (b) Crushed type coarse aggregate of maximum size 20 mm (W_c)
       For   M-15 grade = 210
              M-30 grade = 190
              M-45 grade = 170

Since coarse and fine aggregates of different types are used (crushed and uncrushed respectively), the total free water content is estimated using the expression.

$$W_{FW} = (2/3\ W_F + 1/3\ W_C)$$

Thus by calculation, the free water contents of various mixes are as follows:

Mix	Free water content (kg/m^3)
M-15	206.67
M-30	183.33
M-45	155.67

4. Determination of cement content

The cement content is determined from the relation:

$$\text{Cement content} = \left(\frac{\text{Free water content}}{\text{Free water/cement ratio}} \right)$$

Thus, the cement contents for various mixes are as follow:

Mix	Cement content (kg/m^3)
M-15	240.31
M-30	318.84
M-45	356.07

5. Determination of total aggregate content (T_A)

The total aggregate content is calculated using the formula:

$$T_A = (D - C - W_{FW})$$

where, D = Wet density of the concrete (kg/m³)

C = Cement content (kg/m³)

W_{FW} = Free water content (kg/m³)

From Fig. 20.2, the values of 'D' for various mixes are obtained as:

Mix	D (kg/m³)
M-15	2350
M-30	2365
M-55	2410

Thus the total aggregate contents (T_A) for various mixes are as follows:

Mix	T_A (kg/m³)
M-15	1903
M-30	1863
M-45	1897

6. Selection of fine aggregate content

From Figs 20.3 to 20.5, the proportions of fine aggregate content for mixes of various levels of workability are as follows:

Mix	V-B (secs)	Free water cement ratio	Range of proportions of F.A.	Average value
M-15	3–6	0.86	38–47%	42.5%
M-30	6–125	0.575	31–38%	34.5%
M-45	12	0.440	27–33%	30.0%

The fine aggregate contents of various mixes, calculated using the expression:

Fine aggregate content = (Total aggregate content)
× (Proportion of fines)

The values of fine aggregate contents obtained are as follows:

Mix	Fine aggregate content (kg/m³)
M-15	809
M-30	643
M-45	569

7. Calculation of coarse aggregate content

The coarse aggregate content of various mixes calculated using the expression:

Coarse aggregate content = (Total aggregate content) − (Fine aggregate content)

Mix	Coarse aggregate content (kg/m³)
M-15	1094
M-30	1220
M-45	1328

8. The various ingredients per m³ of concrete for the various mixes are as follows:

Mix	Cement (kg)	Fine aggregate (kg)	Coarse aggregate (kg)	Water (kg)
M-15	240.31	809	1094	206.67
M-30	318.84	643	1220	183.33
M-45	356.07	569	1328	156.67

9. Mix proportions by weight are obtained as follows:

Mix		Proportions by weight		
M-15	1 :	3.37 :	4.55 :	0.86
M-30	1 :	2.02 :	3.83 :	0.58
M-45	1 :	1.60 :	3.73 :	0.44

21

Experimental Review of Indian, British and American Methods of Concrete Mix Design

21.1 GENERAL FEATURES

The main objective of concrete mix design is to select the optimum proportions of the various ingredients of concrete, which will yield fresh concrete of desirable workability and hardened concrete possessing the specified characteristic compressive strength and durability. The mix proportions should also satisfy the additional requirements of the use of minimum possible cement content so that maximum economy is achieved in the unit cost of concrete.

The main purpose of this critical review is to examine by experimental investigations the suitability of the well-established Indian, British, American and other mix design methods outlined in various earlier chapters by designing low, medium and high strength concretes with particular emphasis on economy of cement content. The mix design methods included in the present study by Krishna Raju and Krishna Reddy[183] are the British[182], American[5,79], and Indian Standard[179] methods together with the traditional Road Note No. 4[12] method and the method of Erntroy and Shacklock[88], generally used for high-strength concrete mixes. Concrete mixes corresponding to the grades M-15, M-30 and M-45 were designed using these methods and the control specimens were tested at the end of 28 days to study the theoretical predictions in the light of experimental results as reported by the authors in an earlier research paper[183].

21.2 MATERIALS AND TEST SPECIMENS

Ordinary Portland cement conforming to the India Standard Code IS: 269–1976 was used for all the mixes. The cement conforming to

C-33 Grade had a specific gravity of 3.14 and a surface area of 2250 cm^2 per gm. River sand and crushed granite coarse aggregate of maximum size of 20 mm having the grading requirements shown in Table 21.1, were used for all the concrete mixes. The gradings of the fine and coarse aggregates conformed to the grading requirements specified in the Indian Standard specification IS: 383–1970[20].

Table 21.1: Grading of fine and coarse aggregates

IS sieve size	Percentage passing	
	Fine aggregate	Coarse aggregate
20 mm	–	100
10 mm	–	52
4.75 mm	99.0	8.7
2.36 mm	96.6	–
1.18 mm	74.1	–
600 micron	36.8	–
300 micron	6.3	–
150 micron	1.6	–

The fineness moduli of fine and coarse aggregates were 2.85 and 6.37, respectively. The sand used corresponded to the grading zone II of the IS grading requirements. The specific gravities and bulk densities of the fine and coarse aggregates used were found to be 2.63 and 2.61 and 1598 kg/m^3 and 1564 kg/m^3, respectively. The control specimens comprised of the standard 150 mm cubes and 150 mm by 300 mm cylinders for the determination of compressive strength at 7 and 28 days.

21.3 MIX DESIGN METHODS

Concrete mixes corresponding to M-15, M-30 and M-45 grades were designed according to the procedures specified in the British, American Concrete Institute, Indian Standard, Road Note No. 4 and Erntroy and Shacklock's methods. All the mixes were designed corresponding to the degree of control specified as 'Very good' in the Indian Standard Code. The values of standard deviation and target mean strength computed as recommended in the Indian Standard Code IS: 10262–1982[179] for different grades of concrete are compiled in Table 21.2. For each concrete mix, 8 cubes of 150 mm size and 6 cylinders of 150 by 300 mm size were cast for tests at 7 and 28 days.

Table 21.2: Target mean strength

Grade of concrete f_{ck}	Standard deviations of degree of control 'very good'	Target mean strength $\overline{f_{ak}} = (f_{ck} + 1.65s)$ (N/mm²)
M-15	2.5	$(15 + 1.65 \times 2.5) = 19.13$
M-30	5.0	$(30 + 1.65 \times 5.0) = 38.25$
M-45	6.0	$(45 + 1.65 \times 6.0) = 54.90$

The Road Note No. 4 method was used for the design of mixes of grades M-15 and M-30 only, while Erntroy and Shacklock's method was adopted only for the design of high-strength concrete of grade M-45. The degrees of workability assumed in these methods corresponded to medium, low and very low for M-15, M-30 and M-45 grades, respectively.

In designing concrete mixes by the Road Note No. 4 method, the fine and coarse aggregates were combined to correspond to the standard grading curve no. 3 and no. 1 for M-15 and M-30 grade concrete mixes, respectively. For high strength concrete of grade M-45 designed by Erntroy and Shacklock's method, the aggregates were combined to correspond to the standard grading curve no. 1 which yields the leanest possible mix.

In designing concrete mixes according to the ACI method, a reduction factor of 0.8 was applied to the cube strength to obtain the corresponding cylinder strength. The ACI method is based on the presumption that for a given maximum size of aggregate, the water content in kg/m³ determines the workability of the mix, largely independently of the mix proportions. The determination of the quantity of coarse aggregate in the mix is based on the assumption that the optimum ratio of the bulk volume of dry rolled coarse aggregate to the total volume of concrete depends only on the maximum size of the coarse aggregate and fineness modulus of sand.

The current British mix design method proposed in 1975 by Teychenne[182] to replace the traditional Road Note No. 4 method discards the use of standard grading curves of combined aggregates and the design tables of aggregate/cement ratios for different types of cement and size and type of grading of aggregates used in Road Note No. 4 method.

In the new method, the free water/cement ratio for a target mean strength is determined based on the type of cement and coarse aggregate categorised as crushed and uncrushed. The approximate water content required depends upon the type and maximum size of coarse aggregate used and the desired workability expressed as slump or V-B time. The total aggregate content is determined by subtracting the cement and water content from the wet density of concrete, which is estimated using the water content and relative density of the combined aggregates. The proportions of fine and coarse aggregates are determined from a set of curves involving the various parameters, like water/cement ratio, workability, maximum size and the grading zone of the fine aggregates.

21.4 STANDARD TESTS

Workability of the fresh concrete was tested by the slump, compaction factor and V-B tests conducted according to the procedure specified in IS: 1199–1959. Compressive strength tests on hardened concrete were conducted in accordance with the procedures prescribed in IS: 516–1959. The results of various tests conducted on fresh and hardened concrete are compiled in Table 21.3.

21.5 DISCUSSION OF TEST RESULTS

(a) *Water/cement ratio:* The variation of compressive strength with water/cement ratio is shown in Fig. 21.1 for the mixes designed by the various methods. The general trend of results is in agreement with Duff Abram's water/cement ratio law of decreasing compressive strength with increasing water/cement ratio. For a given grade of concrete, the IS method required the least water/cement ratio while the British method resulted in the higher water/cement ratios compared to the Indian and Road Note No. 4 methods.

(b) *Aggregate/cement ratio:* The lowest aggregate/cement ratios corresponding to the richest mixes were indicated in the IS method while the highest aggregate/cement ratios corresponding to the leanest mixes were obtained for mixes designed by the British method. The aggregate/cement ratio resulting from the different methods varied from 5.7 to 7.9, 4.4 to 5.9 and 2.7 to 5.3 for the M-15, M-30 and M-45 grades of concrete, respectively. The Indian standard method consistently resulted in the richest mix for all the three grades of concrete.

(c) *Workability:* The variation of workability of fresh concrete measured in terms of slump, compaction factor and V-B time

Table 21.3: Details of concrete mixes and results of workability and strength tests

Grade of concrete	Mix design method	Mix proportions (by weight)	W/C ratio	A/C ratio	Cement content (kg/m³)	Slump (mm)	V-B time (secs)	Compaction factor	Cube compressive strength (N/mm²)		28 days cylinder compressive strength (N/mm²)
									7 days	28 days	
M-15	Indian	1 : 2.01 : 3.71	0.60	5.7	315	60	8	0.85	15.0	26.8	18.7
	American	1 : 3.45 : 3.88	0.80	7.3	250	80	4	0.88	7.5	14.1	11.6
	British	1 : 3.37 : 4.55	0.86	7.9	240	90	2	0.90	9.4	13.1	8.2
	Road Note No.4	1 : 2.29 : 3.91	0.77	6.2	290	70	3	0.84	16.2	24.0	14.8
M-30	Indian	1 : 1.09 : 3.32	0.39	4.4	477	20	36	0.80	29.8	43.8	29.7
	American	1 : 2.04 : 3.32	0.55	5.4	336	50	28	0.85	18.9	29.1	22.9
	British	1 : 2.02 : 3.83	0.58	5.9	319	60	24	0.86	17.5	30.1	16.6
	Road Note No. 4	1 : 1.49 : 3.65	0.51	5.1	358	40	25	0.82	18.3	32.36	18.3
M-45	Indian	1 : 0.65 : 2.06	0.30	2.7	591	0	52	0.74	40.9	60.5	31.7
	American	1 : 1.15 : 2.29	0.38	3.4	487	30	46	0.79	34.0	49.9	36.6
	British	1 : 1.60 : 3.73	0.44	5.3	3.56	40	35	0.80	22.0	41.4	20.8
	Erntroy & Shacklock	1 : 1.08 : 3.42	0.40	4.5	410	20	39	0.78	27.0	42.6	26.7

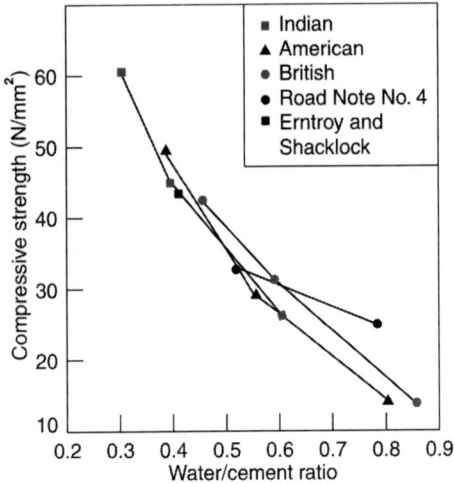

Fig. 21.1: Variation of compressive strength of concrete with water/cement ratio

with water/cement ratio are shown in Figs 21.2, 21.3 and 21.4, respectively. For a given water/cement ratio, highest slumps and compaction factors were recorded for the mixes designed by the British method. For the range of water/cement ratios from 0.35 to 0.60, the mixes designed by the ACI method resulted in the stiffest mixes with high values of V-B time mainly due to the higher fine aggregate content in comparison with the other methods.

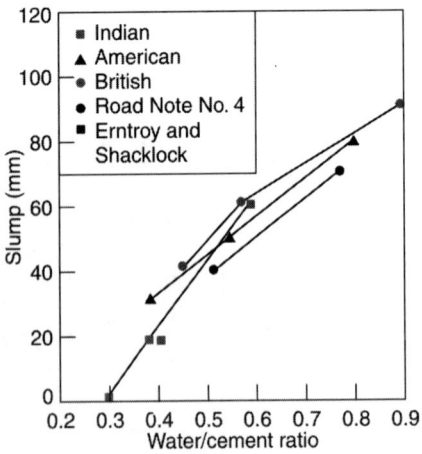

Fig. 21.2: Variation of slump with water/cement ratio

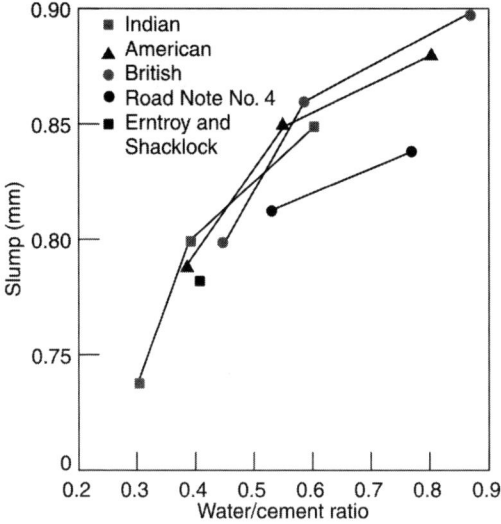

Fig. 21.3: Variation of compaction factor with water/cement ratio

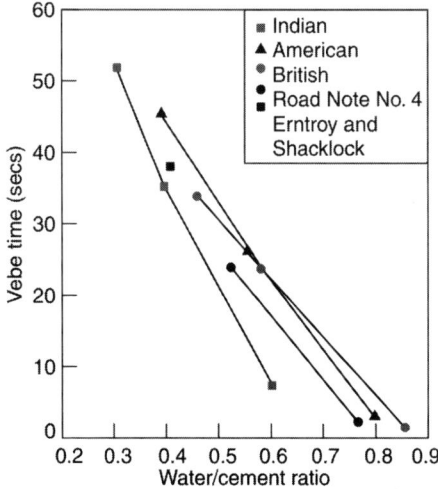

Fig. 21.4: Variation of V-B time with water/cement ratio

(d) *Cement content:* The variation of compressive strength with cement content for the different mixes comprising the three grades of concrete is shown in Fig. 21.5. The trend of results shows that the compressive strength increases more or less linearly with cement content for the mixes designed by the

British method and at the same time results in the most economical cement content for a given compressive strength. The three grades of mixes designed by the Indian method utilised the maximum cement content per unit volume of concrete in comparison with other methods.

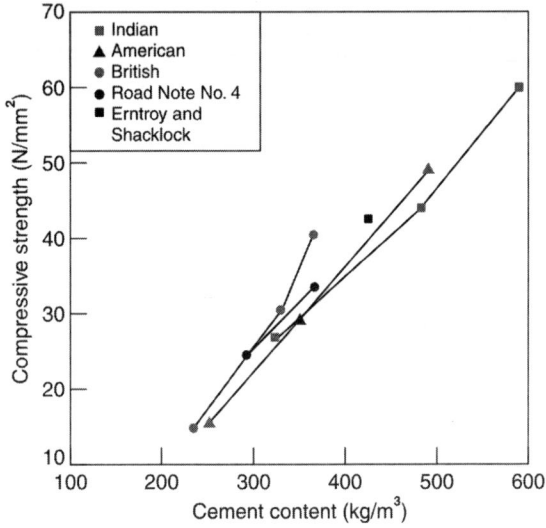

Fig. 21.5: Variation of compressive strength with cement content

(e) *Compressive strength:* The compressive strength results of standard cubes and cylinders are compiled in Table 21.3. The Indian standard method resulted in highly conservative results of compressive strength for all the grades of concrete mainly due to the high content of cement used in conjunction with low aggregate/cement and water/cement ratios in comparison with other methods.

For M-15 grade concrete mix, the marginal lower strengths observed in the mixes designed by the American and British methods are attributed to the very high water/cement ratios used in these mixes. If the maximum water/cement ratios are restricted to a value not exceeding 0.65 from durability considerations, the compressive strength will increase for the M-15 grade concrete mix designed by these methods.

For the medium grade M-30 concrete mix, all the methods yielded the desired characteristic compressive strength except the American method which resulted in marginally lower values.

In the case of high-strength concrete mixes of M-45 grade, the British and Erntroy and Shacklock's methods resulted in marginally lower values, mainly due to the higher aggregate/cement and water/cement ratios in comparison with the other methods.

21.6 CONCLUDING REMARKS

(a) The design of concrete mixes by the British method required the least cement content for all the three grades of concrete. However, the desired strength was not obtained for the lower and high strength ranges mainly due to the higher water/cement ratio.

(b) For all the grades of concrete, mixes designed by the Indian standard method, utilised the highest cement content for unit volume of concrete, in comparison with the other methods.

(c) The ACI method of mix design resulted in concrete of compressive strength very nearly equal to the desired characteristic strength for all the three grades of concrete with economical cement content in the mixes.

(d) The Road Note No. 4 method resulted in the use of higher cement contents than those of the British and American methods. The compressive strength of concrete conformed to the desired characteristic compressive strength.

In the present investigations, ordinary Portland cement of C-33 grade was used. More research work is needed to study the properties of concrete designed by the various mix design methods using C-43 and C-53 grades of cement which are widely used in the building industry.

Silica Fume Concrete

22.1 INTRODUCTION

Condensed silica fume is a byproduct from electric arc furnaces used in the manufacture of silicon metal or silicon alloys. Silica fume contains more than 80–85% SiO_2 in non-crystalline amorphous form with particle size of 0.1 micrometre which is about two orders of magnitude finer than cement particles. Condensed silica fume exhibits excellent pozzolanic characteristics and hence it is ideally suited for use as pozzolanic admixture in concrete leading to several technical advantages, such as reduction in thermal cracking caused by heat of hydration, improved durability and resistance to sulphates and acid waters and significantly improves the compressive strength of concrete. World production of condensed silica fume as a byproduct of silicon alloy and metal industries was of the order of one million metric tonnes in 1981. In the years to come with rapid innovations in steel industry, it is likely that the world production of condensed silica fume will exceed 10 million tonnes by the year 2010 and in the 21st century, condensed silicon fume will be increasingly used as a pozzolanic admixture in the concrete industry. Several investigations over the last four decades by Aitcin[186], Loland and Gjorv[187], Markstead[188], Mehta[189], Garette[190] and Malhotra[191] and various other research workers[192–195] have conclusively proved the advantages of using silica fume to improve the properties of concrete made with Portland cements.

22.2 CHEMICAL AND PHYSICAL PROPERTIES

Depending upon the type of silicon alloy produced and the design of the electric furnace, the physicochemical properties of condensed silica fume vary. Typical chemical and mineralogical composition, density, particle size distribution, water requirement, pozzolanic activity index of condensed silica fume are herewith examined.

(a) *Chemical composition:* Table 22.1 shows the typical chemical composition of silica fume as reported by Aitcin from seven different industrial operations producing various alloys.

Table 22.1: Chemical composition of condensed silica fume

Component	Chemical content of silica fume (per cent)						
	Si	FeSi-75 (heat recovery)	FeSi-75	FeSi-50	FeCrSi	CaSi	SiMn
SiO_2	94	89	90	83	83	53.7	25
Fe_2O_3	0.03	0.6	2.9	2.5	1.0	0.7	1.8
Al_2O_3	0.06	0.4	1.0	2.5	2.5	0.9	2.5
CaO	0.5	0.2	0.1	0.8	0.8	23.2	4.0
MgO	1.1	1.7	0.2	3.0	7.0	3.3	2.7
Na_2O	0.04	0.2	0.9	0.3	1.0	0.6	2.0
K_2O	0.05	1.2	1.3	2.0	1.8	2.4	8.5
C	1.0	1.4	0.6	1.8	1.6	3.4	2.5
S	0.2	–	0.1	–	–	–	2.5
MnO	–	0.06	–	0.2	0.2	–	36.0
Loss on ignition	2.5	2.7	–	3.6	2.2	7.9	10.0

(b) *Physical characteristics:* Table 22.2 shows the typical physical characteristics of condensed silica fume. Fig. 22.1 shows the typical particles size distribution of ordinary Portland cement, fly ash and silica fume as reported by Mehta[196]. The silica fume particles are very fine with average size of 0.1 to 0.2 micrometre which is about two orders of magnitude finer than the ordinary Portland cement and fly ash particles.

(c) *Water requirement and pozzolanic activity index:* Very fine particle size requires high water content to make the slurry with a flowing consistency. The water requirement of a standard Portland cement silica fume mortar is generally far above the 115 per cent permitted for natural pozzolans (ASTM-C-618). The average water requirement of silica fume samples tested by Pistilli[197] was 145%. In spite of the higher water requirement, the pozzolanic activity index (compressive strength ratio between the test mortars having the same consistency) of most silica fume types easily exceeds the minimum 75% requirement to qualify as a pozzolan according to ASTM-C-618.

Table 22.2: Physical characteristics of silica fume

Type of physical characteristic	Typical values
1. Specific gravity	2.2 to 3.1 depending upon the type of alloy produced
2. Loose bulk density	200–250 kg/m^3 (general silica fume)
	500 kg/m^3 (pelletised) (densified)
3. Sieve analysis results	Residue on 45 micron sieve = 3.7 per cent
4. Surface arc (Blaine)	33000 to 77000 cm^2/gm
5. Average particle size	Spherical particle 0.01 to 0.3 micrometre
	Average size = 0.1 to .2 mm

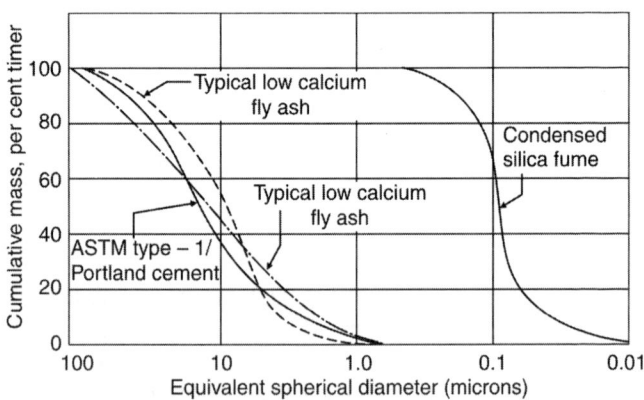

Fig. 22.1: Typical particle size distribution of materials

22.3 PROPERTIES OF CONCRETE CONTAINING CONDENSED SILICA FUME

Pozzolans are defined as siliceous materials of fine particle size that are capable of chemical reactivity with lime at normal temperatures forming cementatious products (calcium silicate hydrates). Thus traditionally the term 'Pozzolanic reaction' is used to designate lime-silica reaction in cementatious systems. When pozzolan is used as an admixture in concrete, it serves as a source of reactive silica.

The lime for the pozzolanic reaction is available from the hydration reactions of tricalcium and dicalcium silicates which are the principal compounds of Portland cement. RILEM technical committee (73 SBC—Siliceous byproducts in concrete) has classified silica fume and low temperature rice husk as highly pozzolanic materials. Both pozzolans contain essentially amorphous silica and possess very high surface area (very fine particles). Hence they are highly reactive and react faster than the other pozzolans, like fly ash which react slowly.

The strength contributions with age of different pozzolans are shown in Fig. 22.2.

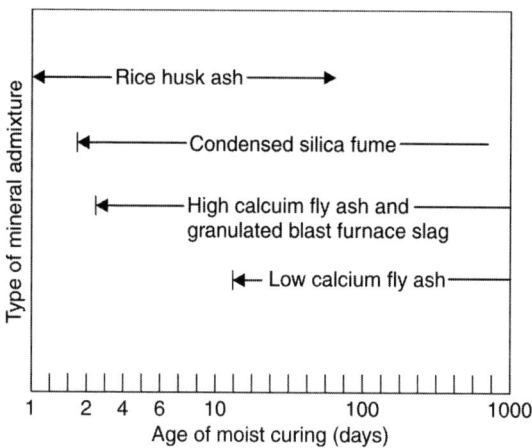

Fig. 22.2: Effective period of strength contribution from various mineral admixtures

1. Durability

Improved durability of silica fume concrete is attributed to the decrease in permeability as a consequence of reduced pore size distribution as shown in Fig. 22.3. Improved durability against chemical attack, such as sulphate waters and reactions causing alkali-aggregate expansion, is mainly due to the decrease in permeability in silica fume concrete which contains smaller pore size distribution in hydrated cement paste.

2. Workability

The effect of silica fume on the rheology of fresh concrete is to improve the cohesiveness of concrete mix due to the finer particle size of silica fume. Segregation and bleeding is reduced and flowability of concrete is increased. For a constant workability, water requirement increases with increase in silica fume content. Figs 22.4 and 22.5 show the water requirement for a constant slump with increasing quantity of silica fume in the mix.

The increased water content helps in better compaction and due to the superior flow characteristics of the concrete, it is ideally suited for pumping and shotcreting.

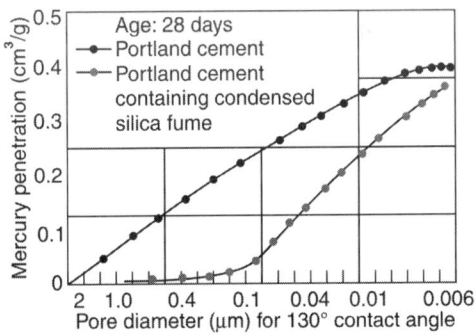

Fig. 22.3: Influence of silica fume addition on pore size distribution of well-hydrated cement pastes

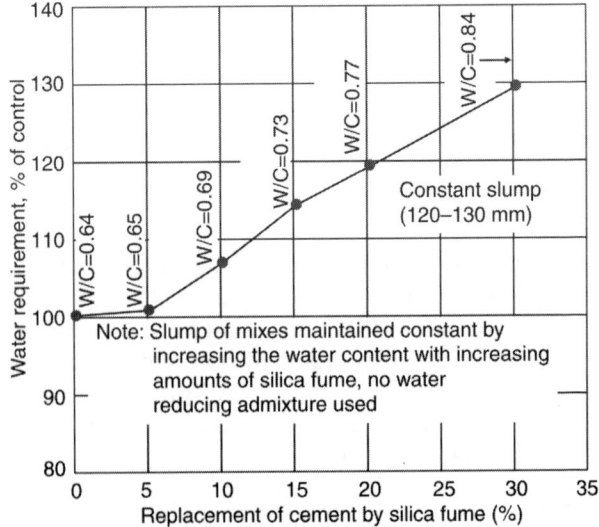

Fig. 22.4: Influence of silica fume addition on water requirement of concrete without water-reducing admixtures

3. Shrinkage

The surface of silica fume concrete must be protected against evaporation of surface water to prevent surface cracking due to plastic shrinkage in hot weather.

The long-term shrinkage of concrete is not affected significantly by the addition of silica fume especially when the water content of concrete mix has not been changed. Research investigations on concrete containing 297 kg/m^3 of ordinary Portland cement and 24 kg/m^3 of silica fume with a water to combined cement and silica fume ratio of 0.6, indicated negligible drying shrinkage even after 64 weeks.

4. Creep

Creep of concrete is inversely proportional to the strength of concrete. Since silica fume concrete has higher ultimate strength, creep is lower than in the corresponding Portland cement concrete. Research investigations on creep using control concrete and silica fume concrete has indicated similar creep after one year with slightly smaller values for silica fume concrete, indicating the beneficial effect of using silica fume in concrete.

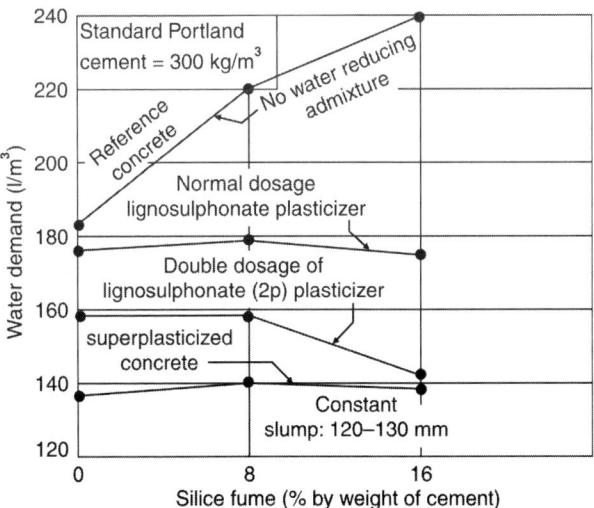

Fig. 22.5: Concrete mixed with and without water-reducing admixtures—Influence of silica fume addition of water demand for constant slump

5. Strength

The variation of compressive strength of silica fume concrete depends upon the cement and silica fume content in the mix, water to combined cement and silica fume ratio and age of concrete. Experimental investigations by Malhotra[198] on air-entrained and non-air-entrained concretes containing different quantities of cement and silica fume and varying water to composite cement ratio indicates significant increase in the compressive strength of concrete with increasing percentages of silica fume as shown in Fig. 22.6. For a given cement content, non-air-entrained concrete exhibited a higher strength than air-entrained concrete.

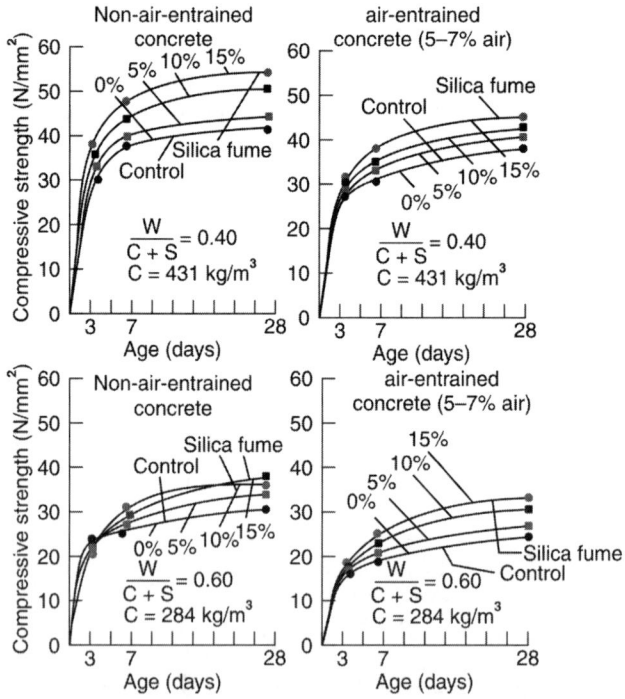

Fig. 22.6: Influence of silica fume on strength of air-entrained and non-air-entrained concretes

Due to high pozzolanic reactivity of silica fume, the strength improving ability of silica fume is nearly 2 to 3 times that of an equivalent mass of Portland cement replaced. Long period of moist curing is more important for silica fume concrete than for ordinary Portland cement concrete.

Wolsiefer[199] has produced very high-strength concrete by using silica fume comprising the following data:

Ordinary Portland cement (Type I) $= 593 \text{ kg/m}^3$
Lime stone aggregate of maximum size $= 10 \text{ mm}$
Condensed silica fume $= 119 \text{ kg/m}^3$

$$\left(\frac{W}{C+S}\right) = 0.22$$

Superplasticizer used
Compressive strength at 14 days $= 100 \text{ N/mm}^2$
Compressive strength at 4 months $= 125 \text{ N/mm}^2$

Bache[200] was able to produce silica fume concrete in laboratory having a compressive strength of 250 N/mm² using 50 per cent each of ordinary Portland cement and silica fume by volume with water/cementatious ratio of 0.13 to 0.15. Calcined bauxite aggregate was used with a large dosage of plasticizer. However, the concrete was more brittle due to the increased strength.

6. Permeability

Investigations by Loland and Hustad[201] indicate that a concrete mixture containing 100 kg/m³ of ordinary Portland cement, 20 per cent of condensed silica fume with superplasticizer showed the same permeability as a concrete containing 250 kg/m³ of Portland cement without silica fume and plasticizer.

The coefficient of permeability measured on concrete containing 250 kg/m³ or ordinary Portland cement with 10 per cent condensed silica fume was around 17.5×10^{-15} m/sec while the coefficient of permeability of concrete containing 250 kg/m³ of ordinary Portland cement was a high as 615×10^{-15} m/sec. Addition of silica fume, which is a highly pozzolanic material, reduces the size of the voids in hydrated cement paste thus making it almost impermeable.

7. Frost Resistance

Durability of concrete containing condensed silica fume to freeze-thaw conditions has been investigated by several research workers[202,203]. Important observations are summarised as follows:

   (i) Non-air-entrained concrete with silica fume shows very low durability factors with very low resistance to frost attack.

(ii) Air-entrained concrete regardless of water/cement ratio (0.4 to 0.6) and containing up to 15 per cent replacement by condensed silica fume performed satisfactorily when tested with ASTM method C-666 for freeze-thaw resistance. Hence it can be concluded that air entrainment is very important rather than using high dosage of silica fume in concrete for frost resistance.

8. Abrasion Resistance

Generally, abrasion resistance improves with strength of concrete. Using special abrasion/erosion test developed at the US Army Corps of Engineers Waterways Experiment Station Laboratory Holland[204] has shown that abrasion resistance of high-strength concrete (90 N/mm^2) containing silica fume is much superior in comparison with control concrete without silica fume as shown in Fig. 22.7. The figure shows the results of abrasion/erosion loss test on concrete used for Kinzua dam stilling basin rehabilitation project.

9. Chemical Resistance

The main reason for high chemical resistance of silica fume concrete is due to the decrease in permeability with small pores and reduction in calcium hydroxide content in cement paste which reacts with sulphates and acids. The reduction in calcium hydroxide content with increasing silica fume content is shown in Fig. 22.8 based on the investigations of Sellevold[205].

Silica fume concrete has excellent resistance against sulphates and acids. More than 20 years of field exposure of concrete to alum shale water (2.89 pH) in Oslo showed that concrete made from normal Portland cement with 15 per cent replacement by silica fume had as much resistance to attack as concrete made from sulphate resisting Portland cement. Laboratory investigations by Mather[206] have shown that cement mortars containing silica fume had superior resistance to sulphate attack than cement mortars without silica fume. Concrete specimens containing 30 per cent silica fume required only 7 days of curing to show high durability to chloride solutions.

10. Resistance to Alkali Aggregate Reaction

In Iceland, Portland cements are of high alkali type (> 0.6% Na$_2$O) and many aggregates contain alkali reactive constituents like silica.

Investigations by Olafasson[207] using mortar prisms made with pyrex glass aggregates and high alkali Portland cement (containing 0.86%, 1% and 1.39% Na_2O equivalent) showed that 5 to 10 per cent cement substitution by silica reduced the 6 months expansion by more than 75 per cent of the expansion in the reference mortar. To overcome the problem of alkali aggregate reaction, Gudmundsson[208], has reported that since 1977, the ordinary Portland cement produced by the Iceland State Cement Works at Akranes (Iceland) has been blended with 5 to 6 per cent silica fume available from a nearby ferro-silicon factory.

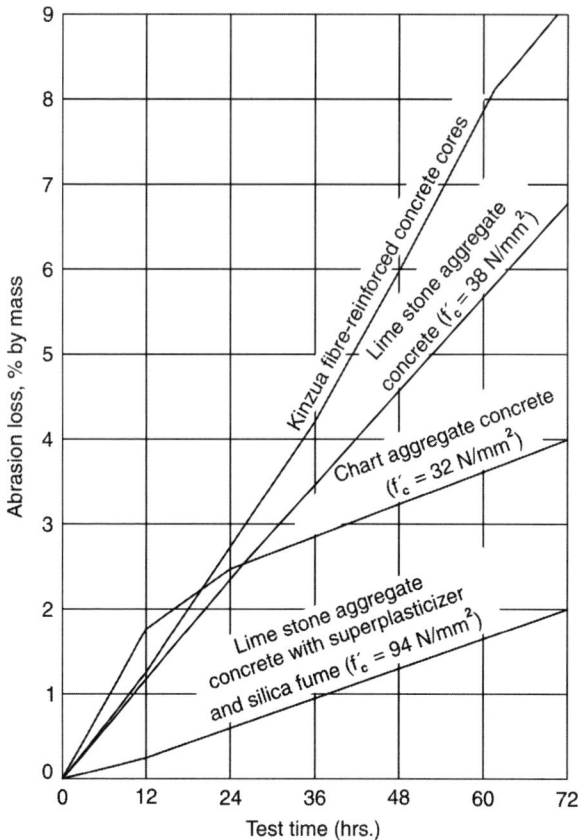

Fig. 22.7: Results of abrasion loss test on concrete for the Kinzua dam stilling basin rehabilitation project

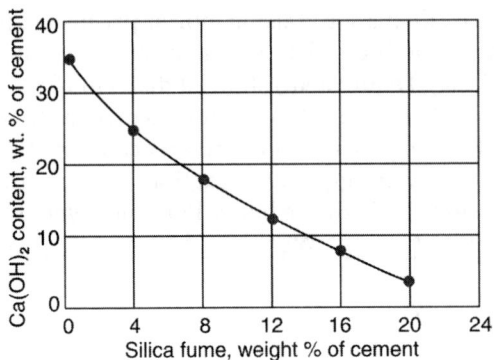

Fig. 22.8: Influence of cement replacement by silica fume on calcium hydroxide content of the cement paste

11. Corrosion of Embedded Steel in Concrete

Corrosion of steel is an electrochemical process requiring the presence of electrolyte, moisture and air. The expansion associated with the formation of rust causes cracking and spalling of concrete. Addition of silica fume increases the ohmic resistance of concrete thus increasing the electrical resistivity and corrosion as shown in Fig. 22.9, which is based on the results of experiments by Gjorv[209]. The increase in electrical resistivity is due to the pore refinement process caused by pozzolanic reaction due to condensed silica fume in the concrete mix.

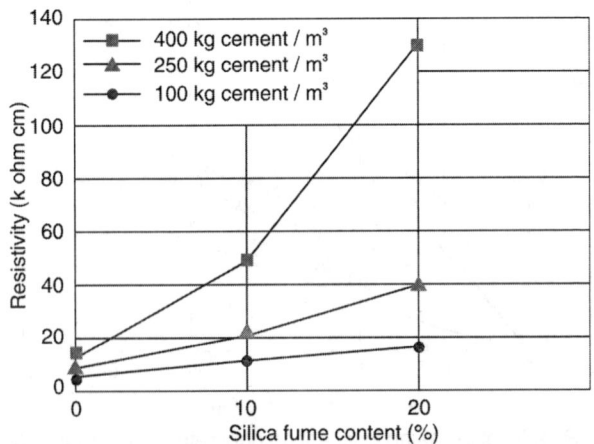

Fig. 22.9: Effect of silica fume on the ohmic resistance of concrete

22.4 EXAMPLES OF INDUSTRIAL USE OF CONDENSED SILICA FUME

(a) *Construction of New Tjorn Bridge in Sweden (1981):* The first industrial use of silica fume on a large scale for obtaining high strength and low heat of hydration was in the construction of the pylons of the New Tjorn Bridge in Sweden in 1981. The compositions of concrete mixes used in the slip formed pylons of the bridge as reported by Rickne and Svensson[210] are as follow:

Portland cement	$= 370 \text{ kg/m}^3$
Condensed silica fume	$= 37 \text{ kg/m}^3$
Fine aggregate	$= 785 \text{ kg/m}^3$
Water	$= 205 \text{ kg/m}^3$

Average compressive strength $(f_{ck}) = 62 \text{ N/mm}^2$
Heat of hydration reduced by the use of condensed silica fume.

(b) *High-rise building in Montreal, Canada (1984):* In 1984, silica fume was used to produce very high-strength concrete for a high-rise building in Montreal, Canada. The average compressive strength of concrete cylinder (f_{cy}) was reported to 60 N/mm^2 at 3 days, 75 N/mm^2 at 7 days and 90 N/mm^2 at 28 days.

(c) *Kinzua dam stilling basin rehabilitation (1983):* Silica fume concrete was first used in concrete where durability to abrasion-erosion was important for the repairs of the stilling basin and the Kinzua dam in Pennsylvania by the US Army Corps of Engineers in 1983. The data from the abrasion-erosion studies by Holland[204] involving different types of aggregates and admixtures showed that a superplasticized concrete containing silica fume had the least loss of material (refer Fig. 22.7). The concrete used for the repairs of the stilling basin of Kinzua dam had the following composition:

Ordinary Portland cement	$= 390 \text{ kg/m}^3$
Condensed silica fume (18% by weight of cement)	$= 70 \text{ kg/m}^3$
Coarse aggregate	$= 975 \text{ kg/m}^3$
Fine aggregate	$= 825 \text{ kg/m}^3$
Water	$= 130 \text{ kg/m}^3$
Cylinder compressive strength f_{cy} (28 days)	$= 90 \text{ N/mm}^2$
Cylinder compressive strength f_{cy} (90 days)	$= 100 \text{ N/mm}^2$

22.5 GUIDELINES FOR MIX PROPORTIONING AND STANDARD SPECIFICATIONS

Condensed silica fume is generally used as an admixture mixed with ordinary Portland cement in the dry state or as suspension in water which contains a water reducing admixture. Hence water/cementatious ratio has to be properly maintained to achieve suitable desirable workability and strength. Silica fume greater then 5 per cent by weight of cement requires the use of superplasticizing or water reducing admixtures to achieve high-strength and durability. For frost resistance concrete, air entraining agent is essential to achieve suitable durability. Dosage of air entraining agent may have to be increased by 25 to 75 per cent of the normal dosage. It is essential to ensure proper curing of silica fume concrete since pozzolanic reaction can take place only in the presence of moisture.

Standard specifications have been evolved only by Norway and Canada at present since these countries are the pioneers in the field of silica fume concrete investigations.

(a) Norwegian Standard-3474 issued in 1978 is the first national standard specification dealing with the use of silica fume in concrete. It has subsequently been revised in 1981. The salient feature of the specification is that the maximum permissible amount of condensed silica fume is 10 per cent by weight of cement. The silica fume should contain at least 85 per cent of silicon dioxide and the silica should be in an amorphous form.

(b) Canadian Standard (CAN 3-A.23.5) also prescribes that the maximum percentage of silica fume should not exceed 10 per cent by weight of cement and the minimum silicon dioxide content should be at least 85 per cent. The pozzolanic activity index should be at least 85 per cent of the control in an accelerated test.

None of these standards contains any guidelines on the effect of silica fume on durability of concrete, probably due to lack of long-term research data on the performance of concrete containing silica fume.

22.6 CONCLUDING REMARKS AND NEEDED RESEARCH

1. The use of condensed silica fume in concrete offers several advantages, like high strength and durability.
2. The cost of condensed silica fume is high due to fine particles and low bulk density. In US and Canada, the material in the

bulk dry state costs as much as cement in the proximity of silicon alloy plants. Condensed silica fume supplied as suspension in water costs almost 2 to 3 times the dry state cost.

3. In spite of the high cost, the demand for condensed silica fume is bound to increase in specific constructions due to its advantages, like very high strength at early ages, durability and abrasion resistance.

Further research data is required in the following areas:

1. Long-term durability of concrete containing more than 10 per cent silica fume.

2. Studies on microstructures of silica fume concrete with 15 to 30 per cent silica fume content which is likely to be significantly different.

3. Research work is required to study the corrosion aspects when silica fume concrete is used in reinforced concrete construction.

4. Research studies on the deformational characteristics of concrete containing 15 to 30 per cent silica fume which may render the concrete more brittle.

23

![Pond Ash Concrete]

Pond Ash Concrete

23.1 INTRODUCTION

Fly ash generated as a by product in the thermal power plants comprises of particles of size varying from one micron to around 600 microns. The finer particles of fly ash collected by electrostatic precipitators are currently being used extensively in the manufacture of blended cements or as partial replacement to cement in concrete mixes. Coarse particles of fly ash which are unutilized are mixed with water and let into large ponds by pumping. This coarser fraction of the fly ash is termed as Pond Ash or Lagoon ash. The world wide present utilization of fly ash used for blending with cements is a meager 16% of the entire production exceeding 1000 million tons.

The design of fly ash concrete mixes using finer particles passing through 45 micron sieve has been presented in detail in section 16.8. Published literature[211,212] exists for the utilization of fly ash as a replacement material to cement in concrete mixes.

Pond ash differs from fly ash collected by electrostatic precipitators since it contains substantial amount of relatively coarser particles (greater than 45 microns and up to 600 microns). Research workers in the past have concentrated largely on the utilization of fly ash in cement or concrete. As far as studies on pond ash utilization are concerned, very limited literature is available and of these works reported, investigations on replacement of sand by pond ash are very few. Ranganath et al[213] have examined the workability and compressive strength of concrete in which pond ash has been incorporated as a part replacement of sand as fine aggregate.

The authors have also stated that most of the thermal power plants in India adopt a wet process of ash disposal. The fly ash collected in hoppers through electrostatic precipitators is mixed with water to form slurry which is transported through pipes for disposal in large ponds. Bottom or grate ash is also mixed for purposes of convenience. It is estimated that presently the coarser particles termed as pond ash amount nearly 1000 million tons in India.

Effective utilization of pond ash is essential from environmental considerations and also from the point of view of minimizing the storage area required for pond ash. Additionally, traditional supply of river and pit sand is becoming scarce and governments in several states of India have imposed ban/restrictions on quarrying of sand.

Any construction industry dealing with the production of precast structural elements and located specifically in the vicinity of thermal stations can profitably utilize the pond ash in large quantities. Characterization is the systematic study to investigate the physical and chemical properties of pond ash obtained from a thermal power station. This will also help in understanding the behaviour of pond ash, thereby paving ways for effective and large scale utilization. The characterization will also help to determine the optimum replacement of sand by pond ash in the production of concrete.

Although extensive published research work is available on the utilization of fly ash as a replacement material, reported work on the utilization of pond ash is very limited. However, recent research publications by Ranganath[214] and Krishna Murthy[215] explore the possibility of using pond ash as fine aggregate in concrete and also its workability characteristics. Cheriaf et al[216] have reported the pozzolanic properties of pulverized coal combustion bottom ash and Kiattikomol et al[217] have investigated the properties of ground coarse fly ash from several sources in Thailand.

Experimental investigations by Pranesh et al[218] have indicated the distinct possibility of replacing sand in concrete by pond ash without sacrificing strength or durability. These investigations reported in National Conferences and Journals over the last few years indicate that pond ash, considered as a useless industrial waste material, occupying precious land in the vicinity of thermal power stations, can be used to replace sand in concrete mixes resulting in cost savings and reduced environmental degradation.

The major findings resulting from the limited investigations on pond ash compiled from literature are listed below:

1. Pond ash is found to contain both reactive small particles (which facilitate pozzolanic reactions) and non-reactive or poorly reactive large particles.
2. In general, pond ash particles vary in size from a minimum of 45 μ to a maximum of 600 μ.
3. Replacement of pond ash as fine aggregate is more beneficial than replacing cement in concrete.
4. Workability of pond ash concrete reduces with increasing percentage of pond ash as replacement to fine aggregate.
5. The reduction in compressive strength generally observed with the addition of pond ash as partial replacement of sand can be restored to a great extent by re-proportioning the fine aggregate to coarse aggregate ratio together with the use of plasticizers.
6. Durability studies have indicated the benefits of using pond ash as part replacement to sand in concrete.

23.2 CHARACTERISTICS OF TYPICAL FLY ASH FROM THERMAL STATIONS

Based on experimental investigations, Kalgal et al[219] have reported the physical and chemical properties of typical fly ash samples procured from Raichur Thermal Power Station (RTPS) located in the state of Karnataka, India. Samples of fly ash collected from four different locations in the ash ponds of RTPS were used in the investigations.

1. Physical Properties

Tests conducted on representative samples of pond ash to determine the specific gravity, bulk density, yielded the following results:
- Specific gravity ...... 2.02 to 2.47
- Bulk density ...... 824 kg/m³ in loose state and 990 kg/m³ in compacted state

The results indicate that pond ash exhibits marginally lower values of specific gravity and hence lighter than that of typical sand used in concrete.

2. X-Ray Diffraction and Electron Micrograph Tests

XRD and SEM analyses were conducted on representative samples of pond ash to determine the type minerals, shape, size and distribution of minute particles. X-ray diffraction test results on a typical sample shown in Fig. 23.1 indicate the predominant presence of quartz (SiO₂) and smaller presence of mullite ($Al_6Si_2O_{13}$). The figure also exhibits large peak heights (counts) indicating higher presence of crystalline particles. The smaller base hump is indicative of very little pozzolanic activity of the pond ash sample.

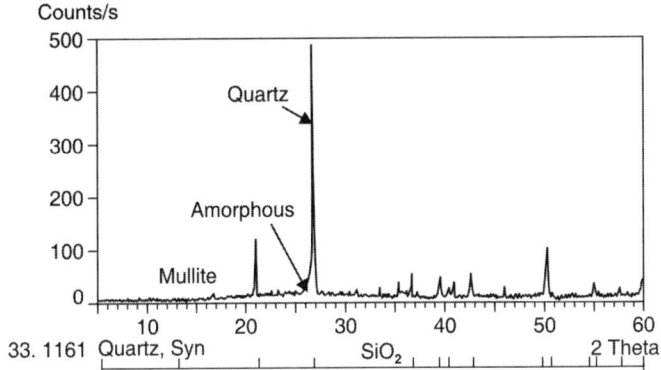

Fig. 23.1: X-ray diffraction test results of pond ash sample

The results of electron micrograph studies conducted on four typical samples are shown in Figs 23.2 to 23.5. The size of particles vary from 2 to 150 µ. Majority of the particles are in the range of 30 to 75 µ. Particles of size in the range of 10 to 40 µ are smooth and spherical in shape.

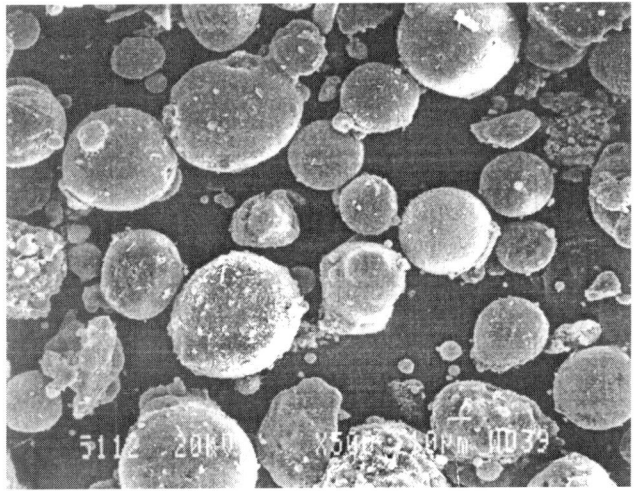

Fig. 23.2: Electron micrograph of pond ash sample P1 (1:500)

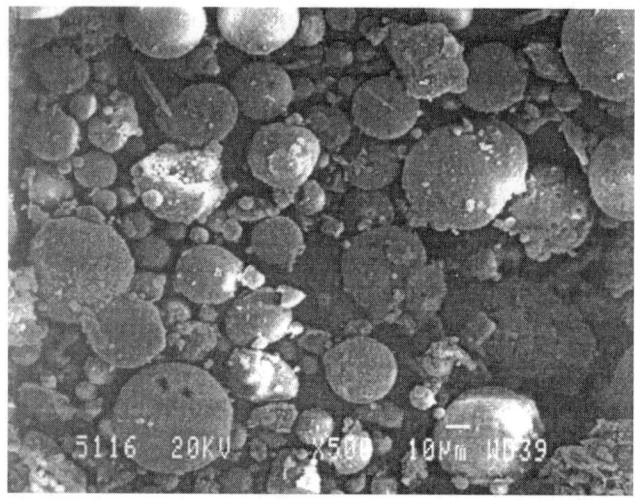

Fig. 23.3: Electron micrograph of pond ash sample P2 (1:500)

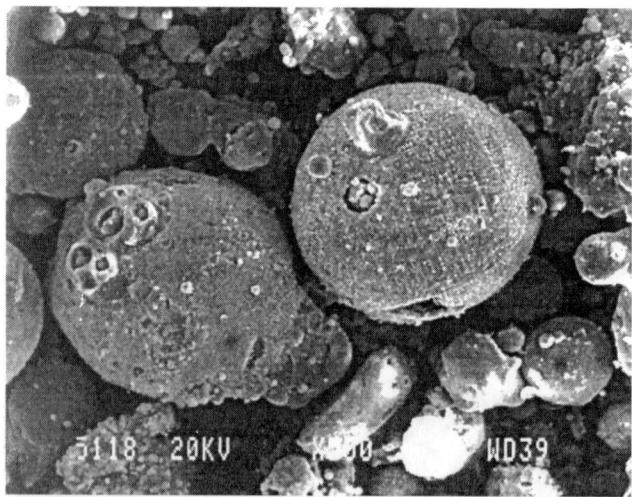

Fig. 23.4: Electron micrograph of pond ash sample P3 (1:500)

Pond ash predominantly contains smooth, glassy and spherical particles. At some locations, sintered particles with surface depositions are also observed. Irregularly shaped particles are relatively coarser with vesicular texture indicating little or no fusion during the combustion process. The presence of large amount of lumped

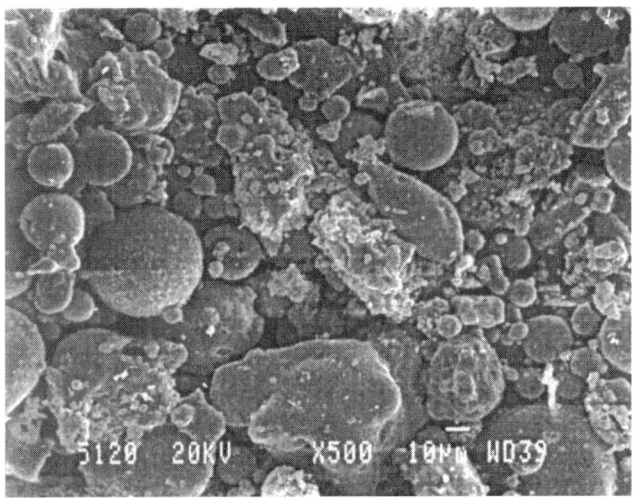

Fig. 23.5: Electron micrograph of pond ash sample P4 (1:500)

vesicular particles indicates that the pond ash comprises of significant amount of bottom ash.

3. Grading Analysis

The particle size distribution of four typical samples designated as P1, P2, P3 and P4 is shown in Table 23.1 and Fig. 23.6. The percentage passing for all the samples were comparable with the grading limits specified in Zone IV of the Indian Standard Code IS: 383[220]. The average value of the fineness modulus of pond ash was found to be around 2 indicating that the average size of the particles as 300 microns.

Table 23.1: Grading analysis of typical pond ash samples

Sieve	Percent passing			
size (mm)	P1	P2	P3	P4
4.75	99.79	98.94	99.88	99.84
2.36	98.83	97.16	98.71	98.98
1.18	95.43	93.52	96.45	97.48
600μ	77.31	76.06	83.41	89.40
300μ	58.18	58.68	69.42	77.65
150μ	29.78	31.40	41.57	44.67
75μ	14.18	13.10	14.17	13.75
45μ	7.23	6.00	7.47	6.58

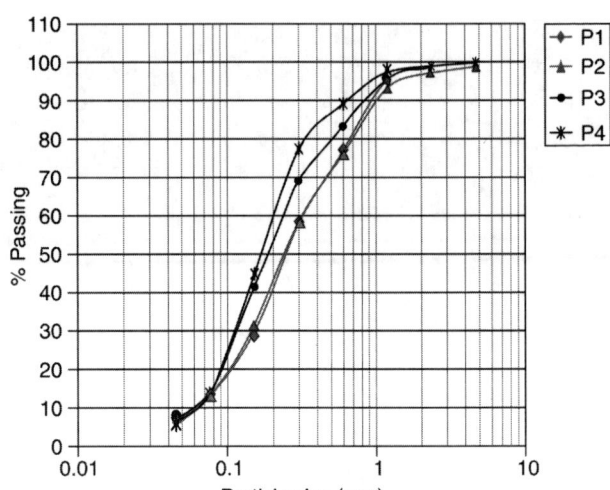

Fig. 23.6: Particle size distribution of pond ash samples

4. Lime Reactivity

Lime reactivity tests conducted as per IS: 1727-1967[221] on four typical samples of pond ash are compiled in Table 23.2. Lime reactivity varying from 0.58 to 0.70 N/mm² indicated negligible pozzolanic activity and suitability of the material as a partial replacement for sand in concrete.

5. Cement

Ordinary Portland cement of 43 grade conforming to IS: 8112-1989[222] was used throughout the investigations. The cement had a specific gravity of 3.15 with initial and final setting times of 120 and 240 minutes, respectively.

6. Aggregates

Locally available river sand and crushed granite passing through 12 mm sieve was used throughout the investigations. The fine aggregate conformed to grading zone II specified as per IS: 383-1970[220].

7. Soundness of Pond Ash–Cement Mix

The soundness of mortar mixes made of cement and typical pond ash samples are shown in Table 23.3. The values are marginally higher when compared with that of normal cement–sand mortar.

Table 23.2: Lime reactivity of pond ash samples

Sample no.	Water (per cent)	Flow (per cent)	Lime reactivity (N/mm²)
P1	20	66.7	0.69
P2	20	68.3	0.58
P3	20	65.0	0.66
P4	22	67.5	0.70

Table 23.3: Soundness of pond ash cement mortar

Sample no.	Soundness (mm)
P1	1.0
P2	1.5
P3	1.5
P4	1.0

8. Normal Consistency and Setting Times of Cement Mortar using Pond Ash

The results of standard tests conducted on cement mortar using pond ash in place of sand to determine the normal consistency and setting times are compiled in Table 23.4. The results conforming to the Indian Standard Code (IS: 269-1989)[14] specifications, are observed to be marginally higher in comparison with cement mortar using standard sand.

Table 23.4: Normal consistency, setting times of cement mortar with pond ash

Sample no.	Normal consistency (per cent)	Initial setting time		Final setting time	
		Hours	Minutes	Hours	Minutes
P1	33	3	00	5	30
P2	33	3	10	5	35
P3	34	3	10	5	20
P4	34	3	00	5	20

9. Compressive Strength of Pond Ash and Standard Sand Cement Mortar Cubes

The variation of compressive strength of cement mortar cubes using standard sand and pond ash replaced as sand is compared in Table 23.5 and Fig. 23.7. The compressive strength values of pond ash mortar cubes were consistently less than that of control cubes without pond ash when tested at ages of 7, 28 and 90 days. However, the rate of increase of compressive strength of pond ash cement mortar with age was more or less similar for the different pond ash samples.

Table 23.5: Compressive strength of pond ash and standard sand cement mortar cubes

Sample no.	7 days strength (N/mm²)	Per cent change in strength	28 days strength (N/mm²)	Per cent change in strength	90 days strength (N/mm²)	Per cent change in strength
Control cube	12.85	0.00	20.25	0.00	32.95	0.00
P1	7.36	42.72	12.57	37.90	22.42	31.96
P2	10.50	18.28	17.06	15.75	28.00	15.12
P3	10.04	21.87	15.79	22.02	25.66	22.12
P4	9.17	28.64	15.46	23.65	23.30	29.28

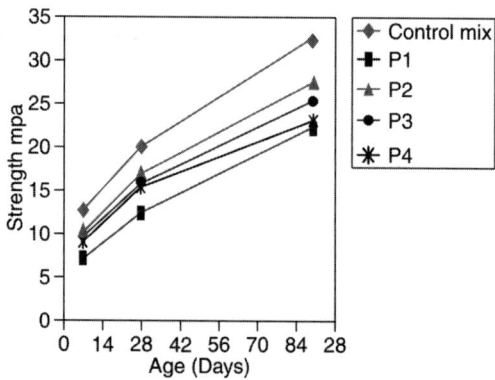

Fig. 23.7: Variation of compressive strength of mortar with age

10. Chemical Composition of RTPS Pond Ash

Chemical tests were conducted on typical samples of RTPS pond ash to determine the salient parameters for comparison with the code limits. The results of tests conducted on these samples are compiled in Table 23.6. The mineral composition and loss on ignition values are well within the limits prescribed in the Indian Standard Code-IS: 3812-1999[223].

23.3 POND ASH CONCRETE MIX PROPORTIONS

Doctoral research investigations were conducted by Pranesh[224] to examine the strength and durability of concrete with pond ash as

Table 23.6: Chemical composition of pond ash (mass percentage)

Sample designation	P1	P2	P3	P4
Loss on ignition	2.61	4.63	2.06	4.55
Total silica (SiO_2)	68.29	67.44	68.10	65.78
Alumina (Al_2O_3)	18.66	18.64	18.90	21.56
Iron oxide (Fe_2O_3)	12.80	6.40	9.00	5.80
Calcium oxide (CaO)	3.00	3.20	2.20	2.40
Magnesium oxide (MgO)	0.40	0.50	0.40	0.50
Sulphuric anhydride (SO_3)	0.40	0.18	0.28	0.35
Insoluble residue	89.90	90.60	90.30	91.50
Soluble salts	0.16	0.18	0.15	0.19

partial and complete replacement to sand and to compare it with normal concrete. The mix proportions were selected based on the criterion of absolute volume with a constant paste volume of 0.29. For all the mix proportions, the degree of packing was measured using a soft ware package entitled LISA based on Modified Andreasson Model[225] for particle packing. This model has been validated for concrete mix design and is very useful for designing optimum mix proportions. The Andreasson equation for solid packing of the mix is expressed as,

$$CPFT = \left[\frac{d^q - d_o^q}{D^q - d_o^q} \right] 100$$

where, $CPFT$ = Cumulative (volume) per cent finer than
$\quad\quad\quad d$ = Particle size
$\quad\quad\quad d_o$ = Minimum particle size
$\quad\quad\quad D$ = Maximum particle size
$\quad\quad\quad q$ = Distribution coefficient

The value of q can be taken between 0.25 and 0.3 for concrete. Practically, it is difficult to compute the exact value of $CPFT$. Hence by varying the proportion of coarse to fine aggregate, we can obtain $CPFT$ value which is close to the theoretical value. Further, in the computation of $CPFT$ only aggregates have been considered and cement paste which is necessary for binding the aggregates into a solid mass has been kept constant.

The author has examined three different criteria in designing the concrete mixes using pond ash as partial and complete replacement to sand.

These mixes are grouped as:
- Type A = Mixes based on workability criteria
- Type B and type C = Mixes based on strength criteria
- Type B and type C = Mixes based on minimum voids concept

Five different mix proportions with 0, 20, 30, 40 and 100 per cent pond ash replacement to sand were considered with mixes designated as BCC1, 20PAB, 30PAB, 40PAB and 100 PAC3. In all these mixes, 2 per cent sulphonated naphthalene polymer based superplasticizer was used to obtain a workability of 8 to 9 Vee bee seconds. Table 23.7 shows the ingredients in these mixes with different percentages of pond ash.

Table 23.7: Pond ash concrete mix proportions

Mix proportions	Cement	Fine aggregate		Coarse aggregate	Water cement	Water volume
		Sand	Pond ash			
	(kg/m^3)	(kg/m^3)	(kg/m^3)	(kg/m^3)	ratio	(l/m^3)
0% Pond ash (BCC1)	330	820	0	1045	0.56	185
20% Pond ash (20PAB)	330	656	134	1045	0.56	185
30% Pond ash (30PAB)	330	574	201	1045	0.56	185
40% Pond ash (40PAB)	330	492	268	1045	0.56	185
100% Pond ash (100PAC3)	344	–	638	1041	0.52	178

23.4 TEST SPECIMENS

Mechanical properties, like density, modulus of elasticity, compressive, flexural and split tensile strengths of hardened concrete at different ages were examined using standard test specimens, such as 100 mm cubes, 100 by 200 mm cylinders and 75 by 75 by 450 mm prisms. Additionally, durability tests were also conducted on specimens of the group BCC1, 40PAB and 100PAC3 to evaluate the resistance of concrete to sulphate, chloride and acid attack and corrosion potential using half cell potentiometer and rapid chloride penetration tests.

23.5 PROPERTIES OF POND ASH CONCRETE

1. Density

The density of pond ash concrete was measured after curing the specimens for 28 days. Table 23.8 shows the density of different mixes used in the investigations. The density of normal concrete is of the order of 2400 kg/m^3. However, the density of pond ash concrete is marginally lesser due to the addition of pond ash replacing sand and gradually decreases with the increase in pond ash content to value of 2200 kg/m³ with 100% replacement of sand with pond ash.

Table 23.8: Density and compressive strength of pond ash concrete

Mix designation	Density (kg/m³) 28 days	Compressive strength (N/mm²)					
		7 days	28 days	56 days	90 days	180 days	360 days
BCC1	23.9	21.6	38.0	46.3	47.2	48.0	51.0
20PAB	23.7	21.6	36.3	46.0	50.0	53.0	60.0
30PAB	23.3	21.3	35.3	43.3	50.6	50.3	56.0
40PAB	22.09	21.0	35.3	43.0	46.6	49.0	56.0
100PAC3	22.0	17.6	29.3	35.0	40.0	40.6	41.0

2. Compressive Strength

The compressive strength of hardened pond ash and normal concrete test specimens were determined using the cube specimens tested at different ages of 7, 28, 56, 90, 180 and 360 days. The results of tests are compiled in Table 23.8. Fig. 23.8 illustrates the variation of compressive strength with age for all mix proportions normalized with respect to 28 days strength. The analysis of strength test results indicates that the compressive strength of pond ash concrete in the

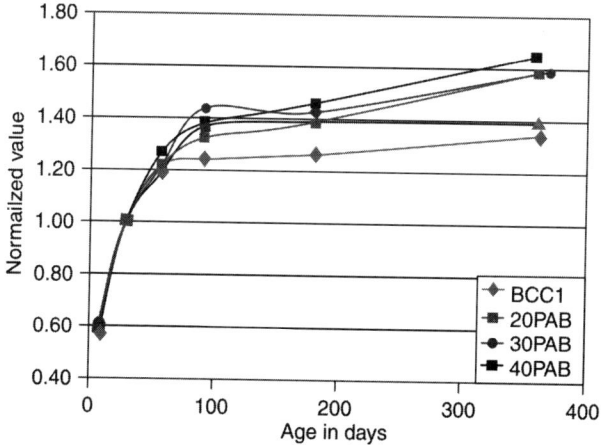

Fig. 23.8: 28 days normalised compressive strength

initial stages of up to 56 days is marginally less than that of normal concrete. However, the compressive strength of pond ash concrete with partial replacement to sand has shown comparatively increased strength beyond 8 weeks and up to one year. This marginal increase in strength is attributed to the pozzolanic activity resulting in strength forming compounds and possible improvement of pose structure.

3. Flexural and Split Tensile Strengths

The flexural and split tensile strengths of normal and pond ash concrete designated as BCC1, 40PAB and 100 PAC3 at various ages are compiled in Table 23.9. Both flexural and split tensile strengths increase gradually with age. The trend is similar in pond ash concrete. Generally, flexural strength is slightly higher than the split tensile strength. As the pond ash content increases in concrete, the flexural and split tensile strengths gradually decreases and the trend is similar to that of compressive strength.

Table 23.9: Flexural and split tensile strengths

Mix designation	Age (days)	Flexural strength (N/mm^2)	Split tensile strength (N/mm^2)
BCC1	28	8.31	3.29
	56	8.53	4.56
	90	9.48	4.88
40PAB	28	8.03	2.87
	56	8.21	4.25
	90	9.80	5.10
100PAC3	28	2.64	2.55
	56	5.20	4.10
	90	4.81	3.40

4. Modulus of Elasticity

Modulus of elasticity of pond ash concrete as determined from standard test specimens of the group 100PAC3 in which sand was completely replaced by pond ash, is compiled in Table 23.10.

The modulus of elasticity of pond ash concrete increases with age in a manner similar to that of normal sand concrete. The increase in modulus of elasticity is commensurate with increase in compressive strength with age.

Table 23.10: Modulus of elasticity of pond ash concrete

Mix designation	Age (days)	Modulus of elasticity (kN/mm^2)
100PAC3	28	20.9
	56	25.9
	90	26.2

5. Durability

The following durability tests were conducted on hardened concrete specimens of the group BCC1, 40PAB and 100PAC3 comprising normal sand and pond ash at different ages.

(a) Exposure to sulphate, chloride and acid solutions
(b) Determination of corrosion potential
(c) Rapid chloride penetration test (RCPT)

(a) *Exposure to sulphate, chloride and acid solutions:* Durability tests comprising of exposure to sulphate, chloride and acid solutions were conducted on different types of hardened concrete specimens. The corrosion potential test and rapid chloride penetration test (RCPT) were conducted on BCC1, 40PAB and 100PAC3 specimen groups to study the resistance of normal and pond ash concrete under exposure to sulphate, chloride and acid solutions. 100 mm cubes were immersed in each of the solutions containing 5% magnesium sulphate, 5% sodium chloride and 10% 2N hydrochloric acid. The loss of weight and compressive strength were determined after 28, 56, 270 and 330 days of exposure. The weight loss in the range of 0.3 to 2.68% was negligible for all the specimens in various solutions. However, significant improvement in strength was observed in sulphate and chloride solutions whereas a slight decrease in strength was noticed in acid solution. The results of durability tests are compiled in Table 23.11.

(b) *Determination of corrosion potential:* The corrosion potential tests were conducted as per ASTM (C 876-77) specifications using half cell potentiometer. The tests were carried out on 75 mm by 75 mm by 450 mm prismatic specimens in which one 8 mm diameter steel rod is placed such that the bar is projecting on either side by a minimum length of 25 mm for connecting the leads of the testing apparatus. These specimens were immersed in 5% sodium chloride solution for 28 days of curing. Corrosion potential was measured in millivolts at 28, 56 and 270 days.

Table 23.12 gives the values of increase in corrosion potential after the curing period. Results show that the corrosion potential values for pond ash concrete is less than that for normal concrete.

Table 23.11: Compressive strength of cubes immersed in $MgSO_4$, NaCl and HCl

Mix designation	Normal 28 days strength (N/mm^2)	Age (days)	Compressive strength (N/mm^2) after immersion in		
			$MgSO_4$	NaCl	HCl
		28	52.0	52.6	32.6
BCC1	38.0	56	54.6	57.5	35.0
		270	56.0	54.0	28.0
		28	52.3	51.3	33.0
40PAB	35.3	56	56.0	57.3	37.0
		270	59.0	58.5	29.0
		28	29.5	30.1	30.0
100PAC3	29.3	150	35.0	34.6	33.9
		330	37.4	39.3	34.8

Table 23.12: Corrosion potential and chloride ion penetration test results

Mix designation	Age (days)	Corrosion potential (mV)	Chloride ion penetration (Coulombs)
	28	−262	3072
BCC1	56	−314	1998
	90	−331	1997
	270	−630	−
	28	−260	2052
40PAB	56	−276	1522
	90	−294	929
	270	−597	−

(c) *Rapid chloride penetration test (RCPT):* The rapid chloride penetration test (RCPT) is a measure of permeability of concrete. The tests were carried out as per the specifications of ASTM (C 1202). Cylindrical specimens of diameter 100 mm and height 200 mm were cast and cured for 28 days before starting the tests. From this cylinder samples, 50 mm thickness were obtained by sawing for conducting the RCPT tests. The testing apparatus consists of a number of cells for housing the specimens. These cells consist of two reservoirs on either side between which the specimen can be placed and sealed with a sealant.

The reservoirs in each cell were filled with 0.3 N sodium hydroxide solution and 3% sodium chloride solution on either side with the specimen in between them. The entire assembly was properly sealed with a sealant.

The leads from the two reservoirs were connected to a power supply and current is passed through the cell maintaining a voltage difference of 60V. For a total period of 6 hours, current readings were obtained in milli amperes (mA) at every hourly intervals. Using these current readings and the time elapsed, charge passed through in coulombs was computed which is a measure of chloride ion penetration through the concrete specimen. Based on the electrical charge passed through the specimen, the degree of permeability of concrete can be interpolated as high, moderate, low, very low and negligible using the following values.

RCPT values > 4000   =   High permeability

> 2000 to 4000   =   Moderate permeability

> 1000 to 2000   =   Low permeability

> 100 to 1000   =   Very low permeability

< 100   =   Negligible permeability

The results of RCPT tests conducted on normal and pond ash concrete at different ages are compiled in Table 23.13. The results show that pond ash concrete has a moderate to very low permeability to chloride ion penetration whereas normal concrete shows moderate to low permeability. Concrete in which sand was completely replaced by pond ash showed the least permeability. This trend is attributed to the possibility of better pore filling effect of pond ash due to pozzolanic activity at later ages. Correspondingly, the compressive strength of pond ash concrete was higher at later ages than at 28 days indicating denser concrete.

Table 23.13: Corrosion potential and chloride ion penetration test results

Mix designation	Age (days)	Corrosion potential (mV)	Chloride ion penetration (Coulombs)
BCC1	28	−262	3072
	56	−314	1998
	90	−331	1997
	270	−630	−
40PAB	28	−260	2052
	56	−276	1522
	90	−294	929
	270	−597	−
100PAC3	28	−273	1410
	56	−354	706
	90	−378	624
	270	−492	−

23.6 PROPORTIONING OF POND ASH CONCRETE MIXES

Based on comprehensive experimental investigations, Pranesh[224] has developed concrete guidelines for designing concrete mixes by partial replacement of normal sand by pond ash from thermal power stations. The parameters investigated comprise the following range of materials:

(a) Cement content varying from 300 to 375 kg/m³ at increasing intervals of 25 kg/m³.

(b) Replacement range of pond ash from 0 to 50% at increasing intervals of 10%.

(c) Water/cement ratio range from 0.45 to 0.60 at increasing intervals of 0.05%.

For different combinations of cement content and water/cement ratio, paste volume was calculated and total aggregate content was obtained based on the criterion of absolute volume.

In the total aggregate content, proportion of sand and coarser fractions was determined based on minimum percentage of voids in the mixture. Based on tests using different percentages of coarse and fine aggregates, it was observed that a mixture of 38% sand and 62% coarse aggregate resulted in maximum bulk density of 18.988 kg/³ with minimum percentage of voids of 27.5%. This critical sand fraction was further replaced by 20, 30, 40 and 50% pond ash successively for designing 80 different mix proportions.

Based on workability tests on trial mixes, the dosage of super-plasticizer required was determined in the range of 0.88 to 19.5 l/m³ to achieve a workability of 8 to 9 Vee Bee seconds for the range of cement contents and water/cement ratios mentioned above. Standard strength tests were conducted on hardened concrete at different ages for all the mixes.

23.7 NOMOGRAMS FOR DESIGNING POND ASH CONCRETE MIXES

Monteiro et al[226] have suggested developing nomograms for designing concrete mixes based on the well-established Abram's law (between strength and water/cement ratio), Lyse's Law (between water/cement ratio and aggregate/cement ratio) and Molinari's law (between cement content and aggregate/cement ratio). The nomograms can serve as a handy tool and are very useful for design office use by mix designers. Using the data from his extensive experimental investigations, the author[224] has developed the following types of nomograms for rapid design of concrete mixes with partial replacement of sand by pond ash and also concrete with pond ash as fine aggregate, completely replacing the conventional sand.

(a) Mix design nomogram (pond ash as partial replacement to sand): Figure 23.9 shows the mix design nomogram for designing concrete mixes using pond ash as partial replacement to sand. The nomogram has been developed based on the 28 days compressive strength results by considering the average values of strength v/s water/cement ratio, water/cement ratio v/s aggregate cement ratio and cement content v/s aggregate/cement ratio for replacement of sand by pond ash in the range of 20 to 50%.

The following simple procedure can be used for designing pond ash concrete.

1. For the design compressive strength required at 28 days, select the value of the water/cement ratio from the curve of strength v/s water/cement ratio (Ex: For M-35 grade concrete, read out w/c ratio as 0.52) (Refer dotted lines in Fig.23.9).

2. From the W/C ratio v/s A/C ratio curve, read out the value of A/C ratio required for the concrete mix (Ex: A/C ratio required is 5.5).

3. For this value of A/C ratio, select the cement content required (Ex: Cement content required is 340 kg/m³).

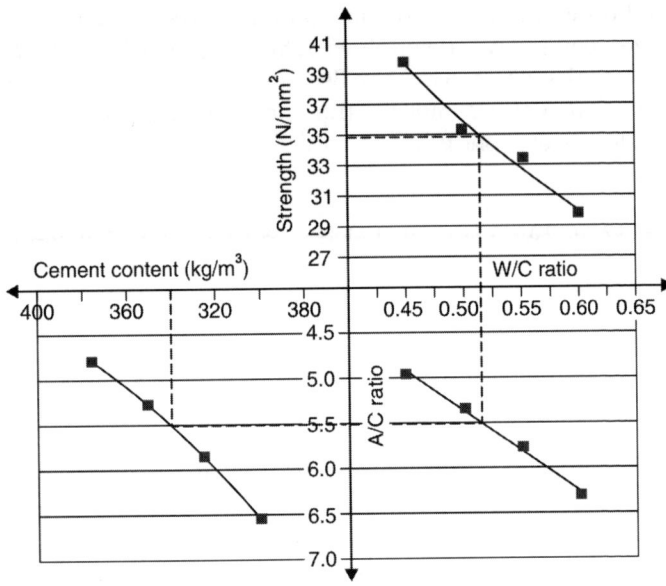

Fig. 23.9: Nomogram for mix design (concrete with pond ash as partial replacement to sand)

4. Determine the percentage of fine and coarse aggregate for minimum percentage of voids. (Ex: 38% F.A. and 62% C.A.)

5. Assuming the level of replacement of sand by pond ash, determine the percentages of sand, pond ash and coarse aggregate.

(Ex: If percentage replacement of sand is assumed as 50%, then the aggregate ingredients are computed as:

Coarse aggregate = 62%
Pond ash (0.5 × 38) = 19%
Sand = 19%

6. Trial mixes are made using the designed mix proportions using the appropriate quantity of superplasticizer to achieve the desired workability.

(b) *Mix design nomogram (pond ash completely replacing sand as fine aggregate)*: Figure 23.10 shows the nomogram for rapid design of concrete mixes with pond ash completely replacing sand as fine aggregate. The procedure to be followed is the same as the various steps followed in case (a).

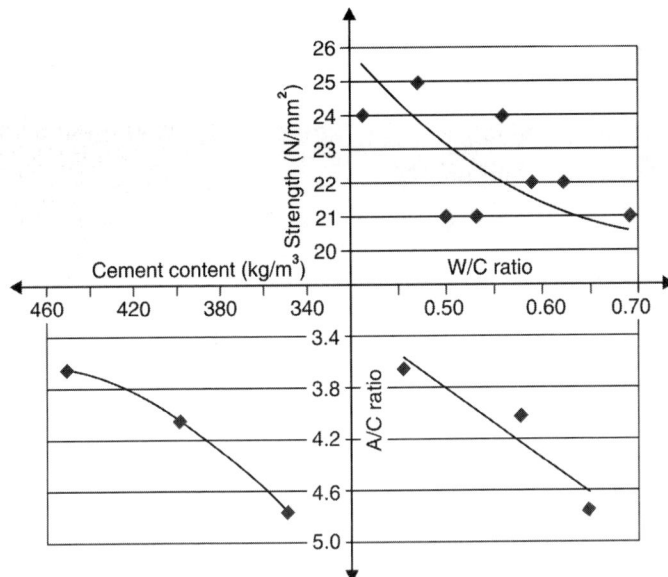

Fig. 23.10: Nomogram for mix design (concrete with pond ash as fine aggregate)

The nomograms illustrated in Figs 23.9 and 23.10 are limited for pond ash having properties similar to that obtained from Raichur Thermal Power Station located in Karnataka state. In addition, they are also limited to the range of strength of concrete covered in the investigation. Partial replacement of sand by pond ash may facilitate the design of concrete of grades in the range of M-30 to M-40, complete replacement may yield concrete of grades in the range of M-20 to M-25 only. However, large scale investigations involving pond ash from different sources are necessary for standardizing and codifying pond ash concrete mix design.

Nanoconcrete

24.1 INTRODUCTION

Nanotechnology dealing with the production and application of physical, chemical and biological systems at scales ranging from a few nanometres to submicron dimensions has found its way into the domain of concrete technology. Nanotechnology was first introduced in the famous lecture of Nobel laureate Richard P. Feynman[227], "There's plenty of Room at the Bottom", delivered in 1959 at the California Institute of Technology. The terminology popularly referred to as Nanotechnology itself was coined by Prof. Nario Tanguchi in 1974. The predictions of Feynman to a large extent have been realized today as we celebrate the Golden Jubilee of nanotechnology on December 29, 2009. Nanotechnology is expected to produce goods and services worth 2.6 trillion dollars in the year 2014 globally. A total of 400,000 research papers and 10000 patents have already come out in this area. Annual research publications worldwide are nearly 60000 in the year 2009. The way we convert food to energy and then to work, very little wastage occurs. If biological machineries were to be as inefficient as our mechanical motors and engines, the food we produce cannot even sustain one-tenth of our population. Thus, biology is nanotechnology in perfection.

According to Konstantin Sobolev[228], nanotechnology has changed and will continue to change our vision, expectations and abilities to control the material world. Significant achievements in this domain comprise the ability to observe the structure of material ingredients at their atomic level and measure the strength and hardness of micro- and nanoscopic phases of composite materials. Among new nanoengineered polymers are highly efficient superplasticizers for concrete and high-strength fibres with exceptional energy absorbing capacity. Nanoparticles, like silicon dioxide, were found to be effective additives to polymers and also to improve the self-compacting capacity, workability and strength of concrete.

Construction industry uses large quantities of Portland cement for infrastructure development throughout the world. Better understanding of the extremely complex structure of cement based materials at the nano-level will apparently result in a new generation of concretes with improved strength and durability. The new types of concretes should not only be sustainable, but also be cost and energy effective and at the same time meet the demands of the modern society. Nanobinders or nanoengineered cement based materials with nanosized cementatious components or other nano-sized particles form the next ground-breaking research domain. At present, the developed countries, like USA, Japan, Germany, USSR and France, are spending billions of dollars per year on nano-technology research funding for the creation of new materials, devices and systems at molecular, nano- and micro-level[229, 230].

Nanotechnology of concrete is set on a path to revolutionize the construction industry by changing the structural properties of concrete to better suit the requirements of structural components. Already several innovative nanoproducts are available in the market which are of immense value in the construction industry. The prominent products comprise novel anti-reflective coatings with increased strength, fire retardant and superior insulation properties at lower maintenance costs. The rapid development in the field of materials science on the nanoscale has opened up a new window of understanding into traditional construction materials like cement and concrete. Nano-cements and concretes with their associated benefits, like overall cost savings and energy consumption coupled with increase in strength and durability, play a significant role in the future of the construction industry and for sustainable development worldwide.

24.2 NANOENGINEERED MATERIALS

Nanotechnology deals with the production and application of physical, chemical and biological systems at scales ranging from a few nanometres to submicron dimensions. Broadly particulate matter less than 100 nm in size is referred to nanomaterial. It also deals with the integration of the resulting nanostructural elements engineered to form larger systems according to Bhushan[231]. Although the science of nanotechnology is new, nanosized devices and objects have existed on this earth as long as life. Biomaterials, like bones or mollusk shells, exhibit exceptional mechanical performance due to the presence of nanocrystals of calcium compounds as reported by Atkinson[232]. Nanomaterials are grouped under the following three broad categories;

(a) *Metallic oxides:* Oxides of metals, such as zinc, iron, titanium, zirconium, cerium and also mixed metal compounds.

(b) *Nanoclay:* Mineral oxides, such as silica, alumina.

(c) *Carbon nanotubes*

Significant achievements in the field of nanoengineered materials include the development of paints and finishing materials with self-cleaning properties, and high scratch and wear resistance. Also nanometre thin coatings have been developed which can protect carbon steel against corrosion and thermal insulation of widow glass.

In the domain of concrete, nanoengineered polymers are highly efficient superplasticizers for concrete and high-strength fibres with exceptional energy absorbing capacity. Nanoparticles, such as silica, are effective additives to polymers and concrete, a development realized in high performance and self-compacting concrete with improved strength and workability. Portland cement being the basic and the largest construction material consumed worldwide, is obviously the product suitable for significant improvements using nanotechnology.

A proper understanding and precise engineering of an extremely complex structure of cement based materials at the nano-level will apparently result in a new generation of concretes which are stronger and more durable with desired stress–strain behaviour and possibly possessing a range of newly introduced and desirable properties, such as electrical conductivity, shrinkage and creep resistance, moisture penetration and stress sensing abilities. At the same time, the new concrete should be sustainable as well as energy and cost-effective meeting the modern societal demands. Nano-binders or nanoengineered cement based materials with a nano-sized cementatious component or other nanosized particles may be the next ground breaking development in the construction industry.

Nanotechnology is currently in its exploration stage and it is emerging from fundamental research to the industrial application. Hence, full scale applications, especially in the construction industry, are limited. However, the tremendous potential of nanotechnology to improve the performance of conventional materials and processes is promising.

24.3 NANOTECHNOLOGY OF CONCRETE

Portland cement is the most widely used material in the construction industry with an estimated production surpassing 6 billion cubic metres per year[233]. The prominent advantages of this material are: Availability of raw materials for production throughout the world, low cost, ease of construction, setting at room temperature and desirable properties. In addition to these advantages, modern day concrete made with cement and aggregates has an excellent performance record of over 180 years. In general, Portland cement is typically used as cementing material with fine and coarse aggregates to create products that are a few mm to several metres thick. The average size of Portland cement particle is in the range of 50 microns. In some applications requiring thinner and stronger products with faster setting time, micro-cement with a maximum particle size of about 5 microns has been used. By reducing the particle size by an order of magnitude, it is possible to obtain nano-Portland Cement.

For decades, major development in the performance of concrete was achieved with applications of super-fine particles of fly ash or silica fume. Nanotechnology has made it possible to introduce nanosilica to improve the properties of concrete. At the micro-level, there is a good analogy between reinforced concrete and fibre reinforced composites. According to Balaguru[234], the lessons learnt from fibre reinforced concrete can also be effectively used for composites made using short discrete fibres, both at micro- and nanolevel. For example, the use of fibre-reinforced concrete containing 0.5% steel fibres is well established in the construction industry to make thin and strong structural elements. The same advantages can be obtained by using 0.5% carbon nanotubes in high performance composites exhibiting superior mechanical and electrical properties[234].

The differences in the particle size distribution and specific area of ingredients in conventional, high-strength/high-performance and nanoengineered concretes are graphically illustrated in Fig. 24.1. The figure clearly indicates the particle size (nm) and specific surface (m^2/kg) of coarse and fine aggregates, Portland cement and fly ash, silica fume and nanosilica.

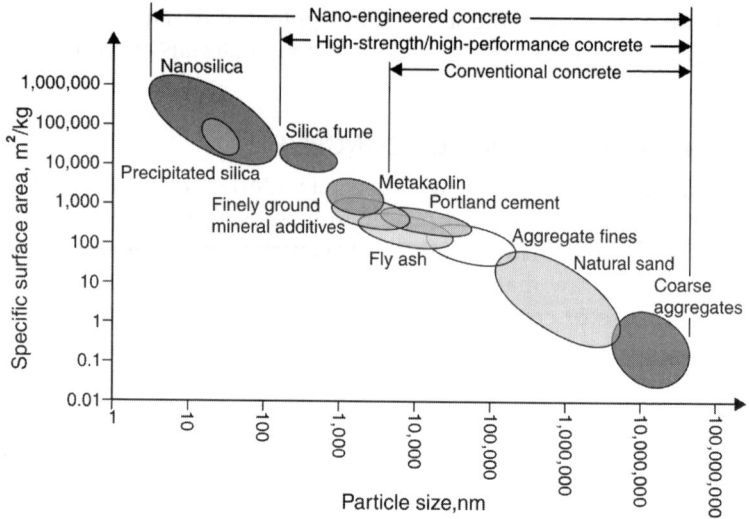

Fig. 24.1: Particle size and specific surface area of different types of concrete materials

In the present state of the art, one can claim that concrete utilizes nanotechnology because it contains nanoparticles as ingredients including nanowater particles and nanoair voids. By using nano-technology, we should be able to control the amount and distribution of the nanoingredients inside the final product to achieve the desired properties. The scalar distribution of the various ingredients of concrete is shown in Fig. 24.2. It is essential to create proper chemical and mechanical tools to control nanoscale pores and the placement of calcium–silicate hydration products to produce concrete which can be termed as a product of nanotechnology. Current research activities concerning nanotechnology in concrete include:

(a) Characterization of cement hydration
(b) Influence of the addition of nanosize silica to concrete
(c) Synthesis of cement using nanoparticles and coatings applied to concrete

24.4 HYDRATION CONCEPTS RELATED TO NORMAL AND NANOCEMENT

Exhaustive data regarding the hydration products of cement are available based on the research supported by the Federal Highway

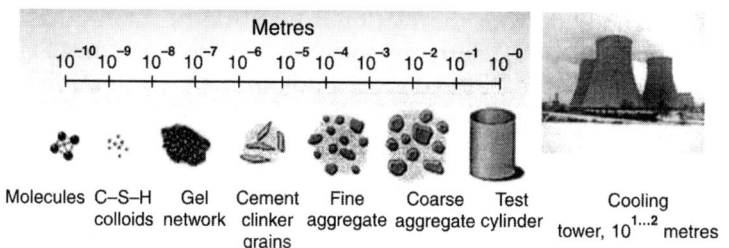

Fig. 24.2: Scales of various constituents of concrete and a typical application

Administration (FHWA) and National Science Foundation (NSF) conducted at the University of Connecticut (USA) [235]. A clear understanding of the hydration process of cement at nanoscale will play an important role in understanding the hydration of nanocement paste. The following facts are noteworthy:

1. One mm^3 of cement occupies about 2 mm^3 of space after complete hydration.
2. The major solid hydration components of cement paste are:
   (a) Calcium silicate hydrate (CSH) which occupies about 50–60 per cent of the volume
   (b) Calcium hydroxide crystals (CH) (20–25 per cent of volume)
   (c) Calcium sulfo-aluminate (CS or etringite) (15–20 per cent of volume)
3. The size of CSH sheets is less than 2 nm and space between sheets = 0.5 to 2.5 nm.
4. Surface area of CSH is large and is of the order of 100 to 700 m^2/g and its strength is attributed to Van der waals forces.
5. CH products are usually large with a width of about 1000 nm.
6. Sizes of capillary voids range from 10 to 1000 nm.
7. Active hydration zone is about 2000 nm thick.
8. For nanocements, through-solution hydration is more suitable since in this mechanism, complete dissolution of anhydrous compounds to their ionic constituents and eventual precipitation of hydrates are possible.
9. Aluminates hydrate much faster than silicates, which make up about 75% of cement with a major role in strength development.

Conventional analytical methods do not provide an accurate model for the rate of cement's reaction with water as a function of temperature, water/cement ratio and grain size because the hydration reactions occur in the nanoscale pores of the cement gel.

Hence scientists from the National Institute for Standards and Technology's (NIST) centre for cold Neutron Research and FHWA are using neutron scattering methods to measure motions and reactions of water at a nanoscale. This study is expected to explain the effects of various factors on the rate of development of hydration products of cement's fractal nanoscale structure according to Balaguru et al[236].

24.5 IMPACT ON CONSTRUCTION MATERIALS

Nanochemistry has revealed the possibilities of new products that can be effectively applied in concrete technology. A simple example being the development of new superplasticizers for improving certain desirable properties of concrete. These nanopolymer based admixtures are ideally suited for the production of rheodynamic concrete[237] generally referred to as self-compacting concrete (SCC). This type of concrete flows under its own weight and completely fills the form work, even in the presence of dense reinforcements without the need for any vibration whilst maintaining homogeneity and resulting in concrete of high early strength and durability. DEGUSSA-MBT Construction Chemicals (India)[238] have developed certain revolutionary type of admixtures using nanopolymer technology which are aimed at targeted performances in concrete. Based on nanoscience, they have developed a range of polymers with the following practical applications:

(a) *Zero energy system:* A system of nanopolymers with longer side chains and shorter main chains to facilitate high early strength in concrete without steam curing and with specific applications in precast reinforced or prestressed concrete structural unit manufacturing industry.

(b) *Glenium sky:* AQ custom made nanopolymer which facilitates long haul concrete mix stability with development of high early strength coupled with high durability of hardened concrete suitable for prestressed concrete flyovers.

(c) *Rheo fit:* A nanopolymer range which meets wider expectations as aesthetics, economics, durability and performance of manufactures concrete products.

The application of nanotechnology in the production of nanopolymers has revolutionized the concrete industry. Self-compacting concrete is commonly used in Europe, Japan, North America, Singapore, Dubai and Taiwan. In India, this technique has gained popularity in some of the major construction projects, such as Delhi Metro Rail Corporation, Nuclear Power Corporation and Indian Space Research Organization.

24.6 NANOTUBES

The most popular nanotubes are carbon nanotubes, discovered by the Japanese Scientist Sumio Iijima in 1991[239]. Arc evaporation techniques have facilitated the production of bulk quantities of nanotubes. Smalley's Group in Texas (USA) developed a method that resulted in high yield of single-walled tubes with unusually uniform diameters having a tendency to form aligned bundles or nanotubes often referred to as nanoropes. Other forms of carbon nanotubes are nanohorns, nano-test tubes and nanofibres possessing very high strength, stiffness, aspect ratio and purity.

Multi-walled nanotubes were first developed by Srivatsava et al[240]. Transmission electron microscopy studies indicate that these tubes appear like nested shells with an interlayer spacing of about 0.34 nm. The equivalent diameter of the tubes is in the range of 10 to 50 nm with typical lengths varying from 100 to 1000 nm. Single layer tubes have a diameter of 1 to 3 nm and a length of 300 nm. Typical stress–strain plots reported by two investigators are shown in Fig. 24.3. The modulus of elasticity varies from 300 to 5000 GPa. In tension mode, the strain at failure is as high as 12% with a tensile strength in the range of 10 to 63 GPa[241].

Carbon nanotubes have excellent potential for use in fuel cells and medical fields. These tubes can be filled with materials including biological molecules. Nanofibres could become the ultimate carbon fibres due to their basic properties, like high strength, stiffness and aspect ratio.

24.7 NANOCOMPOSITES

Nanotubes have been widely used in the fabrication of composite materials inheriting the basic properties of the nanotubes. Mixed with alumino-silicates, carbon nanotubes produce very thin wafers possessing high strength and conductivity. The composite material can also be used as a tough, durable high temperature and low friction coating. At present, alumino-silicate formulations consist of silica particles in the range of 50 to 100 nm[234]. With refined processing, it is possible to reduce the maximum particle size in the matrix to 5 or 10 nm. These matrices reinforced with as low as 0.5%

of nanotubes exhibit extraordinary strength and electrical conductivity. Research investigations conducted by Kashiwagi et al[242] at the university of Pennsylvania have shown that it is possible to build nanoclay-filled polymers possessing excellent mechanical properties and fire resistance. Metal oxide nanoparticles have been used in coatings for protection of ultraviolet light, solar cells, indoor air cleaners and as self-disinfecting surfaces.

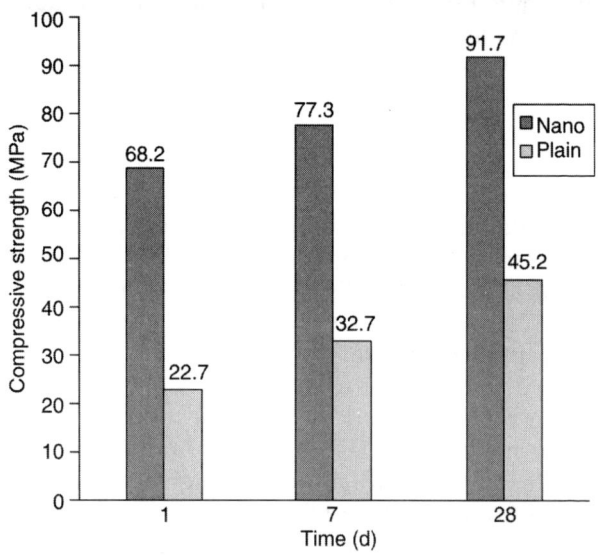

Fig. 24.3: Compressive strength of plain and nanoconcrete

24.8 NANOSENSORS FOR CONCRETE STRUCTURES

Research investigations have been conducted by Song[243] to examine the feasibility of Cyberliths, or smart aggregates embedded in concrete to function as wireless sensors. In future, these micro-sensors might be reduced to dust particle size, with the ability to coat an entire bridge with smart dust for optimum capabilities through a sensor net. These sensors can be used to remotely monitor the condition of the concrete and reinforcement without damaging the structure. Previously strain gauges embedded on concrete and reinforcement at desired locations in the structure helped to study the variations in strain due to various loading conditions using a strain recorder. In future, the load-deformation characteristics can be studied using the microsensors.

24.9 THE CHEMISTRY AND PHYSICS OF NANOCEMENT

Experimental investigations conducted by Maile Aiu and C.P. Huang[244] to determine the advantages of using nanoparticles in building materials, like cement and concrete, have revealed significant improvements in the structural properties of the material, such as strength and durability by the mechanism of mass transfer in nanoscale[245]. Additional investigations by Gengying li[246] have also concluded nanosilica added to fly ash high strength concrete could improve the short- and long-term strength of the concrete.

Nanoparticles are unique because their size affects the behaviour of cement. M.P. Ginebra et al[247] reported that particle size can significantly affect the kinetics of cement pastes. Their findings indicate that a reduction in particle size can lead to a more rapid setting and hardening of cement due to stronger electrostatic attractive forces as a consequence of greater specific surface of the particles. Other studies have shown that the compressive and flexural strengths of cement mortars enhanced with SiO_2 and Fe_2O_3 nanoparticles due to higher modified montmorillonites in comparison with plain cement mortar[248, 249].

Recent studies have shown that nanoparticles used as additives to cement mortar can improve its mechanical properties. However, the behaviour of cements composed exclusively of nanoparticles has not yet been reported. The reader may refer to the experimental studies of Maile Aiu and Huang[244] for detailed information on the structure, morphology, hydration rate, X-ray studies and compressive strength of the cement with and without the addition of nanoparticles.

24.10 NANOCEMENTS

Nanocements are particularly suitable as coatings and repair materials. TX Active™ is a quality label developed by Heideberg cement and Italicementi which shows the durability and photo catalytic functionality of the finished product. This is a self-cleaning cement due to its special formula which is efficient in destroying atmospheric pollutants. In Milan, a commercial building's surface of 3000 m^2 coated with this cement is functioning satisfactorily.

Iglesia Dives (Church) in Misericordia is a well-known building where TX Active cement has been used. In addition to its self-cleaning property, TX Active cement plays a role in reducing pollution. This product has also been used in the construction of two white concrete gateway elements in the 1-35W bridge over the

Mississippi River in Minneapolis (USA). EMACO™ nanocement product from BASF is a concrete repair material with exceptional properties, such as improved bond strength, density and impermeability, reduced shrinkage and cracking tendency coupled with improved tensile strength. It also provides improved compatibility with concrete.

This product has found application in the renovation of office buildings in Brussels, in the waste water plant in France and in the renovation of a bridge structure and cooling towers in Spain.

Chronolia™ developed by Lafarge Company is a quick setting cement used in ready mix concrete made possible by nanotechnology and the understanding of crystalline growth.

Agilia™ developed by Lafarge is considered as the world's first self-compacting and self-leveling concrete. It generated 2.4% of sales volumes and 12% of Lafrage concrete business in 2006. One million cubic metres of Agilia concrete were sold in the year 2006.

Ductal™ also developed by Lafarge is one of the first commercial concretes where steel bars are not used. It exhibits high mechanical strength, durability and self-healing properties. Nanoparticles of titanium oxide (TiO_2), aluminium oxide (Al_2O_3) or zinc oxide (ZnO) are applied as final coating on construction ceramics to bring the following characteristics to the surfaces:

1. TiO_2 has been used for its ability to breakdown the dirt or pollutants when exposed to ultraviolet light and then allowed it to be washed off by rain water on surfaces, like tiles, glass and sanitary ware.
2. ZnO finds application in both coatings and paints used for ultraviolet resistance.
3. Nanosized Al_2O_3 particles are used to make surfaces scratch-resistant and to prevent and decelerate the formation of bad smells, fungus and mould.

Cognoscible technologies estimate that between 5 and 10% of CO_2 emissions in the world are produced by the manufacturing process of cements. At present cognoscible technology is engaged in research activities to achieve three different generations of nanocements:

1. First generation cements with environment friendly nano-particles offering high strength and durability.
2. Second generation cements which reduce CO_2 emissions by 50% of the present values.
3. Third generation liquid nanocements.

The second and third generation cements are expected to achieve 100% of Kyoto Protocol objectives, i.e. to reduce CO_2 emissions to less than 5%. Cognoscible Technologies head quartered at Barcelona (Spain) is working on developing the first generation nanocements through an alliance with Grupo Suner. The news posted by Nano bugle on March 18, 2009 lists the advantages of the nanocement as lower cost of production, higher strength, reduction in CO_2 emissions and with incorporation of Pico catalyst to emit oxygen in place of carbon dioxide.

24.11 NANOADDITIVES TO CONCRETE

Nanotechnology has paved the way for the development of several nanoadditives suitable to enhance the structural properties of concrete. Also nanoadditives of different types have been developed over the years to suit the structural and environmental conditions. Bentz et al[250] have reported the development and use of viscosity enhancing nanoadditives capable of doubling the service life of concrete.

Engineers at the National Institute of Standards and Technology (NIST) have patented a method which is expected to double the service life of concrete in a structure. The key according to the report is a nanosized additive that slows down the penetration of chloride and sulphate ions from road salt, sea water and soils into the concrete. A reduction in ion transport translates to reduction in both maintenance costs and the catastrophic failure of concrete structures. The new technology could save billions of dollars and many lives. The American Society of Civil Engineers (ASCE) estimates that the country spends 54 billion dollars each year to repair the damages caused by poor road conditions.

Infiltrating chloride and sulphate ions during winter cause internal structural damage over a period leading to cracks and weakened concrete. Investigations by Bentz et al have conclusively proved that additive molecules less than 100 nm size increases the viscosity of the concrete mix and also effective in impeding diffusion of ions.

The NIST researchers have also demonstrated that the nano-additives can be blended directly into the concrete with the current chemical admixtures. It has been found that even better performance is achieved when the additives are mixed into the concrete by saturating absorbent lightweight sand. The technology is now available for licensing from the US Government through the NIST office.

24.12 NANOCONCRETE PRODUCTION AND PROPERTIES

1. Gaia Concrete

Gaia nanosilica is the first nanoadditive for concrete developed by Rouzbeh Shahsavari, a doctoral student from MIT School of Engineering. This product has opened the doors for the development of different types of nanoadditives replacing the micro-silica largely used to produce silica fume concrete during the last decades of the 20th century.

The technology was extended by Ulmen S.A. in association with Ferrada et al[251] of Cognoscible Technologies to commercially produce Ga nanosilica for use at ready mixed concrete facilities certified by Environmental Management Systems (ISO14001). Roclano Technologies has obtained the rights for marketing in Spain and Portugal. This product is available in liquid form which facilitates the uniform distribution of nanosilica particles in concrete. According to Ferrada et al, the concrete mixtures with Gaia exhibit perfect workability without segregation or bleeding. Gaia combines the effects of water reduction coupled with increase in slump. This makes the design of self-compacting concrete an extremely easy task.

Table 24.1 shows the salient properties of concrete with and without the addition of Gaia both in the plastic and hardened states.

Table 24.1: Properties of concrete with Gaia nanoadditive

Concrete mix parameters	Plain concrete	Gaia concrete
Cement type (EN-197)	II/A-P 42.5R	II/A-P 42.5R
Cement content (kg/m^3)	460	460
Dosage of additive (%)	–	1.3
Air content (%)	2.7	1.1
Slump (mm)		
After 5 min	60	200
After 30 min	25	210
After 60 min	15	160
After 90 min	–	140
Ambient temperature (°C)	20	20
Compressive strength (N/mm^2)		
At 1 day	22.7	68.2
At 7 days	32.7	77.3
At 28 days	45.2	91.7

The development of strength in plain and Gaia concrete is illustrated in Fig. 24.3.

The application of Gaia at a dosage of 1.3% (by weight of dry cement) results almost twofold increase in the compressive strength at 7 and 28 days.

The early strength of concrete with Gaia is 68.2 N/mm^2 which is nearly three times that of the control concrete. The 28 days compressive strength of concrete made with Gaia demonstrates a classical dependence on water/cement ratio.

As shown in Fig. 24.2, the experimental data has clearly demonstrated the positive action of nanoparticles on the microstructure and properties of cement-based materials. This phenomena can be explained by the following salient factors[252-55]:

1. Nanoparticles well dispersed in concrete mixes increase the viscosity of the liquid phase which helps to suspend the cement grains and aggregates, which in turn improves the segregation resistance and workability of the concrete mix.

2. Nanoparticles fill the voids between the cement grains which results in the immobilization of free water also termed as filler effect.

3. Well-dispersed nanoparticles act as centres of crystallization of cement hydrates which accelerates the hydration process.

4. Nanoparticles favour the formation of small sizes crystals, such as calcium hydroxide and small sized uniform clusters of calcium silicate hydrates (C-S-H).

5. Nanosilica participates in the pozzolanic reactions which results in the consumption of calcium hydroxide and formation of additional dicalcium and tricalcium silicates.

6. Nanoparticles improve the structure of the aggregate contact zone, resulting in a better bond cement paste and aggregates.

7. Crack arrest and interlocking effects between the slip planes provided by nanoparticles improve the toughness, shear, tensile strength and flexural strength of cement based materials.

2. Cuore Concrete

Microsilica has been one of the world's most widely used additives in concrete for over 80 years. Its properties allow high compressive strength, durability and impermeability and they have been part of many important concrete structures. Its main disadvantage being its relatively high cost and contamination which adversely affects the environment and the health of the construction workers.

Microsilica, as a powder, is one-thousandth-fold thinner than cigarette smoke. Hence operators must take special precautions to avoid inhaling microsilica to prevent silicosis, a deadly disease of the lungs. In the middle of 2003, a product which could replace microsilica was developed having better characteristics incurring a lower cost and also fulfilling environmental regulations of ISO-14001. Using tools of physics, chemistry and recent nano-technology, a revolutionary product Cuore nanosilica was developed which had superior advantages in comparison with microsilica. A litre bottle of nanosilica was equivalent to a barrel full of microsilica, extra cement and superplasticizing admixtures.

The Cuore nanosilica product was tested in concrete used for over a year in the world's largest subterranean copper mines to prove its long-term characteristics. Cuore concrete takes care of the environment and also the operator's health. The new product surpassed the expectations of its design and gave concrete not only the high initial and final strength but also increased plasticity of the wet concrete and impermeability to the hardened concrete with cost and cement savings of up to 40%. Also it lowered the levels of environmental contamination.

The nanoparticles of silica turn into nanoparticles of cement (nano-cement) in the chemical reaction that takes place in the concoction of concrete. The advent of Cuore nanosilica concrete in the market has revolutionized the concept of concrete products and their applications in the construction industry.

The salient properties of concrete with Cuore nanosilica (civil engineering portal) are as follows:

1. Cuore concrete is 88% more efficient than microsilica added to concrete and superplasticizers. It exhibits high compressive strength (H-70).
2. The product cost is drastically lower than using traditional production method or formulas.
3. The concrete has an air content of 0 to 1%.
4. The cone test on concrete shows that it preserves the cone shape for more than an hour (with a water/cement ratio of 0.5. Adding 0.5% of nanosilica of the metric volume of the cement used, it conserves the circular shape of 60 cm for two hours with a loss of only 5%). The nanosilica concrete has a plasticity

comparable to that of polycorbilate technology. Hence the use of superplasticizing additive is not necessary.

5. High workability with reduced water/concrete ratio (For Ex. 0.2).

6. Easy homogenization of concrete and the reduction of mixing time allows concrete plants to increase the production.

7. Depending upon the cement and the formulations used for concrete (Tests for strength values in the range of H-30 to H-70) shows that the gain of strength with days are as follows:
   (a) 15 N/mm^2 to 75 N/mm^2 at 1 day
   (b) 45 N/mm^2 to 90 N/mm^2 at 28 days
   (c) 48 N/mm^2 to 120 N/mm^2 at 120 days

8. Cuore nanosilica complies with environmental and health regulations prescribed in ISO-14001. It also protects operators from the danger of being contaminated with silicosis and does not spoil the environment.

9. After successfully passing all the tests, the material is used commercially in different parts of the world.

The following immediate benefits in using the Cuore nanosilica concrete to the user have been identified and listed below as reported by Pascal Maes[256].

1. Cessation of environmental contamination caused by micro-silica solid particles

2. Lower cost of material at site

3. Concrete with higher initial and final compressive and tensile strengths

4. Concrete with good workability

5. Cessation of superplasticizers utilization

6. Cessation of silicosis disease risk to operators

7. High impermeability

8. Reduction in cement content using Cuore nanosilica

9. Cuore nanosilica by itself produces nanocement

10. Compressive strengths of 70 to 100 N/mm^2 have been reported at 28 days

11. Low quantity of nanosilica additive required (1 to 1.5% by weight of cement)

12. Self-compacting characteristics with higher proportions of up to 2.5% nanosilica. Equal or minor raw material cost as in traditional concrete products with superplasticizers and or fibres.

3. High Performance Concrete Using Silica Fume, Fly Ash and Nanosilica

Collepardi et al[257] have investigated the properties of high-performance concretes made by using optimum proportions of silica fume, fly ash and nanosilica to manufacture concrete structural elements suitable for production in precast industry involving steam curing. Among the pozzolanic materials, silica fume appears to be the best performing silicious product for high-performance concretes (HPC). Its behaviour is related to the high content (> 90%) of amorphous silica in the form spherical grains in the range of 0.01 to 1μm. However, silica fume is not available in large amounts and it is also the most expensive mineral additive costing about 0.25 to 0.50 Euros/kg in Europe.

Fly ash is available in large quantities at cheaper rate of 0.02 to 0.03 Euros/kg in Europe and free of cost at Indian thermal stations. However, the performance of fly ash is lower than that of silica fume due to the lower amount of amorphous silica (35 to 40%) and the larger size (0.1 to 04 μm) of its spherical grains. However, a new pozzolanic material developed by Skarp and Sarkar[258] and Collepardi et al[259] and produced synthetically in the form of water emulsion of ultra-fine amorphous colloidal silica (UFACS) is available in the market. The nanosilica is potentially better than silica fume due to its higher content of amorphous silica (> 99%) and the reduced size of its spherical particles (1 to 50 nm). Presently, the price of this nanosilica in Europe is around 0.45 to 0.90 Euros/kg of water emulsion depending on the dry content.

The salient features of the concrete produced using the combination of the ternary system "silica fume — fly ash — nanosilica" are outlined below:

The chemical composition and physical properties of the blended Portland cement and the pozzolanic additives, such as fly ash, silica fume and nanosilica in the form of water emulsion containing 25% colloidal silica are compiled in Table 24.2. The X-ray diffraction patterns and the electron micrographs of the additives are illustrated in Figs 24.4 and 24.5, respectively. The SEM micrograph of the silica fume indicates that the material is of a densified type with some agglomeration of individual grains as usually available in the market suitable for easy transportation and storage. The micrographs of silica fume and fly ash are shown in micrometers while that of nano-silica is shown in nanometers.

Table 24.2: Chemical composition and physical properties

Composition	Cement	Silica fume	Fly ash	Nanosilica (UFACS)
SiO$_2$ (%)	16.7	98.2	60.1	99.1
Al$_2$O$_3$ (%)	3.5	–	22.8	–
Fe$_2$O$_3$ (%)	3.5	0.3	4.7	–
CaO (%)	63.0	0.2	4.6	–
MgO (%)	0.9	–	1.0	–
K$_2$O (%)	0.4	–	2.1	–
Na$_2$O (%)	0.1	–	0.6	–
SO$_3$ (%)	2.5	0.2	0.4	–
CO$_2$ (%)	8.8	–	–	–
Blaine fineness (m^2/g)	0.42	18	0.36	–
Mean size (μm)	15	2	20	0.05

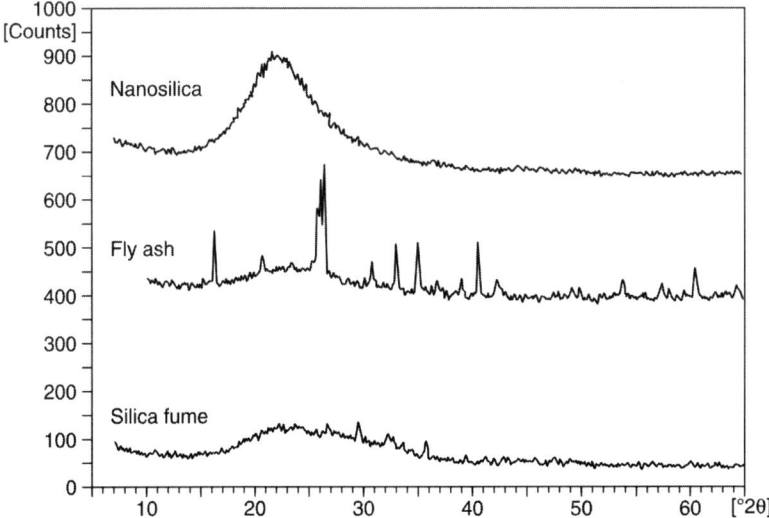

Fig. 24.4: XRD patterns of silica fume, fly ash and nanosilica

Two different types of concretes with properties, such as: (a) Flowing and (b) self-compacting were produced with different combinations of the basic pozzolanic ingredients together with cement, aggregates and superplasticizer as compiled in Tables 24.3 and 24.4, respectively. The mixes designated as SF and FA contained

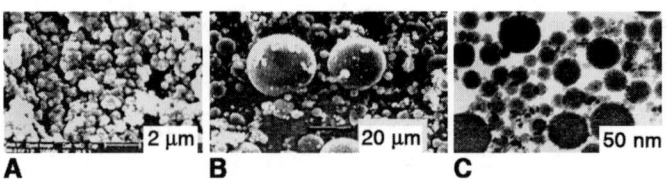

A   **B**   **C**

Fig. 24.5: Electron micrographs showing (A) Silica fume, (B) Fly ash, and (C) Nanosilica

Table 24.3: Flowing concrete mix ingredients

Mix no.	Cement content (kg/m³)	Silica fume (kg/m³)	Fly ash (kg/m³)	Nano-silica (kg/m³)	Aggregate (kg/m³)	Water (kg/m³)	Super-plasti-cizer (%)	W/C ratio	Slump (mm)
SF	396	59	0	0	1800	174	0.87	0.44	230
FA	396	0	59	0	1800	175	0.61	0.44	270
TC1	393	21	29	7.8	1790	173	0.60	0.44	230
TC2	395	15	40	5.1	1800	174	0.55	0.44	230

Table 24.4: Self-compacting concrete (SCC) mix ingredients

Mix no.	Cement content (kg/m³)	Silica fume (kg/m³)	Fly ash (kg/m³)	Nano-silica (kg/m³)	Aggregate (kg/m³)	Water (kg/m³)	Super-plasti-cizer (%)	W/C ratio	Slump flow (mm)
SF	423	60	0	0	1775	186	1.30	0.44	730
FA	424	0	61	0	1780	186	1.20	0.44	740
TC1	425	21	30	7.4	1785	187	1.18	0.44	730
TC2	425	15	40	4.9	1780	187	1.10	0.44	750

only silica fume and fly ash respectively while the mixes TC1 and TC2 contained all the three pozzolanic ingredients. A constant water/cement ratio of 0.44 was maintained in all the concrete mixes. The nanosilica (UFACS) used was in the form of aqueous emulsion having 25% dry content.

The flowing concrete showed a slump of 230 mm while the self-compacting concrete with higher cement content had a slump flow of nearly 730 mm. In both the mixes, the W/C ratio was maintained at 0.44 and the same quantities of aggregates were used. However, the amount of mixing water and superplasticizer was slightly higher in the self-compacting concrete mix. The concrete test specimens were cured at 20°C or steam-cured at 65°C for a period of up to 90 days. The development of compressive strength of flowing and self- compacting concretes with age is shown in Figs 24.6 and 24.7.

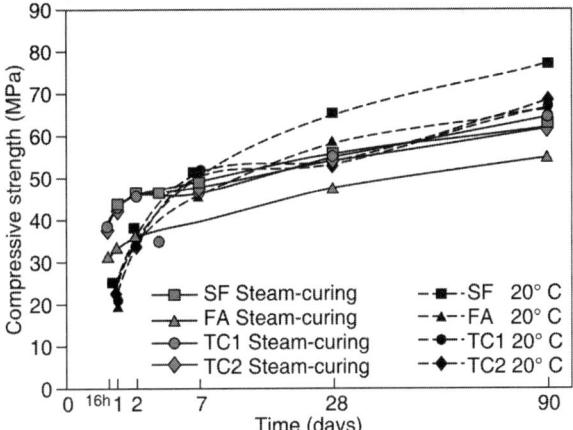

Fig. 24.6: Compressive strength of flowing concretes

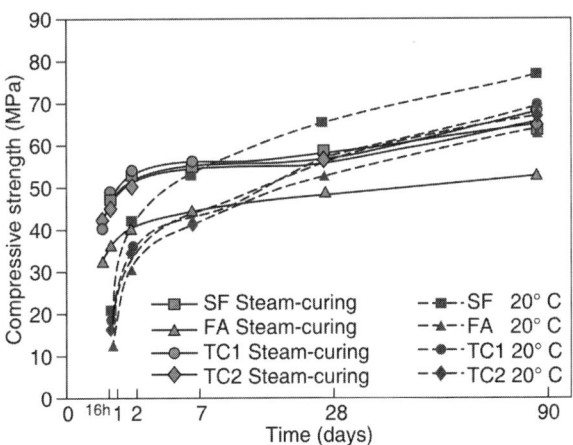

Fig. 24.7: Compressive strength of self-compacting concretes

The results indicate that the strength development of silica fume concrete is better than the fly ash concrete in both flowing and self-compacting mixes regardless of the curing temperature.

However, the difference in the strength development between the silica fume mix (SF) and the ternary combination (TC1 and TC2) appears to be significant only for concretes cured at 20°C, especially at longer ages. In the case of steam-cured concrete, no differences were observed in both types of mixes.

The results of durability[260] studies conducted on test specimens to investigate the penetration of chloride ions and CO_2 is shown in Figs 24.8 and 24.9 for flowing concrete and in Figs 24.9 to 24.11 for self-compacting concrete mixes. In both flowing and self-compacting concretes, the chloride diffusion and carbon dioxide penetration are faster in concretes with fly ash (FA) than in concrete with silica fume (SF) irrespective of the curing temperature. The behaviour of concretes with ternary combinations of silica fume, fly ash and nano-silica (TC1 and (TC2) was very close to that of the corresponding concrete with silica fume (SF) alone.

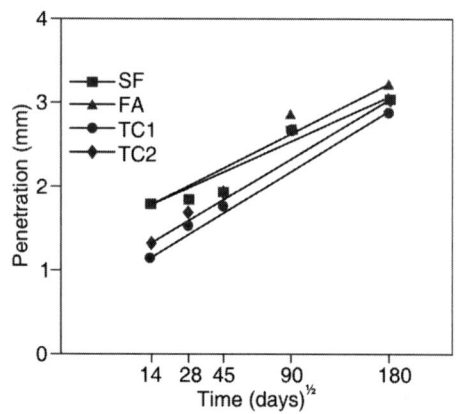

Fig. 24.8: Penetration of chloride ions in flowing concretes

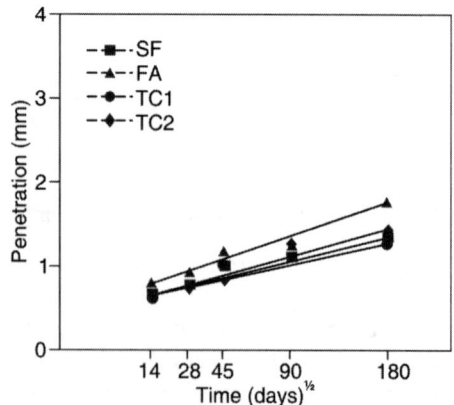

Fig. 24.9: Penetration of carbon dioxide ions in flowing concrete

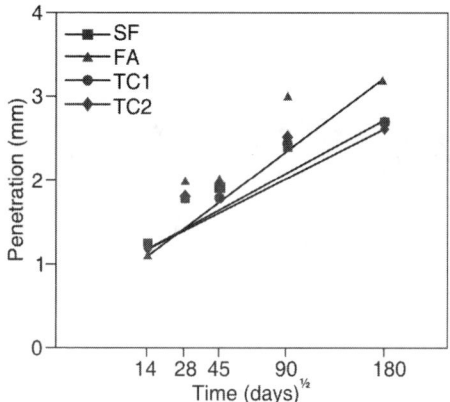

Fig. 24.10: Penetration of chloride ions in self-compacting concretes

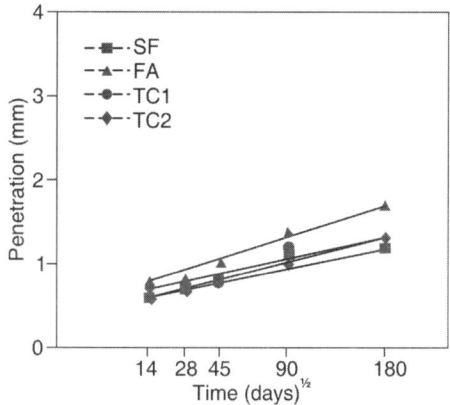

Fig. 24.11: Penetration of carbon dioxide in self-compacting concretes

Based on these investigations, the authors have concluded that use of nanosilica will reduce the quantity of costlier silica fume in concrete mixes. The results of this work also indicate that for steam-cured high-performance concrete (HPC), which is normally used in precast industry, the combined use of silica fume, fly ash and nano-silica will be economical without loss of compressive strength at longer ages.

High-Performance Concrete

25.1 INTRODUCTION

In recent years, the terminology "high-performance concrete" is being increasingly used in the concrete industry and literature dealing with concrete in journals and articles. Concrete is the most widely used building material and any improvement in the design of this material, for example, strength, durability, flowability and cost ripples through the economy. Long-term performance has become vital to the economies of all nations. Concrete has been the major material for providing stable and reliable infrastructure since the days of Greek and Roman Civilizations. At the turn of the 20th century, concrete compressive strength was in the range of 10 to 12 N/mm². However, by 1960, it was in the range of 25 to 40 N/mm². Deterioration, long-term poor performance and inadequate resistance to aggressive environment, coupled with greater demands for more sophisticated architectural form, led to the accelerated research into the microstructure of cements and concretes and more elaborate codes and standards.

According to Henry G. Russel[261], the most significant development in the making of high-performance concrete (HPC) is the virtual elimination of voids in the concrete matrix, which are mainly the cause of most of the ills that generate deterioration.

The American Concrete Institute (ACI)[262] defines high-performance concrete as concrete meeting special combinations of performance and uniformity requirements that cannot always be achieved routinely when using conventional constituents and normal mixing, placing and curing practices. A commentary to the definition states that a high-performance concrete is one in which certain characteristics are developed for a particular application and environment. Examples of characteristics that may be considered critical for an application are grouped as:

1. Ease of placement
2. Compaction without segregation
3. Early age strength
4. Long-term mechanical properties
5. Permeability
6. Density
7. Heat of hydration
8. Toughness
9. Volume stability
10. Long life in severe environments

At present, high-performance concrete is generally replacing the conventional concrete in major construction projects according to Shah[263] and Jensen[264]. The new type of concrete has fast turned from laboratory investigation to practical utilization and became a part of the construction market. Based on recent developments in concrete technology, high-performance concrete is characterized by superior primary properties, such as workability, strength, durability and serviceability. These advantages have resulted in large scale cost savings in various unique projects.

The modern high-performance concrete technology is based on the concept of "densified system with ultra-fine particles" (DSP) which includes the effective use of silica fume (SF) and super-plasticizer (SP) combination to modify the cement system at very low water/cement ratio[265]. The use of silica fume fills space between cement particles and aggregate transition zone, as well as providing a high rate of pozzolanic reaction by forming low compacted calcium silicate hydrates. In essence, high-performance concretes (HPC) can be considered as a logical development of cement concretes in which the ingredients are proportioned and selected to contribute efficiency to the various properties of cement concrete in fresh and hardened states.

25.2 FACTORS INFLUENCING HIGH-PERFORMANCE CONCRETE MIXES

Many characteristics of high-performance concrete are interrelated and hence a change in one may result in changes in one or more of the many characteristics. Consequently, if several characteristics have to be taken into account in producing a concrete for a specific work, each must be clearly specified in the contract documents. A high-performance concrete is something more than is achieved on a routine basis and involves a specification that often requires the concrete to meet several criteria. A specific example being the Lacey V. Murrow Floating Bridge in Washington State (USA), the concrete was specified to meet the specified criteria of compressive strength, shrinkage and permeability. The latter two requirements controlled the mix proportions so that the actual compressive strength was well exceeding the specified strength. This occurred because of the interrelation between the three characteristics.

In situations where concrete sections are reinforced with the maximum permissible percentage of steel bars and also at junctions of beams and columns where reinforcements are lapped together for continuity, flowability of concrete is paramount importance to achieve dense concrete with minimum compacting effort. In such situations, workability criteria are more important than strength. In Marine structures, the primary criterion to be met is the durability of concrete. The concrete should withstand the harmful effects of sea water attack containing sulphates and chlorides and strength is of secondary importance. In workshop floors, the surface strength of concrete and resistance against wear and tear are the primary requirements.

It is pertinent to note that a high-strength concrete is always a high-performance concrete, but a high-performance concrete is not always a high-strength concrete. ACI defines a high-strength concrete as concrete that has a specified compressive design strength exceeding 40 N/mm^2. Some of the European countries specify a higher value of 48 N/mm^2 to be classified as high-strength concrete. The specification of high-strength concrete generally results in a true performance specification, essential for the intended application and the performance can be measured using well-established test procedures, like Schmidt rebound hammer test[266], Ultrasonic pulse velocity test[267] or compressive strength test on a cylindrical core of concrete cut out from the actual structure.

Durability is another prominent requirement specified for concrete used in aggressive environmental conditions, like marine structures, industrial structures manufacturing chemicals, wear-resistant floors of heavy duty machine shops. In such structures, concrete that is used must have a specified minimum strength in addition to other characteristics specified to ensure durability. In the past, durable concrete was obtained by specifying minimum cement content, air content and maximum water/cement ratio. In the present context, performance characteristics may include permeability (water retaining structures), deicer scaling resistance, freeze thaw resistance (highway structures), abrasion resistance (workshop floors) or any combination of these characteristics. Normally, the desired durability characteristics are more difficult to define than strength criteria and hence specifications often use a combination of performance and prescriptive requirements, such as permeability and a maximum water/cement ratio to achieve a durable concrete. The resulting concrete may have a higher strength but this only comes as by-product of requiring a durable concrete.

Most of the high performance concretes produced today contain materials in addition to Portland cement mainly to achieve the compressive strength or durability performance.

The materials generally used include fly ash, silica fume and ground granulated blast furnace slag used separately or in combination. Additionally, chemical admixtures, such as high-range water reducers, are required to ensure easy transportability of concrete. For high-strength concretes, a combination of mineral and chemical admixtures is nearly always essential to ensure achievement of specified strength.

Generally, high-performance concretes have a high cementatious content and a water/cementious material ratio not exceeding a value of 0.40. However, the proportions of the individual constituents vary depending upon local preferences and materials. Mix proportions developed in one part of the country may not necessarily work in a different location. Many trial batches are usually necessary before a successful mix conforming to the specifications is developed. High-performance concretes are also more sensitive to changes in the constituent material properties than conventional concretes. Hence the physical and chemical properties of the cementatious materials and chemical admixtures should be carefully monitored. It is important to note that a greater degree of quality control is necessary for the successful production of high-performance concrete.

25.3 SALIENT FEATURES OF HIGH-PERFORMANCE CONCRETE

High-performance concrete should satisfy certain desirable properties both in the wet and hardened states. Sai Prasad and Kamalesh

Jha[268] have identified the salient features of high-performance concrete and listed them as given below:
 1. Characteristic compressive strength exceeding 80 N/mm²
 2. Water/binder ratio in the range of 0.25 to 0.40
 3. Reduced flocculation of cement grains
 4. Wide range of grain sizes
 5. Densified cement paste
 6. Non-bleeding homogeneous mix
 7. Less capillary porosity
 8. Discontinuous pores
 9. Low free lime content
10. Stronger transition zone at the interface between cement paste and aggregates
11. Endogenous shrinkage
12. Powerful confinement of aggregates
13. Little micro-cracking until about 65–70% of characteristic compressive strength
14. Smooth fracture surface
15. Superior long-term durability under aggressive environmental conditions
16. Good flowability and compactability
17. Good serviceability as indicated by crack and deflection control
18. Good pump ability essential in the construction of high-rise structures

25.4 MIX PROPORTIONING FOR HIGH-PERFORMANCE CONCRETE

Mix proportions for high-performance concrete (HPC) are influenced by many factors, such as:
 1. Specified performance property
 2. Locally available materials
 3. Local experience
 4. Personal preferences
 5. Cost of production
The ingredients of high-performance concretes are almost the same as that of conventional concretes. However, due to the lower water/cementatious ratio, presence of pozzolanic and chemical admixtures is essential. High-performance concretes usually have many features which distinguish them from conventional concretes. From practical considerations, in certain types of concrete constructions,

apart from the final strength, the rate of development of strength is also very important. The high-performance concrete usually contains both pozzolanic and chemical admixtures. Hence the rate of hydration of cementatious materials and the rate of development of strength of high-performance concrete are quite different from that of conventional cement concretes.

The design of mix proportions for conventional or normal concrete is generally based primarily on the well-established water/cement ratio law proposed by Abrams as early as in 1918. In high-strength concrete design, emphasis is mainly on reducing the W/C ratio, increased cement content and use of superplasticizers to improve workability. However, in high-performance concrete mix design, several aspects, like selection of materials, mix design criteria, handling, placing and curing, have to be considered.

The salient interrelated steps involved in the proportioning of high-performance concrete mixtures are listed below:

1. Selection of suitable ingredients, like cement, supplementary cementatious materials, like fly ash, silica fume, coarse and fine aggregates, water and chemical admixtures.
2. Determination of the relative quantities of these ingredients in order to produce as economically as possible, a concrete possessing desirable rheological properties, strength and durability.
3. Effective quality control at all stages of concrete production to achieve the desired properties.

The most commonly used supplementary cementatious materials used for producing high-performance concrete are described below.

1. Silica Fume

Silica fume is a waste by-product in the production of silicon and its alloys. Silica fume is available in different forms, of which the most commonly used is in the densified form. In developed countries, it is already available blended with cement for ready use. Although, it is possible to make high-strength concrete without using silica fume, it is easier to make high-performance, high-strength concrete in the strength range of 60 to 100 N/mm^2 using silica fume.

2. Fly Ash

Fly ash has been used extensively in concrete for a number of years. However, the properties of fly ash vary significantly depending upon the source when compared to silica fume. Most fly ashes will result in strengths not exceeding 70 N/mm^2. For higher strengths, silica fume must be used in conjunction with fly ash. For the production of high-strength concrete, fly ash is used at a dosage of about 15% of cement content.

3. Ground Granulated Blast Furnace Slag

Slags are generally suitable for producing high-strength concrete at a dosage rate between 15 to 30% of cement content. However, for the production of very high-strength concrete exceeding 100 N/mm^2, it is necessary to use the slag in conjunction with silica fume.

Very few formal mix design methods have been developed for high-performance concrete till date. Most commonly, the methods are empirical in nature based mostly on several trial mixtures to decide the final mix proportions to suit the specification requirements, like strength, durability and long-term performance.

25.5 DEVELOPMENT OF MIX DESIGN METHODS FOR HPC USING EXPERIMENTAL INVESTIGATIONS

Most of the high-performance concrete mix proportioning methods are semi-analytical and usually provide aggregate proportioning and calculation of W/C ratio based on compressive strength followed by trial mixes to finalize the desired mix proportions[269, 270]. Many investigators have used the ACI method as the basis for selecting the basic proportions followed by trial mixtures using silica fume, fly ash and superplasticizers.

Exhaustive experimental investigations were undertaken by Sobolev and Soboleva[271] to develop an empirical method of design of high-performance concrete mixes. Their experiments involved three important parameters, i.e.

   (a) Portland cement
   (b) Silica fume
   (c) Superplasticizer termed as PC-SF-SP system

In general, these three materials constitute the basic ingredients for the production of high-performance concrete. Replacement of even small amount of Portland cement by silica fume of 1 to 5% makes the system more denser as the superplasticizer presence increases the thickness of water layers between Portland cement particles resulting in a reduction of viscosity. The best packing and minimum viscosity are reached at the optimum silica fume content of 10 to 20%.

The experimental investigations for developing the mix proportioning data, the following parameters were varied within the ranges specified and the resulting high-performance concrete mixes were tested for their properties in the wet and hardened state. The effect of the rheological parameters on mixes with different water/cement, silica fume and superplasticizer dosages on the concrete slump has been examined.

The authors have developed a flow chart shown in Fig. 25.1 showing the sequence of operations to be followed in designing the final mix proportions of high-performance concrete.

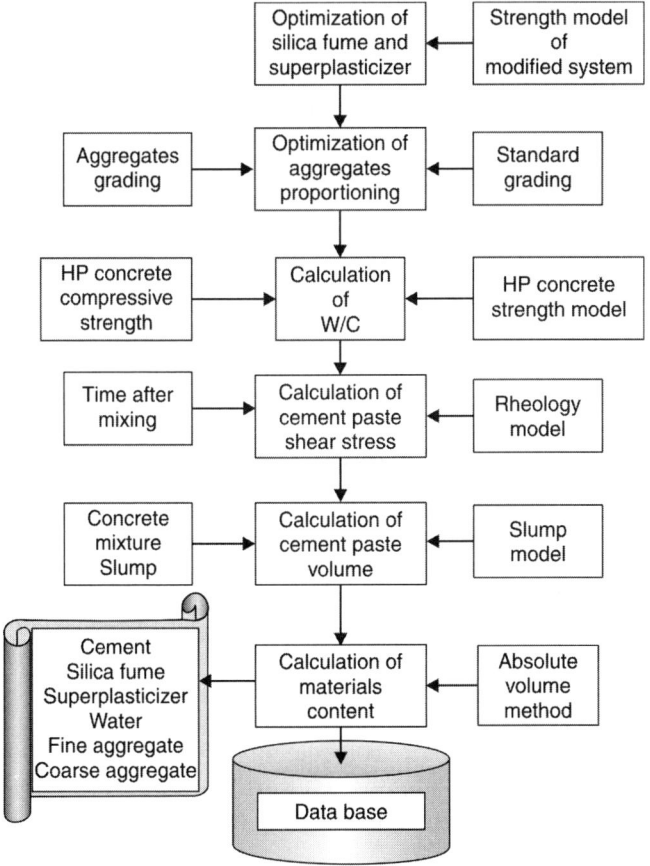

Fig. 25.1: Flow chart for designing mix proportions of high-performance concrete

The computer-based high-performance concrete mix proportioning system based on mathematical models of processed empirical results was established providing a solution for a wide range of design and optimization projects. The results of the computer programme computations are presented using three-dimensional plots showing the inter-relations between the various parameters, like cement content, silica fume and superplasticizer dosage, concrete slump, compressive strength and cost of unit volume of concrete. The authors have presented useful data for the design of high-performance concrete mix design for concretes with a characteristic compressive strength in the range of 60 to 120 N/mm². Table 25.1 contains the basic mix proportioning data which is useful for selecting the ingredients required for HPC concrete of desired 28 days compressive strength along with the required cement and silica fume content, superplasticizer dosage, water content and quantities of coarse and fine aggregates required per unit volume of concrete.

The comprehensive test results clearly show that increasing the superplasticizer dosage from 8 to 18% of silica fume decreases the W/C ratio from 0.31 to 0.26 and increased the 28 days compressive strength from 86 to 97 N/mm². Figure 25.2 shows the relation between the 28 days compressive strength of high-performance concrete and the water/cement ratio which follows an exponential curve together with the experimental results observed in the investigations. The strength model can be conveniently used for estimating the W/C ratio for the target compressive strength of high-performance concrete in the range of 50 to 130 N/mm².

According to Edward G Nawy[271], the ACI Standard is well established for fly ash, silica fume and slag concrete. However, they have not been established in the form of a standard, such as the ACI Manual of Concrete Practice. Nawy has attempted to develop a step by step procedure for designing high-performance concrete containing mixtures using polymers, silica fume and granulated blast furnace slag to comply with the design strength, workability and durability requirements. The following steps are followed to design the high-performance concrete mix proportions:

1. The target design compressive strength F_{ck} is obtained by applying suitable control factors to the characteristic compressive strength, f_{ck}.
2. The slump required for the mix is determined based on the type of structure and the compacting equipment available at site.

Table 25.1: Design table for high-performance concrete mix proportions

	SF 5	SF 10	SF 15	SF 20
Concrete mix proportions (kg/m³)				
Cement	426	449	468	478
Silica fume	22	50	83	120
Total binder	448	499	550	598
Superplasticizer	2.2	5.0	8.3	12.0
Water	153	142	132	121
Coarse aggregate	1169	1155	1136	1119
Fine aggregate	630	622	612	603
Concrete mix parameters				
Silica fume (%)	5	10	15	20
W/C ratio	0.342	0.284	0.239	0.203
Cement paste volume (m³)	0.298	0.307	0.318	0.327
[FA/(Total aggregate)]	0.35	0.35	0.35	0.35
Properties of fresh concrete				
Slump (mm)	100	100	100	100
Density of concrete (kg/m³)	2402	2422	2438	2453
Air content (%)	2.5	2.5	2.5	2.5
Properties of hardened concrete (compressive strength N/mm²)				
1 day	16.8	24.1	34.4	45.1
3 days	28.6	42.2	63.0	84.9
7 days	50.1	67.2	84.8	102.5
28 days	60.0	80.0	100.0	120.0
Cost of concrete per m³				
US (Dollars)	46	54	63	73

3. Maximum size of aggregate to be used is determined based on the minimum thickness of the structural element and the reinforcement used in the member.

4. Mixing water requirement is determined based on the desired slump, entrapped air and the maximum size of aggregates used.

5. Water/cement ratio is determined to suit the required target design compressive strength.

6. Knowing the water content and W/C ratio, the cement content is computed.

7. Silica fume dosage is varied from 5 to 20 per cent of cement content in steps of 5 per cent for various trial mixes.

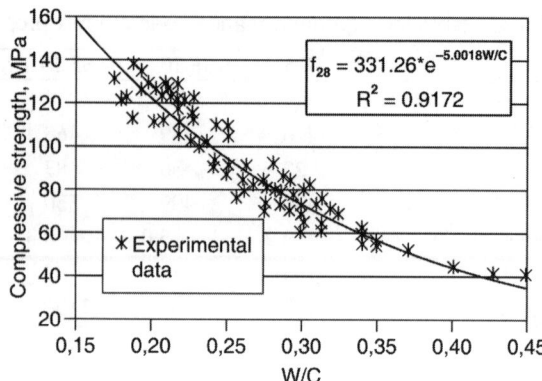

Fig. 25.2: Relation between compressive strength and W/C ratio of high-performance concrete

8. As the silica fume content increases, the degree of workability of the mix reduces.
9. Suitable superplasticizer is used in small dosage to restore the desired slump for each of the trial mix.
10. Test specimens (preferably standard cylinders) are cast using the trial mixes and cured until the time of tests.
11. The mean compressive strength of concrete is determined at different ages, like 28, 56 and 90 days.
12. Based on the test results of fresh and hardened concretes, the optimum trial mix is selected as the design mix for the works.

25.6 CURING OF HIGH-PERFORMANCE CONCRETE

High-performance concrete generally has a very low water/binder ratio, but better particle distribution due to the use of mineral admixtures, which result in significantly less pores per unit volume of cementatious materials in the mix than the conventional cement concrete. In addition, the voids in high-performance concrete are filled up faster with hydration products, because the smaller pores need less hydration products to fill. Also moisture loss due to capillary action stops earlier in the case of HPC compared to CCC under identical curing conditions. The moisture loss from HPC is higher during the first 24 hours.

Due to low water/binder ratio and use of superplasticizers, the early stage hydration rate of HPC is higher than conventional concrete leaving less long-term hydration potential. Also the curing duration after the initial moisture protection has been found to have little effect on long-term chloride permeability of HPC containing micro-silica or fly ash. Due to these reasons, the requirement of curing duration for HPC is less compared to CCC. Method of curing has similar effect on HPC both for creep and shrinkage of concrete, which are again influenced by the type and duration of curing.

Proper curing of HPC is important since early age loss of moisture from concrete leads to detrimental effects on the soundness and long-term properties of concrete. Protection against moisture loss from fresh HPC is crucial for the development of strength, prevention of plastic shrinkage cracks and also for durability. Also wet curing cannot be adopted for HPC, because this will increase the water/binder ratio adjacent to the exposed surface causing deterioration of the quality of concrete. In the case of HPC, the loss of moisture is maximum during the first 24 hours after placement. Fresh concrete mix of HPC is more cohesive and bleeding is less compared to CCC. However, evaporation of bleed water takes place rapidly rendering HPC more prone to plastic shrinkage cracks especially around the initial setting time. Therefore, to overcome this problem, curing process should start immediately after the placement of fresh HPC.

Wet curing cannot be adopted on fresh concrete in view of the harmful effects on the quality of surface layers of the hardened concrete. Applying wet curing before final setting of the concrete will dilute the cement paste near surface, thereby increasing the W/C ratio. Consequently, the strength and impermeability of concrete will be seriously hampered. Hence the curing procedure of HPC is divided into two stages comprising: (a) Initial curing where water is not directly used and (b) Wet curing. Loss of moisture from fresh HPC depends upon the ambient conditions, wind velocity, temperature and humidity and also exposed surface area to volume ratio (S/V). In addition, structural geometry, reinforcement layout and construction methods have a direct bearing on the initial curing procedure.

The concrete surfaces that are generally encountered for initial curing can be grouped under two types, such as:

1. *Type-1 surface*: The exposed surface of fresh concrete of structural elements, like slabs, shell roofs which have a high ratio of surface to volume are grouped under this category. This group is subcategorized as:

  (a) Type-1A, where the concrete surface on which the finishing work is to be carried out at the time of construction, viz. shell segments of domes and plate elements of folded plate roofs used in factories and aircraft hangers.

(b) Type-1B, where the concrete surface for which no finishing work is specified, such as concrete pavements.

2. *Type-2 surface*: It comprises beams of large depth, such as ring beams of large domes and water tanks. This type of surface has lower exposed area to volume ratio. Also the exposed surface may have reinforcements protruding from it and stoppage work for a period may develop laitance on the exposed surface which has to cleaned before starting the next pour of fresh concrete.

Moisture loss from type-1 surface is generally more than that from type-2 surface considering the ratio of surface area to volume. Hence, separate methods for initial curing of each of these surfaces are important.

Initial Curing for Type-1 Surfaces

Soon after placing of wet concrete, it has been observed that bleed water collects at the top of the concrete surface. If curing compound is spread before the bleeding water dries up, local ponding of the curing compound mixed with the bleeding water affects the long-term properties of surface concrete layers especially in dry and hot climate. In many cases, cracks with random orientation develop on the membrane formed by the curing compound when it dries up. These cracks are not plastic shrinkage cracks. Hence the cracks render the curing compound ineffective in preventing the moisture loss from the exposed surface of fresh high-performance concrete.

Surface finish for this type is to be done during the period of initial curing. However, it cannot be started immediately after the placement of fresh concrete. In constructions, like Kaiga and Rajasthan Nuclear Power Plants, it is reported that initial curing was started about 2 hours after depositing the fresh concrete. The surface finishing work has to be completed prior to the commencement of initial setting process for high-performance concretes.

The surface needs to be covered prior to starting surface finishing work. Also the duration of surface finishing work should be as minimum as possible. Many investigators have reported that covering of fresh high-performance concrete is an efficient method for initial curing.

Initial Curing for Type-2 Surfaces

It is not feasible to spread any cover sheet over this type of surface in contrast to type 1 surface. The surface must be cleaned of laitance to achieve a good construction joint. Green cutting seems to be the best method for removing the laitance from the concrete surface. However, this has to be done after final setting time so that the additional water available from water-air jet of green cutting on the surface does not cause harm on the quality of concrete.

The rate of strength gain at early ages is faster in HPC compared to CCC. Hence to overcome the contradictory situations, the best procedure is to use retarders to delay the setting process of HPC adjacent to the exposed surface to a depth 4 to 5 mm even 12 hours after the placement of concrete.

The formation of plastic shrinkage cracks up to the time of final setting can be prevented by applying retarding agents to the surface concrete. Normally, moisture loss from the surface can be prevented by using covers on fresh concrete. A case study conducted on HPC of grade 60 used at Kaiga, nuclear power plant located in Karnataka, it was found that the potential for the development of plastic shrinkage cracks was higher in finished surfaces than those which were left unfinished.

Final Curing Method

In the case of conventional concrete, wet curing, such as ponding of water on the exposed face or covering the exposed surface by wet burlap and keeping it wet by periodical sprinkling of water, has been found to be effective.

Initial curing of HPC is started almost immediately after the placement of fresh concrete and continued till the final setting of the concrete. A better proposition will be to extend the initial curing for an hour after the final setting time. Normally, the initial curing is followed by wet curing.

The total curing period of HPC is the sum of the initial curing and the final wet curing which is longer and forms the second part. It is reported that the method of curing has a similar effect on HPC both for creep and shrinkage of concrete, which are again influenced by the type and duration of curing. In general, a curing period of 7 to 10 days will be required for HPC. Although 2–3 days curing is sufficient for gain of strength, the longer period is essential from considerations of durability.

25.7 ADVANTAGES OF USING HIGH-PERFORMANCE CONCRETE

After an exhaustive study of the production, properties and uses of high-performance concrete, Sai Prasad and Kamalesh Jha[268] have summarized the various advantages of using high-performance concrete in structural concrete elements as detailed below:

1. Reduction of overall cross-sectional dimensions of the structural member, resulting in direct savings in the volume of concrete and increase in plinth/useable area.
2. Reduction in self-weight and superimposed dead load resulting in smaller foundations.
3. Reduction in form work area and cost with the accompanying reduction in shoring and stripping time due to high early age gain in strength.
4. Overall savings in real estate costs in high-rise constructions located in congested areas.
5. Fewer beam elements of longer spans to support the same magnitude of loading.
6. Reduced axial shortening of heavily loaded compression members.
7. Reduction in the number of supports and the supporting foundations due to the increase in span lengths.
8. Reduction in the thickness of floor slabs and supporting beam sections which form the major component of the weight and cost of the majority of structures.
9. Superior long-term service performance under static, dynamic and fatigue lading.
10. HPC structures experience low creep and shrinkage.
11. Reduced permeability making it suitable for water retaining structures
12. Service load deformations are comparatively less due to higher stiffness of members as a consequence of higher modulus of elasticity of concrete.
13. Higher resistance to freezing and thawing, chemical attack and significantly improved long-term durability.
14. Increased wear resistance making it suitable for pavements.
15. Reduced maintenance and repair costs.
16. Depreciation costs are comparatively less.

25.8 APPLICATIONS OF HIGH-PERFORMANCE CONCRETE

High-performance concrete is rapidly emerging as the basic structural material of the 21st century for innumerable number of applications, such as pavements, high-rise structures, large building complexes, storage structures, towers, trusses, marine structures, aircraft hangers, silos, highway structures, rehabilitation projects, bridge deck overlays, railway bridges, prestressed concrete tanks, pipes piles and sleepers, nuclear power stations, grid floors, folded plate and shell roofs for industrial structures. Special types of high-performance concretes which gain strength rapidly in a few hours are being increasingly used in advanced countries for pavement repairs to facilitate early opening to traffic.

A brief review of several important applications of high-performance concrete reported by a working group of European Concrete Committee/International Federation for Prestressed Concrete (CEB-FIP) and other sources is presented in the following paragraphs.

1. Fast Track Concrete

Fast track concrete is a special type of high-performance concrete designed to give high strength at a very early age without using special materials or techniques. The early strength is controlled by the water/cement ratio, cement content and characteristics. Fast track concrete paving (FTCP) was developed originally by the concrete paving industry in Iowa by Grove et al[272]. This technology is well established and widely used for rapid repairs and also adapted in UK according to Walker[273]. The application of fast track concrete technology is well suited in the following five areas:

1. Complete pavement reconstruction
2. Partial replacement by an inlay of at least one lane
3. Strengthening of existing bituminous or concrete pavements by a concrete overlay
4. Rapid reconstruction process and maintenance
5. Airfield pavements

The additional benefits of using FTCP technology are: (i) a reduced contract period, resulting in savings of contract overhead costs, (ii) early opening of the pavement to traffic, (iii) minimizing the use of expensive concrete paving plant and traffic management systems and (iv) reduced traffic delay costs.

Further investigations by Nagi et al[274, 275] have shown that it is possible to perform full depth pavement repairs at faster rate using a variety of concrete mixtures as demonstrated under the SHARP C-205 and Sharp C-206 projects. They have shown that opening of repaired pavements to traffic is possible in as short a time as 2 to

4 hours after repairs, if special rapid strength gain cements are used in association with chemical accelerators and high-range water reducers. Durability studies conducted under this project showed very good freeze-thaw resistance of the repaired concrete overlays. Field studies by Seehra et al[276] and Ozyildirim[277] using special rapid-strength-gain cements, such as magnesium phosphate cement (MPC) and pyrament blended cement (PBC), have shown that complete pavement replacement work can be satisfactorily done at a faster rate than conventional repairs.

2. **Bridge Deck Overlays in USA**

(a) *Washington overlays:* Rehabilitation and/or protection using latex-modified concrete (LMC) and low-slump dense concrete overlays was undertaken for 12 concrete bridge decks in the state of Washington. These decks were evaluated by Babaei and Hawkins[278] to identify the factors affecting the serviceability of the overlaid bridge decks. The evaluation included overlay freeze-thaw scaling, surface wear and skid resistance, surface cracking, bond with the underlying deck, chloride and water intrusion and the overlay's ability to retard continued corrosion of the reinforcing steel bars. The results of the evaluation indicate that, regardless of concrete deterioration caused by corrosion of reinforcing steel, concrete overlaid bridge decks will require resurfacing after about 25 years of service, as a result of traffic action and deterioration due to adverse environmental conditions.

(b) *Virginia overlays:* In Virginia, a two-lane four-span bridge deck was overlaid with dense concrete containing silica fume at 7 or 10% by weight of cement instead of the latex modified concrete (LMC) often used by the departments of transportation. Post work test results reported by Ozylidirim[279] indicated that the overlaid concrete bonded well with the base concrete with the finished overlay exhibiting very low permeability, high strength and satisfactory freeze-thaw resistance. To avoid the potential problem of plastic shrinkage cracks in LMC and silica fume concrete, proper curing after the placement of concrete was recommended as an essential step.

(c) *Polymer concrete overlays:* The status of polymer concrete overlays for concrete bridge decks has been reviewed by

Sprinkel[280]. The report provides valuable information on the properties, application and performance record of the concretes overlays. Polymer overlays constructed using epoxy methacrylate and polyester styrene binders and graded silica and basalt aggregates can provide skid resistance and protection against chloride intrusion for 1 to 20 years. They are an economical technique for extending service life of reinforced concrete bridge decks, especially when the overlays must be constructed during off-peak traffic periods to minimize inconvenience to the traffic.

(d) *High-strength–high-performance concrete pavements:* High-performance – high-strength concrete pavements have been used in Norway because of the need to provide increased wear resistance to steel studed tyres as reported by Gjorv et al[281]. Highways E-6 and E-18 were paved with high-strength concrete having a thickness of 180 mm with concrete volumes of the order of 22,000 m^3. The concrete used had a characteristic compressive strength of 85 to 90 N/mm^2. The fresh concrete had a slump of 20 to 60 mm. After four years of service, the wearing resistance of the pavements was found to be satisfactory except for some longitudinal cracks which developed close to the joints which was attributed to fatigue.

3. Typical Bridge Structures

The benefits of using high-strength–high-performance concrete for bridges are well known to bridge engineers. Based on an extensive survey of published literature, Zia[282] et al, Adelman and Cousins[283], Schemmel and Zia[284], Taerwe[285] and Russel[286] have concluded that the use of high-strength–high-performance concrete would enable the standard prestressed concrete girders to span longer distances with slender cross-sections and support heavier live loads and have better durability under aggressive environmental conditions.

In an earlier report of the National Research Council, Zia[287] claims that the use of high-strength–high-performance concrete for bridges has received much wider and earlier acceptance in Europe and Japan than in US and lists several bridges built in those countries in the chronological order from 1968 to 1990 in Table 25.2. The location of the bridge along with maximum span and the grade of concrete used are also included in the table.

Table 25.2: Data of bridges built in Europe, USA and Japan using high-strength concrete

Sl no.	Name of bridge	Location	Year	Maximum span (m)	Max. design concrete strength (N/mm²)
1	Nitta highway bridge	Japan	1968	30	59
2	Kaminoshima highway bridge	Japan	1970	86	59
3	2nd Ayaragigawa bridge	Japan	1973	50	60
4	Iwahana bridge	Japan	1973	45	89
5	Ootanable railway bridge	Japan	1973	24	79
6	Fukamitsu highway bridge	Japan	1974	26	69
7	Akkagawa railway bridge	Japan	1976	46	79
8	Kylesku bridge	Scotland	–	79	53
9	Deutzer bridge[+]	Germany	1978	185	69
10	Tower road bridge	Washington	1981	49	62
11	East Huntington bridge	W.Virginia	1984	274	55
12	Annacis bridge	Vancuvor	–	–	55
13	Sylans viaduct	France	1986	–	60
14	Re island bridge	France	1987	–	60
15	Braker lane bridge	Texas	1987	26	66
16	Pont du joigny	France	1988	46	60
17	Pont du pertuiset	France	1989	88	110
18	Arc sur la rance	France	1989	–	60
19	Giske bridge	Norway	1989	52	55
20	Sandhomoya[+]	Norway	1989	154	55
21	Boknasundet[+]	Norway	1990	190	60
22	Helgelandsbrua	Norway	1990	425	65
23	Kwung tong by pass	Hong Kong	1990	–	65

[+] Lightweight concrete

(a) *Bridges in Japan:* Japan National Railways built three high-strength concrete bridges in 1973 which are of historical importance. The high-strength–high-performance concrete was adopted to minimize the dead loads and deflections under total design service rolling loads of the rail coaches. Additional advantage accrued by using high-strength concrete was significant reduction in maintenance costs and these first generation of high-strength concrete bridges have performed according to all the expectations without any major repairs and disruption of traffic.

The 2nd Ayaragigawa bridge was the first high-strength concrete bridge built using post-tensioned bulb T-beams. The bridge with a 60° skew and the design concrete strength of 60 N/mm² was chosen to reduce the self-weight of T-girders to less than 150 t for lifting.

Iwahana railway bridge was the first medium span prestressed concrete trussed bridge in Japan made with high-strength concrete with a compressive strength 89 N/mm². The 45 m single span Warren truss bridge was selected to satisfy the clearance beneath the bridge and to reduce deflections. The truss elements were prefabricated and jointed at site using concrete bridge was selected in preference to steel to overcome the problem of noise and vibrations and to reduce the maintenance costs under adverse environmental conditions.

The Akkagawa railway bridge spanning 305 m was built using prestressed concrete Warren trusses of 45 m maximum span lengths. The compressive strength of concrete at site was 96 N/mm². After casting, the members were steam cured at 65°C for 12 hours and then were autoclave cured for at 180°C at a pressure of 10 atmospheres for an additional period of 20 hours. The different parts were assembled into 45 m span units and lifted into position. The joints were cast in situ with high-strength concrete.

(b) *Bridges in France:* High-strength self-compacting concrete (flowable concrete) with characteristic compressive strength of 80 N/mm² at 28 days was adopted for the construction of Pertuiset cable stayed bridge over Loire river in France. The concrete was designed to have slump of more than 200 mm to facilitate easy deposition of concrete in the towers. A water/cement ratio of 0.33 with suitable superplasticizers was adopted for casting the deck slab of 180 mm thickness and the pylons supporting the cables at the top.

Mailer et al[288, 289] and his team of researchers have reported the construction of Joigny Bridge by using high-strength concrete from a ready mixed commercial concrete plant without using silica fume. It is a three span bridge with a central span of 46 m. The superstructure comprises sections of double I-girders prestressed externally[290]. The average concrete strength achieved was 78 N/mm². By using HSC, the authors claim to have achieved 30% savings in concrete and 24% load reduction on the pier abutments and foundations. The reduction in weight also resulted in some savings in the number of prestressing strands. The team has also monitored the long-term performance record of the bridge by periodical observations of temperature, humidity, permeability and deformations under severe loading conditions which satisfied the design criteria.

High-strength concrete of characteristic strength 97 N/mm² was used in the construction of Elom bridge spanning 400 m. High-strength was achieved for concrete used in this cable stayed bridge by using silica fume to attain structural efficiency and durability. For the same reason, high-strength concrete of 60 N/mm² was chosen for the Normandie bridge, which was also a cable stayed bridge with a long span of 624 m, constructed during 1990–95.

(c) *Bridges in Norway:* The majority of all concrete highway structures built in Norway since 1989 have followed a general requirement of using a water/binder ratio of less than 0.40 combined with the use of silica fume mainly to improve the chloride resistance due to deicing agents and marine environment. According to a CEB-FIP Report of 1994, the annual consumption of such concrete ranged from 150,000 to 200,000 m³. Sandhomoya bridge was built in 1989 using light-weight high-strength concrete (LWHSC) of 56 N/mm². The three-span cantilever bridge with a central span of 154 m, derives its advantage of reduced weight and increased strength due to the use of LWHSC.

Strongsundet bridge built using four 65 m long precast girders in 1990 was post-tensioned.

High-strength concrete of 75 N/mm² with a water/cement ratio of 0.35 was used. The Stovset bridge built in 1992–93 is a prestressed concrete cantilever bridge was cast using light-weight high-strength concrete (LWHSC) of 74 N/mm². The bridge covering a central span of 220 m had the advantages of reduced self-weight and increased strength due to adoption of LWHSC.

(d) *Bridges in Denmark:* The islands of Sporogoe and Funen were connected by the West bridge, a combined road and railway bridge with a total length of 6600 m. The structure consists mainly of 110 m long precast concrete girders. The bridge has been designed for a service life of requirement of 100 years. Hence durability is a major consideration in the design of the concrete mix. Table 25.3 gives the limits established for the concrete proportions selected for the project.

Table 25.3: Summary of concrete mix design and test results

Grade of concrete	Target mean strength (N/mm^2)	W/C ratio	A/C ratio	Total cement content (kg/m^3)	Admixture content (kg/m^3)	Compressive strength at (N/mm^2)			
						1 day	3 days	7 days	28 days
M-60 with micro-silica	83	0.24	3.70	521	10.42	25.32	41.27	71.42	85.64

Note: Slump observed on wet concrete = collapse

(e) *Bridges in Germany:* Post-war years accelerated the bridge construction activity in Germany, well ahead of the other countries. Among the 500 and odd bridges built in post-war Germany, during 1949–53. Seventy per cent of them used prestressed concrete. Deutzer bridge crossing the Rhine river close to Cologne is an excellent example of a free cantilever construction with three spans of 132, 185 and 121 m. Sixty-one metres of the central span were cast with lightweight concrete and the rest of the bridge with normal weight concrete. The mean strength obtained in the field was 69 N/mm^2 for the normal weight concrete and 73 N/mm^2 for the lightweight concrete according to the state of the art report released by the CEB-FIP in 1990.

(f) *Bridges in Canada:* Portneuf bridge constructed in Quebec in 1992, uses precast post-tensioned girders of 24.8 m span. The average concrete strength was 75 N/mm^2 with a water/cement ratio of 0.29 and an air content of 5 to 7.5%. By using high-strength concrete, smaller loss of prestress and consequently larger permissible stresses and smaller cross-section were achieved. The service life of the structure improved with enhanced durability.

St. Eustache Bridge in Quebec, built as a replacement for a 17 m span bridge, uses precast pretensioned channel-shaped girders cast with 60 N/mm² concrete. The concrete mix had a water/cement ratio of 0.26 and an air content of 4.5%. High-strength and high-performance concrete was chosen not for strength but for durability. The initial cost of the design proved to be more economical than the steel-concrete composite girder.

High-strength concrete was selected for a Highway-50 overpass in Mirabel, Canada. The HSC design was selected based on economy and prolonged service life. Although the specified concrete strength was 60 N/mm², the actual field strength of the cylinders was reported as 80.7 N/mm² with an air content of 6.2% according to the CEB-FIP 1994 reports.

(g) *Bridges in United States of America:* The application of high-strength–high-performance concrete in the United Sates was more centred towards high-rise buildings in the early period than in bridges. However, for the last 30 years, the use of prestressed concrete is well established for the construction of long-span bridges. The first prestressed concrete bridge built in USA was the Magnel's Walnut Bridge in 1949. During the last 30 years, prestressed concrete has been the choice for long-span girder bridges like the Lin Cove Viaduct of 180 m span built in 1976. The Dames Point bridge of 396 m span and the Sunshine Sky Way bridge[291] of 365 m span are excellent examples of prestressed concrete cable stayed bridges built in USA towards the end of 20th century. The trend is clear that more bridges will be built with higher and higher concrete strength and superior properties in the foreseeable future as the industry becomes more familiar with advances in concrete technology.

An overview of the Federal Outlook for high-strength concrete bridges has been presented by Lane[292] and Podolny as early as in 1993. Several examples of high-strength concrete bridges have been presented along with research and design studies and issues of codes and specifications and future research needs in the report. Standardized pretensioned girders were first used in USA for the construction of the Brake Lane Bridge over I-35 in Austin. Type C girders, each 26 m long and spaced at 2.6 m, were designed for a specified compressive strength of 66 N/mm² but the field strength achieved was of the order of 92 N/mm² at 28 days and 51 N/mm² in 17 hours necessary for the release of high-tensile wires from the pretensioning bed. Durning and Rear[293] have reported the extensive materials development programme undertaken to produce the high-strength concrete for the bridge, utilizing the locally available materials.

The report of static and fatigue load tests conducted on three full sized Texas Type C prestressed girders of 14.6 m span, made with 69 N/mm² concrete was presented by Russel and Burns[294] in 1993. Two of the girders contained debonded strands and the third girder contained draped strands. The girders were tested for in flexure by a combination of static overloads and repeated service loads which varied from a minimum of 225,000 cycles to a maximum of 700,000 cycles. The test results confirmed that the behaviour of girders made with high-strength concrete can be adequately predicted by the current design procedures and that use of debonded strands is a viable alternative to fully bonded draped strands. The advantages of using high-strength concrete in the construction of highway bridges were investigated by actual testing of the pretensioned tee girder under fatigue or repetitive loads by Bruce et al[295], Roller et al[296,]. Four full size 1370 mm deep pretensioned bulb tee girders were cast with 68 N/mm² concrete. The girders with a span length of 21.3 m contained the same number and configuration of longitudinal prestressing strands and same amount of web reinforcement.

A deck slab was cast on two of the four girders. One girder with a deck slab and one without the deck slab were tested for flexure and shear strengths. The second girder with a deck slab was used to determine the long-term behaviour under full design dead load over an 18-month period, and the remaining girder without a deck slab was subjected to fatigue test under 5 million cycles of repeated loading. The girders tested in flexure and shear satisfied the design and specification requirements. The long-term sustained load test indicated that losses of prestress were significantly less than the predicted values. Also the camber and deflections were well within the values calculated by conventional analysis.

The girder tested under fatigue loading of 5 million cycles satisfied all the strength and serviceability requirements[297].

25.9 TYPICAL EXAMPLES OF HIGH-PERFORMANCE CONCRETE MIXES USED IN MAJOR STRUCTURES

1. Flyovers at University Circle and Agricultural College, Pune

(a) Design specifications

Grade of concrete	= M-60
Characteristic compressive strength	= M-75
Assumed standard deviation as per Table 8 of IS: 456-2000	= 5 N/mm^2
Target mean strength	= 83 N/mm^2
Desired workability	= 75 to 125 mm
Type and size of coarse aggregate	= Angular, crushed, 20 mm maximum size
Type of fine aggregate	= River and crushed sand
Type of quality control	= Very good
Type of exposure	= Normal

(b) Ingredients of design concrete

mix cement	= 425 kg (Birla super 53 grade)
Specific gravity	= 3.15
Fly ash	= 60 kg (specific gravity = 2.40)
Micro-silica	= 36 kg (Astech brand, grade 920D)
Specific gravity	= 2.20
Cementatious content	= (4225 + 60 + 36)
	= 521 kg
Water/ binder ratio	= 0.24
Water content	= 125 kg
Admixture type	= Sulphonated Naphthlene [Rheobuild 4134 (M)]
Admixture dosage	= 2% by weight of cementatious material
Coarse aggregate content	= 60% (1184 kg)
Fine aggregate	= 40% (945 kg)

The details of mix ingredients are shown in Table 25.3.

2. Two Union Square Building, Seattle, Washington

A good example of the use of high-performance and high-strength concrete in the range of 138 N/mm² at 56 days is the two Union Square Building located in Seattle, Washington. The high-strength concrete had a modulus of elasticity of 53.8 kN/mm². The details of design and actual mix proportions of high-strength and high-performance concrete are compiled in Table 25.4.

Table 25.4: Typical mix details of high-performance and high-strength concrete

Mix details	Cement content (kg)	Silica fume content	Fine aggregate (kg)	Coarse aggregate (kg)	Water (kg)	Superplasticizer (W.R. Grace)	
						Dartard –40 (kg)	Mighty –150 (kg)
Actual (1)	433	45.2	528	848	98	0.507	4.45
Actual (2)	433	45.2	528	858	98	0.507	7.45
Design Mix	430	32 kg⁺	498	818	95	1.42	(up to 11)

Note: The first and second rows of values represent actual mix proportions
The third row indicates design mix proportions
⁺ Weight of solid silica fume only. Water contained as part of the emulsion must be subtracted from the total water allowed.

25.10 ULTRA-HIGH-PERFORMANCE CONCRETE

1. Introduction

Ultra-high-performance concrete (UHPC) is also referred to as reactive powder concrete[298] (RPC) is a high strength, ductile material formulated by combining the various ingredients, like Portland cement, silica fume, quartz floor, fine silica sand, high range water reducer, steel or organic fibres and water. The concrete made with all these materials is capable of providing compressive strengths of up to 200 N/mm² and flexural strength of up to 50 N/mm².

The ingredients of ultra-high-strength concrete are usually supplied in a three component premix comprising of:

(a) Reactive powder (Portland cement, silica fume, quartz flour and fine silica sand) preblended in bulk bags.

(b) Superplasticizers (high-range water reducers)

(c) Steel or organic fibres, like polypropylene fibres

Water is added before mixing the ingredients to achieve the desired workability before depositing in the structural form work. The fibres impart ductile behaviour to the high-strength performance concrete with the capacity to deform and support flexural and tensile loads even after initial cracking. The use of this material for construction is simplified by the elimination of reinforcing steel bars and the material can be virtually self-placed or dry cast to the desired shape.

The superior durability characteristics are due to combination of fine powders selected for their grain size not exceeding 600 µm and chemical reactivity. The unique combination of these various ingredients results in maximum compactness of the resulting concrete with a small disconnected pore structure.

2. Mix Ingredients for UHPC

The following materials are required for the preparation of ultra-high-performance concrete.

    (a) Portland cement
    (b) Silica fume
    (c) Coarse aggregate
    (d) Fine aggregate
    (e) Steel or polypropylene fibres
    (f) Superplasticizers
    (g) Water

The recommended percentages of ingredients according to Ductal Lafarge Company[299] who have specialized in the commercial production of ultra-high-performance and high-strength concrete are compiled in Table 25.5.

3. Procedure for Production of UHPC Mix

    (a) First mix the silica fume and coarse aggregate with a little quantity of water just enough to make the coarse aggregate surface damp. Time of mixing = 1.5 to 2 minutes.

    (b) Add Portland cement to the mix and continue the dry mixing for 1.5 to 2 minutes. Cement will not become active until the water is added.

    (c) Add fine aggregate (sand) and polypropylene fibres and mix the ingredients for another 1.5 to 2 minutes.

    (d) Add the required quantity of water and superplasticizer, mix all the ingredients for 5 minutes.

    (e) Allow the concrete to rest for 3 minutes after initial mixing so that water has a chance to penetrate all the ingredients.

    (f) Mix again all the ingredients for another 5 minutes and then the concrete is ready for use.

Table 25.5: Recommended percentages of ultra-high-performance concrete ingredients

Sl no.	Mix ingredient	Percentage by weight
1	Portland cement	10 to 15
2	Coarse aggregate (gravel or crushed rock)	35 to 45 –
3	Fine aggregate (sand)	24 to 28
4	Silica fume	4 to 5 % by weight of C.A.
5	Superplasticizer	3 to 5 % by weight of cement
6	Polypropylene fibres	Equal to the amount of silica fume
7	Water	8 to 10% of overall volume of the mix

For site use, Ductal Lafarge Company sells ready made bags of UHPC mix and at site, the required quantity of water and super-plasticizer is added to make the concrete.

4. Structural Properties

(a) Strength

Compressive strength: 120 to 200 N/mm^2

Flexural strength: 15 to 50 N/mm^2

Modulus of elasticity: 45 to 50 kN/mm^2

(b) Durability

Freeze/thaw (after 300 cycles):   100%

Salt scaling (loss of residue):   < 60 g/m^2

Abrasion (relative volume loss index):   1.7

Oxygen permeability: < 10^{-20} m^2

Carbonation depth: < 0.5 mm

5. Applications of UHPC

Ultra-high-performance concrete has been used for a number of innovative solutions including projects, such as transportation structures, acoustic sound panels, seal walls, bridge anchor plates and beams for power plant cooling towers. In Seoul, Korea, a famous pedestrian bridge named as the "Foot Bridge of Peace" has a 130 m span arch constructed entirely with a proprietary product of UHPC named as DUCTAL is considered as an structural and architectural wonder. The arch bridge has no middle supports and has a platform thickness of just 45 mm of UHPC.

Perry and Zakariasen[300] have reported the first use of UHPC for the construction of a train station canopy. The Shawnessy Light Rail

Transit (LRT) station constructed during 2003–2004, forming a part of the southern expansion to Calgary's LRT system is considered as the first structure to use UHPC. The station platform roof is made up of 24 thin-shelled canopies 5.1 m by 6 m having thickness of just 20 mm, supported on single columns.

UHPC technology has a unique combination of superior technical characteristics including ductility, strength, and durability while providing highly moldable products with a high quality of surface finish. The design strength of the concrete used in the canopies was 130 N/mm^2. In addition to the canopies, the other components include struts, columns, beams and gutters. The total volume of concrete used for the station totaled 80 m^3.

References

1. TALBOT, A.N. and RICHART, F.E. The strength of concrete– its relations to the cement, aggregates and water. *University of Illinois, Engineering* Experimental Stations, Bulletin No. 137, 1922.
2. ABRAMS D. A. Design of concrete mixtures. *Structural Materials Research Laboratory, Lewis Institute,* Chicago, 1918, Bulletin No. 1.
3. FERET, R. Sur La compacite Mortiers Hydrauliques. (On the compaction of hydraulic mortars), *Annales does Ponts et Chanssees,* Paris, Vol. 4, No. 21, 1892, Memories Serie 7e.
4. GONNERMAN, H.F. and LERCH, W. Changes in characteristics of Portland cement as exhibited by laboratory tests over the period 1904 to 1950. ASTM *Special Publication.* No. 127, 1951.
5. ACI Committee–212, 1999 (reapproved 2009). Standard practice for selecting proportions for normal, heavy weight and mass concrete, ACI–211.1–1991, American Concrete Institute, Formington Hills, Michigan, USA, 2002.
6. IS: 456–2000, Indian Standard *Code of Practice for Plain and Reinforced Concrete* (Fourth Revision) B.I.S. New Delhi.
7. BLOEM, D.L. Effect of maximum size of aggregate on strength of concrete. *National Sand and Gravel Association Circular No. 74,* Washington DC, February 1959.
8. National Sand and Gravel Association, Joint Technical Information Letter No. 155, Washington DC, April 1959.
9. SINGH, B.G. Specific surface of aggregates related to compressive and flexural strength of concrete. *Journal of the American Concrete Institute,* Vol. 54, April 1958, pp. 897–907.
10. AKROYD, T.N.W., CONCRETE (Properties and Manufacture) Pergamon Press, London, 1962, p. 267.
11. NEVILLE, A.M. *Properties of Concrete,* Pitman, London, 1963, p. 465.
12. ROAD RESEARCH LABORATORY, *Design of Concrete Mixes* DSIR, Road Note No. 4, London, HMSO, 1950.

13. MURDOCK, L.J. The workability of concrete. *Magazine of Concrete Research*, Vol. 12, No. 36, November 1960, Cement and Concrete Association, London, pp. 135–44.

14. IS: 269–1989, (*Fourth revision*), *Specification for 33 Grade Ordinary Portland Cement*, (With Amendment No. 3), Bureau of Indian Standards, New Delhi.

15. BS: 8500–2:2006, *Concrete–Complementary British Standard* to BS EN 206–1, specification for constituent materials and concrete, British Standards Institution, London, 2006.

16. ASTM C 150–11/C 150M–11 Standard Specification for Portland Cement 2009.

17. LEA, F.M. *The Chemistry of Cement and Concrete*, London, Arnold, 1956.

18. LEA, F.M. Modern developments in cements in relation to concrete practice. *Journal of the Institution of Civil Engineers*, London, February 1943.

19. NEVILLE, A.M. *Properties of Concrete*, Pitman, London, 1963.

20. IS: 383–1963, (re-approved in 1997) *Indian Standard Specification for Coarse and f aggregates from Natural sources for Concrete*, Bureau of Indian Standards, New Delhi, 1997.

21. BS: 882–1992, *British Standard Specification for Concrete Aggregates from Natural source for Concrete*, British Standards Institution, London, 1992.

22. ASTM C 33/C33M–11a Standard Specification for Concrete Aggregates, 2011.

23. McINTOSH J.D. and ERNTROY, H.C. The workability of concrete mixes with 3/8 inch aggregates, *cement and concrete Association, Research Report No. 2*, London, June 1955.

24. IS: 2386–1963, Part III, *Methods of Test for Aggregates for Concrete Specific Gravity, Density Voids, Absorption and Bulking*, Bureau of Indian Standards, New Delhi, 1963.

25. *Concrete Hand Book*, The Concrete Association of India, Cement House, Bombay, 1969.

26. GLANVILLE, W.H., COLLINS, A.R. and MATTHEWS, D.D. The Grading of Aggregates and Workability of Concrete, 2nd Edition, *Road Research Technical, Paper No. 5*. HMSO 1947, pp. 38.

27. NEWMAN, K. *Properties of Concrete, Structural Concrete*, Vol. 2, No. 11 September 1965, pp. 451–82.

28. CUSENS, A.R. The measurement of the workability of dry concrete mixes. *Magazine of Concrete Research*, Vol. 8, No. 22, March 1956, pp. 23–30.

29. KELLY, J.W., and POLIVKA, M. Ball test for field control of concrete consistency. *Journal of American Concrete Institute*, Vol. 51, May 1955, pp. 881–8.

30. ASTM Standard C-124–39 (1996). Test for flow of Portland cement concrete by use of the flow table, part 10. *Concrete and Mineral Aggregates*. October 1969, pp. 77–79.

31. BAHRNER, V., Vibrotekniska Undersokningen, (Vibration technique Investigation). Rapport, 1. *Svenska Cement foreningen, Tehniska Meddelanden och Undersoknings rapporter Nr. 1*, Malmo-cStockholm 1940, pp. 23.

32. POWERS, T.C. Studies of workability of concrete. *Journal of the American Concrete Institute*, Vol. 28, 1932. pp. 419–48.

33. TAYLOR, W.H. *Concrete Technology and Practice*, Angus and Robertson Ltd., Third Edition, London, 1969.

34. HUGHES, B.P. The rational design of high quality concrete mixes, concrete. *Journal of the Concrete Society*, London, Vol. 2, No. 5 May 1968, pp. 212–222.

35. WRIGHT, P.J.F. Entrained air in concrete, *Proceedings of the Institution of Civil Engineers*, Part I, Vol. 2. No. 3, May 1953, pp. 337–58.

36. Cardon, W.A. Size and number of samples and statistical considerations in sampling. *Significance of tests and properties of concrete and concrete aggregates*, ASTM, Special technical Publication No. 169 Philadelphia, 1956.

37. HIMSWORTH, F.R. The variability of concrete and its effect on mix design. *Proceedings of the Institution of Civil Engineers*, London, Vol. 3, March 1954, p.163.

38. STANTON WALKER. Application of theory of probability to design of concrete for strength specifications. *Paper presented at 14th Annual Meeting of National Ready Mixed Concrete Association at New York* 27th January 1944 (Washington, NRMCA, 1955).

39. Graham, G., and Martin, F.R., 'Heathrow. The construction of high grade quality concrete having for modern transport aircraft'. *Journal of the Institution of Civil Engineers*, London, Vol. 26, p. 117.

40. SPARKES, F.N. Control of variations in quality of concrete and its effect on mix proportions. *The Reinforced Concrete Review*, Vol. 1, 1949, p. 543.

41. *Proceedings of a Symposium on Mix Design and Quality Control of Concrete*. Cement and Concrete Association, London, May 1954.

42. MURDOCK, L.J. The control of concrete quality. *Proceedings of the Institution of Civil Engineers*, London, Part I, Vol. 2, No. 4, 193, p. 426.

43. ERNTROY, H.C. Contribution to discussion on the paper. 'The Control of Concrete Quality'. *Proceedings of the Institution of Civil Engineers*, Part I, Vol. 3, 1954, p. 236.

44. ERNTROY, H.C. The variation of works test cubes. *Cement an Concrete Association*, London, Research Report No. 10 1960, pp. 28.

45. ERNTROY, H.C. The variation of works concrete set cubes. *International Association for Bridge and Structural Engineering*, Sixth Congress, Stockholm, June 1960, pp. 679–92.

46. Anon. 'The Vibration of Concrete', Joint Committee. *Institution of Civil Engineers and Institution of Structural Engineers*, 1956, pp. 64.

47. Dudding, B.P., and JENNETT, W.J. Control chart technique when manufacturing to a specification. *British Standard Institution*, London, BS: 2564–1955, pp. 77.

48. PEARSON, E.S., The application of statistical methods to industrial standardisation and quality control. *British Standards Institution*, London, BS: 600–1935, Reprinted 1960 (with minor amendments).

49. Introduction to statistical methods for quality control of concrete. *Cement and Concrete Association*, London, Advisory Note Number 8, May 1965, pp. 9.

50. BS EN 1992–1–1:2004, Euro Code–2, Design of Concrete Structures-General Rules and Rules for Buildings, 2004.

51. ACI: 214–77, American Concrete Institute Standards Recommended Practice for evaluation of strength test results of concrete. *ACI, Manual of concrete Practice*, Part–I ACI 214R–02, 2011.

52. ROWE, R.E., CRANSTON, W.B., and BEST, B.C. New concepts in the design of structural concrete. *The Structural Engineer*, Vol. 43 December 1965, pp. 399–403.

53. KRISHNA RAJU, N. Limit state design for structural concrete. *Journal of the Institution of Engineers* (India), CI, Vol. 51, January 1971, pp. 138–143.

54. PLOWMAN, J.M. Maturity and the strength of concrete. *Magazine of Concrete Research*, Vol. No. 22, March 1956, pp. 13–22.

55. TAYLOR, W.H. *Concrete Technology and Practice*, Angus and Robertson Ltd. London, Third Edition, p. 92.

56. NEVILLE, A.M. Some aspects of the strength of concrete. *Civil Engineering,* London, Vol. 54, Part I, October 1959, pp. 1153–56, Part 2. November 1959, pp. 1308-II, part 3, December 1959, pp. 1435–39.

57. GRIFFITH, A.A. The phenomena of rupture and flow of solids. *Philosophical Transaction, Royal Society, Series A*, Vol. 221, 1921. pp. 163–198.

58. EPSTEIN, B. Statistical aspects of fracture problems. *Journal of Applied Physics* Vol, 19, February 1948, pp. 140–147.

59. GONNERMAN, H.F. Effect of size and shape of test specimen on compressive strength of concrete. *Proceedings of American Society of Testing Materials* Vol. 25, Part II, 1925, pp. 237–230.

60. Rilem Commission for Concrete, 'Correlation Factors between the strength of different specimen types', *International Union of Testing and Research Laboratories for Materials and Structures* (RILEM), Bulletin, No. 39, 1957, pp. 81–105; No. 12, New series, September 1961, pp. 155–156.

61. KESLER, C.E. Effect of length to diameter ratio on compressive strength. *Proceedings of the American Society for Testing and Materials,* Vol. 59, 1959, pp. 1216–1228.

62. NEWMAN K, and LACHANCE, L. The testing of brittle materials under uniform uniaxial compressive stress. *Proceedings of the American Society far Testing and Materials*, Vol. 64,1964, pp. 1044–1067.

63. SIGVALDASON. O.T. The influence of the testing machine on the compressive strength of concrete. *Proceeding of a Symposium on Concrete Quality, Cement and Concrete Association*, London, November 1964, pp. 162–171.

64. EVANS, R.H. The plastic theories for the ultimate strength of reinforced concrete beams. *Journal of the Institution of Civil Engineers*, London, Vol. 21, 1943–44, pp. 98–121.

65. BS EN: 12390–3 and 4–2009. Testing hardened concrete, compressive strength and flexural strength of Test Specimens, *British standards Institution*, London, 2009.

66. TAYLOR, W.H., *Concrete Technology and Practice,* Angus and Robertson Ltd., London, Third Edition, p. 115.

67. ACI 211.1–1991, (reapproved in 2002) *Standard Practice for selecting proportion for normal, heavy weight and man concrete, ACI,* Farmington Hills, Michigan, USA, 2002.
68. AKROYD, T.N.W. *Concrete, Properties and Manufacture,* Pergamon Press, London, 1962, p. 111.
69. ASTM, Standard C 231–10/C 231 M–10, Standard Test Method for Air Content of Freshly mixed Concrete by the Pressure Method.
70. IS: 1199–1959, Indian Standard, *Methods of Sampling and Analysis of Concrete.* Bureau of Indian Standards, New Delhi, 1959.
71. WRIGHT, P.J.F. The effect of the method of test on the flexural strength of concrete. *Magazine of Concrete Research,* Vol. 3, No. 11, pp. 67–76.
72. IS: 516–1959, Indian Standard Code of Practice. *Methods of Test for strength of concrete,* Bureau of Indian Standards, New Delhi, 1959.
73. C, 78–10; ASTM Standard Test for flexural strength of concrete (using simple beam with third point loading). Part 10.
74. PRICE, W.H. Factors influencing concrete strength. *Journal of the American Concrete Institute,* Vol. 47, February 1951, pp. 417–32.
75. JONES, F.E. The physical structure of cement products and its effect on durability. *Proceedings of the Third International Symposium on the Chemistry of Cement,* London 1952, Cement and Concrete Association, 1954 p. 368.
76. SHESTOPYROV, S.V. Durability of concrete. *Proceedings of the Fourth International Symposium on the Chemistry of Cement,* Washington 1961. US National Bureau of Standards, 1962. Monograph 43. Vol. 2, pp. 889–907.
77. LEA, F.M., and DAVEY, N. The durability of concrete in structures. *Journal of the Institution of Civil Engineers,* London. Vol. 32. No. 7 1149, pp. 248–275.
78. NURSE, R.W. Assement of concrete durability. *Proceedings of a Symposium on Concrete Quality,* Cement and concrete Association, London, November 1964, pp. 71–77.
79. ACI Committee 211, Recommended Practice for selecting proportions for No slump concrete (ACI: 211.3–75), (Revised in 1987 and Reapproved in 1992).
80. THAULOW, SVEN, Field Testing of Concrete' Published by *Norsk Cement forening,* GSLO, 1952, Resume published in *ACI. Journal Proceedings,* Vol. 50, No. 7, March 1954, News letter pp. 10–11, 24–26.

81. SHACKLOCK, B.W. and WALKER, W.R. The specific surface of concrete aggregates, and its relation to the workability of concrete, Research Report No. 4, *Cement and Concrete Association,* London, July 1958.

82. BS: 812–1960, *Methods for sampling and testing of mineral aggregates, sand and fillers.* British Standards Institution, London, 1960.

83. MURDOCK, L.J., and BLANKLEDGE, G.F. *Concrete Materials and Practice,* Edward Arnold, London, 1968, pp. 110–224.

84. SAUL, A.G.A. Principle underlying the steam curing of concrete at atmospheric pressure. *Cement and Concrete Association,* Technical Report, TRA/196, London, July 1955.

85. COLLINS, A.R. *The principles of making high strength concrete*– Report of Eleven Lectures on Prestressed Concrete given at the Building Exhibition, Cement and Concrete Association, London, November 1949.

86. PARROTT, L.J. Selection of constituents and proportions for producing workable concrete with a compressive cube strength of 80–110 N/mm². *Cement and Concrete Association,* Technical Report, TRA/416, May 1969.

87. PARROTT, L.J. *The Production and Properties of High Strength Concrete,* Vol. 3 No. 11, November 1969, pp. 443–448.

88. ERNTROY, H.C., and SHACKLOCK, B.W. Design of high strength concrete mixes. *Proceedings of a Symposium on Mix Design and Quality Control of Concrete,* Cement and Concrete Association, London, May 1954, pp. 55–65.

89. NEVILLE, A.M. Tests on the strength of high alumina cement concrete. *Journal of New Zealand Institution of Engineers,* Vol. 14, No. 3, March 1959, pp. 73–76.

90. BS: 915–1947, *High Alumina Cement British Standards Institution,* London, 1947.

91. NEWMAN K. The design of concrete mixes with high alumina cement. *Reinforced Concrete Review,* Vol. 5, No. 5, March 1960.

92. HUSSEY, A.V., and ROBSON, T.D. High alumina cement as a constructional material in the chemical industry. *Symposium on Materials of Construction in the Chemical Industry,* Birmingham, April 1950, Published by the Society of Chemical Industry.

93. GOTTLIEB, S. The hardening of aluminous cement at high and low temperatures, *Cement and Lime Manufactures,* Vol. 13, No. 4 April 1940.

94. NEVILLE, A.M. The effect of warm storage conditions on the strength of concrete made with high alumina cement. *Proceeding of the Institution of Civil Engineers,* London. Vol. 10, June 1958, pp. 185–192.

95. NEVILLE, A.M. Deterioration of high alumina cement concrete. *Proceedings of the Institution of Civil Engineers,* London, 1963.

96. MARTIN, F.R. Concrete runways. *Proceedings of Meeting of the Pavings Development Group (Concrete and Soil Cement),* 17th October 1956, Cement and Concrete Association, Loudodf 1657, pp. 1–52.

97. ACI 617–58. Standard specifications for concrete pavements and concrete bases. *Journal of the American Concrete Institute,* Vol. 30, No. 1, July 1958.

98. KAPLAN, M.F. Flexural and compressive strength of concrete as affected by the properties of coarse aggregates. *Journal of the American Concrete Institute,* Vol. 55, May 1959, pp. 1193–208.

99. SHACKLOCK, B.W., and KEENE, P.W. Comparison of the compressive and flexural strengths of concrete with and without entrained air, *Civil Engineering,* London, January 1959, pp. 77–80.

100. KAPLAN, M.F., Effects of incomplete consolidation on compressive and flexural strength, ultrasonic pulse velocity and dynamic modulus of elasticity of concrete. *Journal of the American Concrete Institute,* Vol. 56, March 1960, pp. 853–67.

101. WRIGHT, P.J.F. The design of concrete mixes on the basis of flexural strength. *Proceedings of a Symposium on Mix Design and Quality control of concrete,* Cement and Concrete Association. London, May 1954, pp. 74–76.

102. SHORT, A and KINNIBURGH, W. *Lightweight Concrete,* Asistat Publishing House, 1963, p. 3.

103. BLAKE, L.S. The development of concrete blocks in Great Britain. *Fifth International Congress of the Precast Concrete Industry,* London. British Precast concrete Federation, 1966, pp. 61–71.

104. LEA, F.M. Lightweight concrete aggregates. *Building Research Station Bullitin No. 15,* HMSO 1936, pp. 14.

105. GERWICK, B.C. Effective utilization of prestressed lightweight concrete. *Proceeding of the Fist International Congress on Lightweight Concrete,* Vol. 1, Cement and Concrete Association. London, 1968, pp. 243–50.

106. NEWMAN, A.J. and TEYCHENNE, D.C. A classification of natural sands and its use in concrete mix design. *Proceedings of a Symposium on Mix Design and Quality Control of Concrete.* May 1954, Cement and Concrete Association, London, pp. 175–194.

107. LLEWELLIN, J.D. Lightweight aggregates in blocks, screeds and panels. *Chemistry and Industry,* No. 15, April 1964, pp. 601–609.

108. BS: 3797:1990. Specification for lightweight aggregates for concrete, B.S.I, London, 1990.

109. BS EN 13055–1:2002(E), Part-1, Lightweight aggregates for concrete mortar and grout, British-Euro Standards, 2002.

110. C330–64T, *Tentative Specifications for lightweight aggregates for structural concrete,* ASTM. Standards, Part 10, Concrete and Mineral aggregates. October 1967, pp. 251–258.

111. TEYCHENNE, D.C. Lightweight aggregates, their properties and used in concrete in the United Kingdom, *Proceeding of the First International Congress on Lightweight concrete,* Cement and Concrete Association, London, may 1968, pp. 23–37.

112. HOBBS, C. The physical properties of lightweight aggregates and concretes. *Chemistry and Industry,* Vol. 11, 1964, pp. 504–600.

113. LANDGREN, R, HANSON, J.A. and PFEIFER, D.W. An improved procedure for proportioning mixes of structural lightweight concrete. *Journal of the Portland Cement Association Research and Development Laboratories,* Vol. 7, No. 2 May 1965, pp. 47–65.

114. LLEWELLIN. J.D. Handling, mixing, transporting and placing lightweight aggregates concrete. *Proceedings of the First International Congress on Lightweight concrete,* Cement and Concrete Association, London, Vol. 1, May 1968, pp. 55–62.

115. SHORT, A. and KINNIBURGH, W. Lightweight concrete. Asia Publishing House, 1968, p. 29.

116. SHORT, A. The use of lightweight concrete for reinforced concrete construction. *The Reinforced Concrete Review,* Vol, 5, No. 3, September 1959, pp. 141–148.

117. HANSON, J.A. Replacement of lightweight aggregate fines with natural concrete. *Journal of the American Concrete Institute* Proceedings Vol. 61, No. 7,. July 1964, pp. 779–793.

118. TEYCHENNE, D.C. Structural concrete made with lightweight aggregates. *Concrete,* Vol. 1, No. 4, April 1967, pp. 111–112.

119. TEYCHENNE, D.C. Crushing, transverse and tensile strength and the elastic modulus of compacted lightweight aggregates

concrete. *Proceedings of the Symposium on Testing and Design Methods of Lightweight Aggregate Concretes*. Budapest, 1967.

120. ACI Standard. *Recommended Practice for Selecting Proportions Structural lightweight concrete*; ACI 211.2–98 (R 2004) American Concrete Institute, Jan. 1998 pp. 20.

121. KLIEGER, P and HANSON, J.A. Freezing and thawing tests of lightweight aggregate concrete. *Journal of the American Concrete Institute*, Proceedings Vol. 57, No. 7 January 1961, pp, 779–796.

122. KLUGE, R.W. Structural lightweight aggregate concrete. *Journal of the American Concrete Institute*, Proceedings, Vol. 53, No. 4 October 1956. pp. 383–402.

123. ACI 213R–03; Guide for Structural Lightweight-Aggregate Concrete, American Concrete Institute, 2003.

124. KLUGE, R.W., SPARKS, M.M. and TUMA, E.C. Lightweight aggregate concrete. *Journal of the American Concrete Institute*, Proceedings, Vol. 45, No. 9, May 1949, pp. 625–644.

125. SHIDELER, J.J. Lightweight aggregate concrete for structural use. *Journal of the American Concrete Institute*, Proceedings, Vol. 54, No. 4, October 1957, pp. 299–328.

126. *Design of Concrete Mixtures*, Concrete Information Sheet ST-100, Portland Cement Association, 1966.

127. ORCHARD, D.F. *Concrete Technology*, Contractors Record Limited, Vol. 1, 1962, pp. 154–55.

128. McINTOSH, R.H. BOTTON, J.D. and MUIR, CHD. NO. Fines concrete as a structural material. *Proceedings of the Institution of Civil Engineers*. London, Part 1, Vol. 5, No. 6, November 1959, pp. 677–94.

129. Report of ACI. Committee-207. Mass concrete for dams and other massive structures. *Manual of Concrete Practice*, 1970, Part 1, pp. 207–1 to 37.

130. ASTM Standard C150–11. *Portland Cement*, American Society for Testing Materials, Philadelphia, 2011.

131. BS 8500–2:2006, Concrete, Complimentary British Standard to BS EN 206–1, Specification for Constituent Materials and Concrete, BSI, London, 2006.

132. Properties of Mass Concrete in United States and Foreign Deams, Report No. C-880, Concrete Laboratory, United States Bureau of Reclamation, July 1958, 3 pp.

133. McINTOSH, J.D. The use in mass concrete of aggregates of large maximum size. *Civil Engineering and Public Works Review,* Vol. 52, London, September 1957, pp. 1011–15.

134. McINTOSH, J.D. *Concrete Mix Design,* Cement and Concrete Association, London, 1966, pp. 93–96.

135. ORCHARD, D.F. *Concrete Technology,* Vol. 1, Contractors Record Limited, 1962, pp. 208–264.

136. HIGGINSON, E.C. WALLACE, G.B., and ORE, E.L. Effect of maximum size of aggregate on compressive strength of mass concrete. *Symposium on Mass Concrete,* SP-6, ACI-Publications, Detroit, 1963, pp. 219–256.

137. *Symposium on Mass Concrete,* Special Publication, SP-6 American Concrete Institute, Detroit, Michigan, 1963.

138. HARBOE, E.M. Properties of mass concrete in Bureau of Reclamation Dams, *Report No. C-1009, Concrete Laboratory* USSR, December 1961, 6 pp.

139. AKROYD, T.N.W. *Concrete, Properties and manufacture,* Pergamon Press, 1962, pp. 302–310.

140. DAVIS, H.S. High density concrete for shielding atomic energy plants. *Journal of the American Concrete Institute,* Proceedings, Vol. 54, May 1958, pp. 965–77.

141. HENRIE JAMES, O., Magnetite iron ore concrete for nuclear shielding. *Journal of the American Concrete Institute,* Proceedings, Vol. 51, No. 6, February 1955, pp. 547–48.

142. WITTE, L.P. and BACKSTROM, J.E., Properties of heavy concrete made with baryte aggregates. *Journal of the American Concrete Institute,* Proceedings, Vol. 51, No. 1, September 1954, pp. 65–88.

143. DAVIS, H.S. BROWNE, F.L. and WITTER, H.C. Properties of high density concrete made with iron aggregate. *Journal of the American Concrete Institute,* Proceedings, Vol. 52, 1959 pp. 705–26.

144. JORDAN, J.P.R. Further properties of barytes, concrete, *Technical Report-TRA/176.* Cement and Concrete Association, London, January 1955, 9 pp.

145. FIESENAEISER, E.I. and WASIL, B.A. Heavy steel aggregates concrete. *Journal of the American Concrete Institute,* Proceedings Vol. 52, No. 1, September 1955, pp. 73–82.

146. Anon. An investigation of pozzolanic nature of coal ashes. *Engineering News,* 71(24), 1914, pp. 1334–1335.

147. McMILLAN, F.R. and POWERS, T.C. A method for evaluating admixtures. *Proceedings of the American Concrete Institute,* Vol. 30, 1934, pp. 325–344.

148. DAVIES, R.E., CARLSON, R.W., KELLY, J. WARD and DAVIS, H.E. Properties of cements and concrete containing fly ash. *Proceedings of the American Concrete Institute,* Vol. 33, 1937, pp. 577–612.

149. USBR. Report on physical and chemical properties of fly ash, Hungry Horse Dam, Laboratory Report CH-95, 1948.

150. BERRY, E.E. and MALHOTRA, V.M. Fly ash for use in concrete. Part 2, A critical review of the effects of fly ash on the properties of concrete. Report 78–16, CAMMET, Energy, Mines and Resources, Canada, 1978.

151. SMITH, I.A. Design of fly ash concretes. *Proceedings of the Institution of Civil Engineers,* London, Vol. 36, April 1967, pp. 769–90.

152. CANNON, R.W. Proportioning fly ash concrete mixes for strength and economy. *Journal of the American Concrete Institute,* Proceedings, Vol. 65, No. 11, November 1968, pp. 969–79.

153. DAVIES, R.E. et al Properties of cements and concretes containing fly ash. *Journal of the American Concrete Institute,* Proceedings, Vol. 33, May-June 1937, pp. 577–612.

154. DAVIES, R.E. et al Weathering resistance of concretes containing fly ash cements. *Journal of the American Concrete Institute,* Proceedings, Vol. 37, January 1941, pp. 281–96.

155. DHIR, R.K., MUNDAY, J.G.L. and ONG, L.T. Mix proportioning of concrete with pulverized fuel ash. A Critical Review, ACI., Special publication SP; 79, 1983, pp. 267–288.

156. KRISHNA RAJU, N., VALSA IPE, T and SRINATH, N. Mix proportioning and strength characteristics of Portland cement and pulverized fly ash concrete. *Journal of the Indian Concrete Institute,* Vol. 49, Oct-Dec 1994, pp. 27–32.

157. Methods of achieving high strength concrete; FIPV. Congress, Paris, *Journal of the American Concrete Institute,* Proceedings, Vol. 64, No. 1, January 1967, pp. 45–48.

158. ALEXANDER, K.M. WARDLAW, J and GILBERT, J.D. Aggregate cement bond. Cement paste strength and the strength of concrete. *Proceedings of the International Conference on the Structure of concrete.* Cement and Concrete Association, London, 1968, pp. 59–81.

159. HARRIS, A.J. High strength concrete: Manufacture and properties. *The Structural Engineer*, Vol. 47, No. 11, November 1969, pp. 441–446.

160. FREYSSINET, E. Prestressed concrete: Principles and applications. *Journal of the Institution of Civil Engineers*. London, Vol. 33, 1950, pp. 331–380.

161. LAWRENCE, C.D. The properties of cement paste compacted under high pressure. *Cement and Concrete Association*, London, Research Report, October 1968.

162. BENNETT. E.W. and GOKHALE, V.G. Some experiments on the compaction of cement paste, mortar and concrete by vibration of different frequencies. *Indian Concrete Journal*, Vol. 41, November 1967, pp. 421–428.

163. PARROTT. L.J. Selection of constituents and proportions for producing workable concrete with a compressive cube strength of 80–110 N/mm². *Cement and Concrete Association*, *TRA-316*, London, May 1969.

164. PARROTT, L.J. The production and properties of high strength concrete. *Concrete*, Vol. 3, No. 11, November 1969, pp. 43–448.

165. GERWICK, Jr. B.C. *Construction of Prestressed Concrete Structures*, John Wiley and Sons Inc. Interscience, New York, 1971.

166. KUKACKA, L.E., STEINBERG, M. and MANOWITZ, B. Preliminary cost estimate for the radiation induced plastic impregnation of concrete. BNL Report-11263, *Brookhaven National Laboratory*, Upton, New York, April 1067.

167. CONCRETE POLYMER MATERIALS–First Topical Report– BNL Report 53134 (T-509), *Brookhaven National Laboratory*, Upton New York, Also USBR General Report No. 31, Denver, December, 1968.

168. DIKEOU, J.T., KUKACKA, L.E., BACKSTROM, J.E. and STEINBERG, M. Polymerisation makes tougher concrete. *Journal of the American Concrete Institute*, Proceedings, Vol. 66, No. 10, October 1969, pp. 829–839.

169. HADLEY, G. *Linear Programming*, Addison-Wesley Publishing Co., Reading, Massachusetts, 1962.

170. FOX RICHARD, L. *Optimisation Methods for Engineering Design*, Addisonwesley Publishing Co., Reading, Massachusetts, 1971.

171. DANTZIG, G.B. Linear Programming and Extension, Princeton University Press, Princeton, New Jersey, 1963.

172. HADLEY, G. Non-linear and dynamic programming, Addison Wesley Publishing Co., Reading, Massachusetts, 1964.

173. ROSEN, J.B. The gradient projection method for non linear Programming. *Journal of the Society for Industrial and Applied Mathematics*, Part I, Linear Constraints, Vol. 8, No. 1, 1960, pp. 181–217. Part II Non Linear Constraints, Vol. 9, No. 4, 1961, pp. 514–532.

174. ZOUTENDIJK, G. *Methods of Feasible Directions*, American Elessvicr Publishing Company inc. New York, 1960.

175. CARDON, W.A. and GILLESPIE, A. Variables in concrete aggregates and Portland cement paste which influence the strength of concrete. *Journal of the American Concrete Institute*, Proceedings, Vol. 60, No. 8, August 1963, pp. 2029–1050.

176. WALKER, S., and BLOEM, D.L. Effects of aggregates size on properties of concrete. *Journal of the American Concrete Institutes*, Proceedings, Vol. 57, No. 9, September 1960, pp. 283–298.

177. BLOEM, D.L. and GARNOR, R.D. Effects of aggregates properties on strength of concrete. *Journal of the American Concrete Institute*, Proceedings, Vol. 60. No. 10, October 1963. pp. 1429–1454.

178. DHANANJAYA, H.R. A Software for concrete mix design. *The Indian Concrete Journal*, September 1996, pp. 489–493.

179. IS: 10262–1982, (reaffirmed in 1999) Indian Standard Recommended Guide lines for concrete mix design Bureau of Indian Standards, New Delhi, 1980, pp. 1–21.

180. VISVESVARAYA, H.C. and MALLICK, A.K. Relation between water content in concrete mixes and compressive strength. Second International CTB/RILEM Symposium on Moisture Problems in Buildings, Rotterdam, Netherlands, 1974.

181. Handbook on Concrete Mixes Based on Bureau of Indian Standards. New Delhi, 1982.

182. BS EN:206–1/BS 8500–1:2006 Specifications for Concrete, British-Euro Standards, 2006.

183. KRISHNA RAJU, N. and KRISHNA REDDY, Y. A critical review of the Indian, British and American methods of concrete mix design. *The Indian Concrete Journal*, Vol. 63, No. 4, April 1989, pp. 196–201.

184. KRISHNA RAJU, N. Properties of high density concrete. *The Indian Concrete Journal*, Vol. 58. No. 5, May 1884, pp. 130–133.

185. KRISHNA RAJU, N., DWARAKANATH, H.V. and GAURI SHANKAR SINGH. Production and properties of high-

density concrete using haematite aggregates. *Proceedings of the International Symposium on Innovative World of Concrete (ICI-IWC)*, August-September, 1993, Proceedings Vol. II, pp. 3–305 to 314.

186. AITCIN, P.C. Condensed silica fume. University of Sherbrooke, Quebec, Canada, 1983.

187. LOLAND, K.E. and GJORV. O. Silica fume in concrete, Research Institute for Cement and Concrete at the Norwegian Institute of Technology, Trondheim, Norway, June 1981, 8 Reports.

188. MARKSTEAD. A. Addition of silica fume to concrete. *Proceedings of 6th Nordic Concrete Research Congress*, August 1969.

189. MEHTA, P.K. and GJORV. O.E., Properties of Portland cement and concrete containing fly ash and condensed silica fume. *Cement and Concrete Research*, Vol. 12, No. 5, pp. 587–595.

190. GARETTE, C.G. and MALHOTRA, V.M. Silica fume in concrete. Preliminary Investigations, Canadian Centre for Mineral and Energy Technology Report No. 82-1E, February 1982, pp. 1–15.

191. MALHOTRA, V.M. and GARETTE, C.G. Silica fume concrete: Properties, application and limitations. *Proceedings of Annual meeting of Institute of Concrete Technology*. Fulmer Grange, Slough, England, June 1982, pp. 1–23.

192. SELLEVOLD, E.J. and RADJY, F.F. Condensed silica fume in concrete: Water demand and strength development. *Proceedings of First International Symposium on Fly Ash, Silica Fume and Slag*, ACI, Special Publication SP-79, Montebello, Canada, August 1983, pp. 677–694.

193. BUIL, M. and ACKER, P. Creep in silica fume concrete. *Cement and Concrete Research*, Vol. 15, No. 3, 1983, pp. 486.

194. WOLSIEFER, J. Ultra-high-strength field placeable concrete with silica fume admixture. *Concrete International*, Vol. 6, No. 4, pp. 25–31.

195. SKRASTINIS, J.I. and ZOLDERNS, N.G. Ready mixed concretes incorporating silica fume. *First International Conference on Fly Ash, Silica Fume and Slag, ACI*, Special Publication SP-79, Montebello, Canada, 1983, pp. 813–830.

196. MEHTA, P.K. Typical particle size distribution of materials. *Proceedings of the First International Symposium on Fly Ash, Silica Fume and Slag*, ACI Special Publication SP-79, Montebello Canada, 1983, pp. 1–46.

197. PISTILLI, M.F. WINTERSTEEN, R and CECHMNER, R. The uniformity and influence of silica fume source on the properties of Portland cement concrete. *Cement, Concrete and Aggregate,* Vol. 6 No. 2, 1984, pp. 120–124.

198. MALHOTRA, V.M. Mechanical properties of freezing and thawing resistance of non-air-entrained and air-entrained condensed silica fume concrete using ASTM TEST C-666. *Cement, Concrete and Aggregates.* 1986.

199. WOLSIEFER, J. Ultra-high-strength field placeable concrete with silica fume admixture. *Concrete International,* Vol. 6, no. 4, 1984, pp. 25–31.

200. BACHE, H.H. Densified cement ultra-fine particle based materials. Second International Conference on Super plasticizers in Concrete, Ottawa, Canada, June 1981, Paper-9, pp. 1–35.

201. LOLAND K.E. and HUSTAD, T. Silica in concrete permeability. Cement and Concrete Institute at the Norwegian Institute of Technology. Trodenheim, Norway, Report No. STF 65–A–81031, June 1981.

202. TRAETTBERG. A frost action in mortar of blended cement with silica dust. *Proceedings of Conference on Durability of Building Materials and Component,* ASTM-STP 691, 1980, pp. 536–548.

203. OPSAHL, O.A., Silica in concrete frost resistance. Cement and Concrete Institute at the Norwegian Institute of Technology, Trondheim, Norway, Report no. STF-65 A 81301, June 1981.

204. HOLLAND, T.C. Abrasion-erosion evaluation of concrete mixtures for Kinzua dam stilling basin repairs. US Army Engineers Waterways Experiment Stations, Misc. Paper SL-83–16, Sept. 1983.

205. SELLEVOLD, E.J., BAGER, D.H. KLITGAARD JENSON, E. and KNUDSEN, T. Silica fume cement pastes: Hydration and pore structure. Proceedings of Nordisk Mini Seminar on Silica Concrete, Norwegian Institute of Technology, Trondheim, Norway, December 1981, pp. 1–32.

206. MATHER, K. Factors affecting sulphate resistance of mortars. *Proceedings of 7th International Congress on the Chemistry of Cements,* Paris, July 1980, Vol. IV, pp. 580–585.

207. OLAFASSON, H. Effect of silica fume on alkali silica reactivity. *Proceedings of Nordic Research, Seminar on Condensed Silica Fume in Concrete,* Trondheim, Norway, December 1981, pp. 141–149.

208. GUDMUNDSSON, G. Production of blended cement based on silica fume. *Proceedings of Nordic Research, Seminar on Condensed Silica Fume in Concrete,* Trondheim, Norway, Dec. 1981, pp. 135–140.

209. GJORV, O.E. Durability of concrete containing condensed silica fume. *Proceedings of First International Conference on Fly Ash, Silica Fume and Slag,* ACI, Special publication SP-79, Montbello, Canada, July 1983, pp. 695–708.

210. RICKNE, S and SVENSSON, C. The New Tjorn bridge, Nordisk Betong (Stockholm), Vol. 2, No. 4, 1982, pp. 213–216.

211. MALHOTRA, V.M., Use of Fly Ash in Concrete, ACI Materials Journal Committee Report, Vol. 84, 1987, pp. 381–403.

212. NEVILLE, A.M., Properties of Concrete, British English language Society, Fourth and Fourth and Final edition, London, 2000.

213. RANGANATH, R.V, BHATTACHARJEE, and KRISHNA MOORTHY, S. Reproportioning of aggregate mixes for optimal workability with pond ash as fine aggregate in concrete. *The Indian Concrete Journal,* July 1999, pp. 441-449.

214. RANGANATH, V., Study on the Characterization and Use of Ponded Fly ash as Fine Aggregate in Cement Concrete, Ph.D Thesis, Civil Engineering Department, I.I.T, New Delhi, August 1995.

215. RANGANATH, R.V and KRISHNA MURTHY, S. Influence of ponded fly ash mineral admixture on the strength and workability of cement concrete. *Proceedings of the National Conference on Mineral Admixtures,* I.I.T, Madras, 1992.

216. CHERIAF. M, CAVALCANTE ROCHA. J and PERA. J. Pozzolanic properties of pulverized coal combustion bottom ash. *Cement and Concrete Research,* Vol. 29, 1999, pp. 1387-1391.

217. KIATTIKOMOL. K, JATURPITAKUL. C, SONGPIRIYAKIZ. S, CHUTUBITIM.S. A study of ground coarse fly ashes with different finenesses from various sources as pozzolanic materials. *Cement and Concrete Composites,* Vol. 23, 2001, pp. 335-343.

218. PRANESH, R.N et al., Strength and durability of concrete with pond ash as fine aggregate. *Indian Concrete Journal,* April 2007.

219. KALGAL, M.R. et al. Characterization of RTPS pond ash. *Proceedings of the International Congress on Fly Ash,* New Delhi, Dec. 2005.
220. IS: 383-1970, Specifications for Coarse and Fine aggregates from Natural Sources for Concrete, Second Revision, BIS, New Delhi, 1970.
221. IS: 1727-1967, Specifications for Method of Test for Pozzolonic materials, BIS, New Delhi 1967.
222. IS: 8112-1980, Specifications for 43 Grade Ordinary Portland Cement, BIS, New Delhi 1980.
223. IS: 3812-1999, Specifications for Fly Ash for Use as Pozzolana and Admixture, BIS, New Delhi, 1999.
224. PRANESH, R.N., Pond Ash in cement Concrete–Some Studies on its Feasibility, Doctoral Thesis submitted to the Visvesvaraya Technological University, Belgaum, 2007-2008 pp. 1-165.
225. SENTHIL KUMAR, V and MANU SANTHANAM. Particle packing theories and their application in concrete mixture proportioning – A review. *The Indian Concrete Journal,* 2003 pp. 1324-1331.
226. MONTEIRO, P.J.M et al. Designing concrete mixtures for strength, elastic modulus and fracture energy. *RILEM Journal,* Vol. 26, No. 162, 1993.
227. FEYNMAN, R. There is plenty of room at the bottom (Reprint from speech delivered at the annual meeting of the West of the American Physical Society). *Engineering Science Journal,* Vol. 23, 1960, pp. 22–36.
228. SOBOLOV, K and GUTIERREZ, M.F. How nanotechnology can change the concrete world, American Ceramic Society Bulletin, Vol. 84, No.10, October 2005, pp. 14–18.
229. CHONG, K.P., Nano-science and engineering in mechanics and materials. *Journal of Physics and Chemistry of Solids,* Vol. 65, 2004, pp.1501–1506.
230. COLLEPARDI, M. et al. Combination of silica fume, fly ash and amorphous nanosilica in superplasticized high performance concretes. *Proceedings of the First International Conference on Innovative Materials and Technologies for Construction and Rehabilitation,* Lecce, Italy, 2004, pp.459–468.
231. BHUSHAN, B., Hand Book of Nano Technology, Springer Publications, 2004.

232. ATKINSON, W.I., Nanocosm-Nanotechnology and the Big Changes Coming from The inconceivably small, AMACOM, pp. 36–39.

233. BALAGURU, P and SHAH, S.P. Fiber Reinforced cement Composites, McGraw-Hill Publications, New York, 1992, pp. 530.

234. BALAGURU, P and CHANG, P. High strength composites for repair, rehabilitation and strengthening of concrete. *Indian Concrete Institute Journal*, No.3, 2003, pp.7–18.

235. MEHTA, P.K. Concrete, Structure, Properties and Materials, Prentice-Hall, New Jersey, 1986, pp. 449.

236. PERUMALA SWAMY BALAGURU and KEN CHONG. Nanotechnology and Concrete Proceedings of the ACI Session on Nano technology of Concrete, recent developments and Future Prospectives, Nov. 7, 2006, Denver, USA, pp. 15–28.

237. KRISHNA RAJU, N. Prestressed Concrete (Fifth Edition), Tata McGraw-Hill Publishing Company, New Delhi, 2012, pp. 699–700.

238. DEGUSSA CONSTRUCTION CHEMICALS (India) Pvt. Ltd. Product Promotional Pamphlet-C, 68, MIDC, Thane, Belapur Road, Yurbhe, Nava Mumbai, July 2006.

239. IIJAIMA, S. Helical Micro-Tubes of Graphite Carbon, *Journal of Nature*, 354 (6348), 1991, pp. 56–58.

240. SRIVASTAVA. D, WEI. C and CHO. K. Nano-Mechanics of Carbon Nano-Tubes and Composites, Applied Mechanics Review, Vol. 56, 2003, pp. 215–230.

241. QIAN. D, WAGNER. G.J, LIU. W.K, YU. M and RUOFF. R.S. Mechanics of Carbon Nano-Tubes, Applied Mechanics Review, Vol. 55, 2002, pp. 495–533.

242. KASHIWAGI. T, DU. F, DOUGLAS. J.F, WINEY. K.I, HARRIS. R.H and SHIELDS. J.R., Nano-particle net works reduce the flamability of polymer nano composites. *Journal of Nature Materials*, Vol. 4, 2005, pp. 928–933.

243. SONG. G. Smart Aggregates: A distributed Intelligent Multi-Purpose Sensor Network (DIMSN) For Civil Structures, IEEE International Conference on Networking, Sensing and Control, 2007, pp.775–780.

244. MAILE AIU and HUANG. C.P. The Chemistry and Physics of Nanocement, University of Delaware, USA, August, 2006.

245. HANEHARA. S and ICHIKAWA. M. Nanotechnology of cement and concrete. *Journal Taiheiyo Cement Corporation*, 2001, pp. 141: 47–58.

246. GENGYING LI. Properties of high volume fly ash concrete incorporating nanosilica. *Cement and Concrete Research*, Vol. 34, 2004, pp. 1043–1049.

247. GINEBRA. M.P, DRIESSENS. F.C.M and PLANNEL. J.A. Effect of the particle size on the micro- and nano structural features of calcium sulphate cement; A kinetic analysis. *Biomaterials*, V.25, 2004, pp. 3453–3462.

248. HUI LI, HUI GANG, XIAO, JIE YUAN and JINPING OU. Micro-Structure of Cement Mortar with Nano Particles, Composites, Part-B, V.35, 2004, pp. 1185–1189.

249. WEN YIH KUO, JONG SHIN HUANG and CHI HSIEN LIN. Effect of organo modified monmorillonite on strength and permeability of cement mortars. *Cement and Concrete Research*, Vol. 36, 2006, pp. 886–895.

250. BENTZ. D.P, PELTZ. M.A, SNYDER. K.A and DAVIS. J.M, VERDICT: Viscosity Enhancers Reducing Diffusion in Concrete Technology, Concrete International, 31(1), pp. 31–36, January 2009.

251. FERRADA. M.G., ESCOBAR. M., DOMINGUEZ, FERRADA. R.V. and AVALOS. P.I., U-Silice ISO-14001: Silice, A Favor del Medio Ambiente, XIV Jornadas Chilenas del Hormigon, Valdivia, Chile, 2003.

252. COLLEPARDI. M, OGOUMAH-OLAGOT, SKARP.U and TROLI. R. Influence of Amorphous Colloidal Silica on the Properties of Self-Compacting Concretes, Proceedings Of the International Conference: Challenges in Concrete Construction–Innovations and Developments in Concrete Materials and Construction, Dundee, U.K., 2002, pp. 473–83.

253. H. LI, XIAO. H-G, YUAN. J, OU. J., Micro Structure of Cement Mortar with Nano-Particles, Compos, Part-B, Vol. 35, 2004, pp. 185–189.

254. BATRAKOV. V and SOBOLEV. K. Multicomponent Cement Based Superplasticized High-Strength Concretes; Design, Properties and Optimization, 5th CANMET/ACI International Conference on Fly Ash, Silica Fume, Slag and Natural Pozzolans in Concrete 1995, pp. 695–710.

255. KONSTANTIN SOBOLEV and MIGUEL FERRADA GUTIERREZ. How Nano Technology Can Change the Concrete World, American Ceramic Society Bulletin, Vol.84. No.10, Nov. 2005, pp. 16–20.

256. PASCAL MAES., Cuore Nano Silica Concrete- A Report published in Civil Engineering Portal, 2009.
257. COLLEPARDI. M, COLLEPARDI. S, SKARP. U and TROLI. Optimization of Silica Fume, Fly Ash and Amorphous Nano-Silica in Super plasticized High performance Concretes, Proceedings of 8th CANMET/ ACI International Conference on Fly Ash, Silica Fume, Slag and natural Pozzolans in Concrete, SP-221, Las Vegas, 2004, pp. 495–506.
258. SKARP, U and SARKAR, S.L., Enhanced Properties of Concrete Containing Colloidal Silica, Concrete producer, December 2000, pp. 1–4.
259. COLLEPARADI, M et al. Influence of amorphous colloidal silica on the properties of self-compacting concretes. *Proceedings of the International Conference in Concrete Construction–Innovations and developments in Concrete Materials and Constructions*, Dundee, Scotland, UK, September 2002, pp. 473–483.
260. COLLEPARADI, M. Quick method to determine free and bound chlorides in concrete. *Proceedings of RILEM Workshop on Chloride Penetration into Concrete*, Saint Remyles Chevreuse, 1995, pp. 10–16.
261. RUSSEL HENRY, G. What is high performance concrete?, *Journal of Concrete Products*, January, 1999.
262. ACI Committee 211, Guide Lines for selecting Proportions for High Strength Concrete With Portland cement and Fly Ash (ACI 211-4R), American Concrete Institute, Detroit, USA, 1993.
263. SHAH S.P. Recent Trends in the Science and Technology of Concrete, Concrete Technology, New Trends, Industrial Applications, Proceedings of the International RILEM Workshop, E and FN Spon, London, 1993, pp. 1–18.
264. JENSEN, J.J. Structural Aspects of High Strength Concretes, Concrete Technology, New Trends, Industrial Applications, Proceedings of the International RILEM Workshop, Eand FN Spon, London, 1993, pp. 197–212.
265. BACHE, H.H. Densified cement/ultra-fine particle based materials. *Proceedings of the Second International Conference on Superplasticizers in Concrete*, Ottawa, 1981.
266. KRISHNA RAJU, N. A Critical review of testing of Concrete and Concrete Structures, Short term Training Programme on Advances in Cementatious Materials and Testing, M.S. Ramaiah Institute of Technology, Bangalore, March 1998, pp. NKR 2–1 to 2–11.

267. KRISHNA RAJU, N. Prestressed Concrete (Fourth Edition), Tata McGraw-Hill Co, New Delhi, 2007, p. 718.

268. SAI PRASAD, P.V and KAMALESH JHA. High Performance Concrete, A Project Work Undertaken for Course No. 624 at IRICE, Pune, 2009.

269. DEWAR, J.D. A Concrete Laboratory in a Computer- Case Studies of Simulation of Laboratory Trial Mixes, ERMCO-95, proceedings of the XIth European Ready Mixed Concrete Congress, Istanbul, 1995, pp. 185–193.

270. FRANCOIS de LARRARD. A Method For Proportioning High Strength Concrete Mixtures, Cements, Concretes, and Aggregates, Vol. 12, No.2, 1990, pp. 47–52.

271. SOBOLEV, K.G and SOBOLEVA, S.V. High Performance Concrete Mix Proportioning, Special Publication, SP: 179-26, American Concrete Institute.

271. NAWY, EDWARD, G. Fundamentals of High Performance Concrete, (Second Edition) John Wiley and Sons Inc, 2000, pp. 89–121.

272. GROVE, J.D, JONES, K, BHARIL, K.S , ABDUL SHAFI, A and CALDERWOOD, W. Fast Track and Fast Track II, Cedar Rapids, Iowa Transportation Research Record No 1282, pp. 1–7.

273. WALKER, B. Fast Curing concrete overlays to Challenge Black top Supremacy, Highways, Vol. 58, No.1966, 1990, pp. 16–17.

274. NAGI, M, JANSSEN, D and WHITING, D.A. Durability of Concrete for Early Opening of Required High Ways- Field Evaluation, Durability of Concrete, proceedings Of the Third International Conference, Nice, France, Edited by V.M Malhotra, American Concrete Institute, Detroit, Michigan, (ACI : SP-145), 1994, pp. 811–833.

275. NAGI, M and WHITING, D.A. Field Studies of New Test procedures and Materials For Concrete Pavement Rehabilitation, Proceedings of the Conference-Workshop on the Repair and Rehabilitation of the Infrastructure of the Americas, University of Puerto Rico, Mayaquez, Puerto Rico, Edited by Houssam A. Tountanji, 1994, pp. 223–235.

276. SEEHRA, S.S, GUPTA, S, and KUMAR, S. Rapid Setting Magnesium Phosphate Cement for Quick Repair of concrete Pavements – Characterization and Durability Aspects, Cement and Concrete Research, Vol. 23, No.2, pp. 254–266.

277. OZYILDIRIM, C. A Field Investigation of Patches Containing Pyrament Blended Cement, Virginia Transportation Research Council, Charlottesville, VA, 1994, 16 pp.

278. BABAEI, G and HAWKINS, N.M. Performance of Bridge Deck Concrete Overlays: Extending the Life of Bridges, ASTM Special publication No.1100, 1990, pp. 95–108.

279. OZYILDIRIM, C. A Field Investigation of a Concrete Overlay Containing Silica Fume On Route 50 over Opequon Creek, Final report, Virginia transportation Council, Charlotsville, VA, 1993, 22 pp.

280. SPRINKEL, M.M. Polymer Concrete Bridge Overlays, Transportation Research Record, No. 1392, 1993, pp. 107–116.

281. GJORV, O.E, BAERLAND, T and RONNING, H.R. Abrasion Resistance of High Strength Concrete Pavements, Concrete International: Design and Construction, Vol.12 No.1, 1990, pp. 45–48.

282. ZIA, P, SCHEMMEL, J.J and TALLMAN, T.E. Structural Applications of High Strength Concrete, Final Report, Center for Transportation Engineering Studies, Department of Civil Engineering, North Carolina State University, NC, June 1989, pp. 330.

283. ADELMAN, D and COUSINS, T.E. Evaluation of the use of high strength concrete bridge girders in Louisiana, *Prestressed Concrete Institute Journal*, Vol. 35, No.5, Sept-Oct 1990, pp. 70–78.

284. SCHEMMEL, J.J and ZIA, P. Use of High Strength Concrete in Prestressed Concrete Box Beams for High Way Bridges, Transportation Research Record No.1275, 1990, pp. 12–18.

285. TAERWE, L. High Strength Concrete for Prestressed Concrete Girders, IABSE Reports Vol. 64, 1991, pp. 355–360.

286. RUSSEL, B.W. Impact of high strength concrete on the design and construction of prestressed girder bridges, *P.C.I. Journal*, 1994, Vol. 39, No. 4, pp. 76–89.

287. ZIA, P, LEMING, M.L and AHMAD, S.H. High performance concrete. A State of the Art Report, Strategic Highway Research Program, National Research Council, Washington, D.C, 1991, pp. 251.

288. MAILER, Y, PLISKIN, L, MILLAN, A, HAGOLLE, D, HIGUET, P, CADORET, G LEBOULICAUT, J.P, REGNIER, J.M. de LARRARDD, I, SCHALLER, I, BONNET, G, SUDRET, J.P and BRAZILLIER, D. High Performance Bridge: The Experimental

Structure of Joigny (Yonne Region of France), Travaux, April 1989, No. 642 pp. 57–65.

289. MAILER, Y, PLISKIN, L. Bridge at Joigny; High Strength Concrete Experimental Bridge, Transportation Research Record No. 1276, 1990, pp. 45–48.

290. MAILER, Y, and BIZILLIER, D and ROI, S. The Bridge of Joigny, Concrete International, Vol. 13, No.5, 1991, pp. 40–42.

291. KRISHNA RAJU, N., Design of Bridges (Fourth Edition), Oxford and IBH, New Delhi, 2009, pp. 437–471.

292. LANE, S.N and PODOLNY. Jr. The federal Outlook for High Strength Concrete Bridges, Journal of the Prestressed Concrete Institute, Vol. 38, No. 3, May-June 1993, pp. 20–33.

293. DURNING, T.A and REAR, K.B. Braker Lane Bridge–High Strength Concrete in Prestressed Bridge Girders, PCI Journal, Vol. 38, May-June 1993, pp. 46–51.

294. RUSSEL, B.W and BURNS, N.H. Static and Fatigue behavior of Pretensioned Composite bridge Girders Made With High Strength Concrete, PCI Journal, Vol. 38, No. 3, May-June 1993, pp. 116–128.

295. BRUCE, R.N, RUSSEL, H.G, ROLLER, J.J, and MARTIN, B.T. Feasibility Evaluation of Utilizing High Strength Concrete in Design and Construction of High way Bridge Structures, Interim Report, Louisiana Transportation Research Center, Baton Rouge, Los Angeles, 1992, xxii, 228 pp.

296. ROLLER, J.J, and MARTIN, B.T, RUSSEL, H.G, and BRUCE, Jr. R.N. Performance of Prestressed High Strength Concrete Bridge Girders, P.C.I. Journal, Vol. 38, No.3, 1993, pp. 34–45.

297. KRISHNA RAJU, N, et al. Limit State Behavior of Pretensioned I-Beams with Limited Prestress, Indian Concrete Journal, Vol. 49, No.7, July 1975, pp. 213–218.

298. PERRY, V.H. "What is reactive Powder Concrete?", High performance Concrete Bridge Reviews, No. 16, July-August 2001.

299. LAFARGE NORTH AMERICA. Technical Characteristics; UHPC with organic fibers National Building Code of Canada, 1995.

300. PERRY, V.H, ZAKARIASAN, D. First use of High performance Concrete for an Innovative Train Station canopies, Concrete Technology- Today, August 2004, pp. 1–8.

Subject Index

Author Index